BIM Teaching and Learning Handbook

T0256278

This book is the essential guide to the pedagogical and industry-inspired considerations that must shape how BIM is taught and learned. It will help academics and professional educators to develop programmes that meet the competences required by professional bodies and prepare both graduates and existing practitioners to advance the industry towards higher efficiency and quality.

To date, systematic efforts to integrate pedagogical considerations into the way BIM is learned and taught remain non-existent. This book lays the foundation for forming a benchmark around which such an effort is made. It offers principles, best practices, and expected outcomes necessary to BIM curriculum and teaching development for construction-related programmes across universities and professional training programmes. The aim of the book is to:

- Highlight BIM skill requirements, threshold concepts, and dimensions for practice;
- Showcase and introduce tried-and-tested practices and lessons learned in developing BIM-related curricula from leading educators;
- Recognise and introduce the baseline requirements for BIM education from a pedagogical perspective;
- Explore the challenges, as well as remedial solutions, pertaining to BIM education at tertiary education;
- Form a comprehensive point of reference, covering the essential concepts of BIM, for students;
- Promote and integrate pedagogical consideration into BIM education.

This book is essential reading for anyone involved in BIM education, digital construction, architecture, and engineering, and for professionals looking for guidance on what the industry expects when it comes to BIM competency.

M. Reza Hosseini is currently the associate head of school (research) at the School of Architecture and Built Environment, Deakin University; a research fellow of the Centre for Research in Assessment and Digital Learning (CRADLE); and the founder and leader of the Australian BIM Academic Forum (ABAF). He had 12 years of experience working in various areas of the construction industry prior to joining the academic world. His main research and teaching areas are building information modelling (BIM) and digital engineering, and he has written around 190 published papers and book chapters.

Farzad Khosrowshahi is dean of the College of Engineering and Science at Victoria University, Melbourne, and Fellow of the Chartered Institute of Building. He is the founder of the BIM Academic Forum, Information Visualisation Society, and Uniting Construction Information. Farzad has previously served as the director of Construct IT, chair of ARCOM (Association of Researchers in Construction Management), and chair of IV Society. He has served on a number of committees including The European Council for Construction Research, Development and Innovation, buildingSMART UK, and the Construction Industry Council Research & Innovation.

Ajibade A. Aibinu is currently an associate professor at the Faculty of Architecture, Building and Planning, The University of Melbourne, Australia, where he was the Assistant Dean, Research Training. His research interests cut across the built environment project and asset management, from design and construction to operations, particularly data-driven processes to ensure value for money. He is the founder of the Intelligent Cost Manager, a cloud-based cost-management solution that leverages deep learning and predictive modelling to generate cost estimates with greater accuracy using historical data. Ajibade is a member of the Australian Institute of Quantity Surveyors and was previously an associate of the Chartered Institute of Arbitrators (2008–2013).

Sepehr Abrishami is a BIM programme leader and senior lecturer in the School of Civil Engineering and Surveying at the University of Portsmouth, UK. He has carried out extensive research in the field of BIM, automation in design and construction, and environmental design. His areas of expertise include building information modelling (BIM), generative and parametric evolutionary design, artificial intelligence (AI), digital design and construction, integrated design, design for manufacturing and assembly (DfMA), offsite manufacturing, Big Data, blockchain, Internet of Things (IoT), and IT integrated architectural design.

BIM Teaching and Learning Handbook

Implementation for Students and Educators

Edited by
**M. Reza Hosseini, Farzad Khosrowshahi,
Ajibade A. Aibinu, and Sepehr Abrishami**

Routledge
Taylor & Francis Group

LONDON AND NEW YORK

First published 2022
by Routledge
2 Park Square, Milton Park, Abingdon, Oxon OX14 4RN

and by Routledge
605 Third Avenue, New York, NY 10158

Routledge is an imprint of the Taylor & Francis Group, an informa business

© 2022 selection and editorial matter, M. Reza Hosseini, Farzad Khosrowshahi, Ajibade A. Aibinu, and Sepehr Abrishami; individual chapters, the contributors

The right of M. Reza Hosseini, Farzad Khosrowshahi, Ajibade A. Aibinu, and Sepehr Abrishami to be identified as the authors of the editorial material, and of the authors for their individual chapters, has been asserted in accordance with sections 77 and 78 of the Copyright, Designs and Patents Act 1988.

All rights reserved. No part of this book may be reprinted or reproduced or utilised in any form or by any electronic, mechanical, or other means, now known or hereafter invented, including photocopying and recording, or in any information storage or retrieval system, without permission in writing from the publishers.

Trademark notice: Product or corporate names may be trademarks or registered trademarks, and are used only for identification and explanation without intent to infringe.

British Library Cataloguing-in-Publication Data
A catalogue record for this book is available from the British Library

Library of Congress Cataloging-in-Publication Data
Names: Hosseini, M. Reza, editor.
Title: BIM teaching and learning handbook : implementation for students and educators / edited by M. Reza Hosseini, Farzad Khosrowshahi, Ajibade A. Aibinu, Sepehr Abrishami.
Description: New York : Routledge, 2021. | Includes bibliographical references and index.
Identifiers: LCCN 2021001439 (print) | LCCN 2021001440 (ebook) | ISBN 9780367427955 (hbk) | ISBN 9780367855192 (ebk)
Subjects: LCSH: Building information modeling--Study and teaching.
Classification: LCC TH438.13 .B63 2021 (print) | LCC TH438.13 (ebook) | DDC 720.285--dc23
LC record available at https://lccn.loc.gov/2021001439
LC ebook record available at https://lccn.loc.gov/2021001440

ISBN: 978-0-367-42795-5 (hbk)
ISBN: 978-1-032-03472-0 (pbk)
ISBN: 978-0-367-85519-2 (ebk)

Typeset in Baskerville
by Deanta Global Publishing Services, Chennai, India

Contents

Contributors

Hamid Abdirad earned his PhD in the Built Environment at the University of Washington (UW) with a concentration in technology and project delivery. With a multi-disciplinary background in architecture, engineering, and construction, Hamid currently practices with Magnusson Klemencic Associates (MKA) to research and develop BIM automation workflows in a diverse portfolio of domestic and international projects. Before joining MKA, he taught BIM and VDC courses at the graduate and undergraduate levels at the UW. He is a published author in the field with research on digital project delivery, information exchange, BIM process mining, BIM software customization, and BIM curriculum design.

Tayyab Ahmad is a freelance architectural engineer and an academic. He has degrees in architectural engineering and construction management. He completed his PhD at the University of Melbourne in 2020. As an academic, Tayyab is strongly driven to play his role in sustainable development of construction projects. His expertise is in the areas of built environment sustainability, building information modelling, life-cycle assessment, project success, and decision-making frameworks. For the last six years, he has considered green building projects from design, performance-modelling, and project-development perspectives. He aspires to bridge the theory and practice of built environment to achieve sustainable outcomes.

Abiola Akanmu is an associate professor of Smart Design and Construction in the Myers Lawson School of Construction at Virginia Tech. Her research has been focused on the design of cyberlearning and educational technologies with applications to workforce health, safety and technical training, cyber-physical construction systems, building information modelling, and wearable robots. Her research has been funded by private and federal agencies including National Science Foundation.

Hande Aladağ has been working as an assistant professor in the Construction Management Division of the Civil Engineering Department in Yildiz Technical University (YTU), Istanbul, since 2018. Previously, she worked as research assistant in the Construction Management Division of the Civil Engineering Department between 2008 and 2018. She completed a doctoral degree in architecture from YTU in 2017. Her research interest areas are construction management, project management, sustainable architecture, contract administration, risk management, urban regeneration, and BIM.

Salam Al-Bizri has more than 30 years' experience in higher education and the construction industry. She has been exploring issues surrounding the integration of design and construction processes and how, through education, a better understanding of issues can be achieved and best practice delivered. By acquiring and exploiting the knowledge and skills of experts in the field, students can be educated in the cause and effect relationships that result from decision making. This is demonstrated as best practice as well as illustrating the weaknesses in the way activities are currently carried out.

Mohammad Alhusban has a BSc in Civil Engineering, an MSc in Quantity Surveying, and a PhD in Construction Management from the University of Portsmouth, UK. He is currently the head of the civil engineering department in the Middle East University, Lebanon. He has 5 years of practical engineering experience in civil engineering, construction management, and health and safety. His research interests revolve around integrating building information modelling/management into procurement arrangements and how such integration can affect the delivery of sustainable construction projects.

Yusuf Arayici is a professor in construction project management at Northumbria University. With an international outlook, he previously fulfilled an academic management role as a dean. He also successfully completed his research fellowship TUBITAK (Research Council of Turkey) in Digital Construction.

Since 2000, his research projects have ranged from building information modelling to sustainability. He has led substantial research groups over a prolonged period of time through continuous cycles of research with funded research projects, has graduated many PhD and MSc students, and has published more than 100 papers and five books.

Currently, he is researching on AI-supported heritage BIM.

Abbas Mehrabi Boshrabadi currently works at Centre for Research in Assessment and Digital Learning, Deakin University, Melbourne, Australia. Over the last decade, Abbas has been involved in conducting research and development in assessment and feedback in higher education, teaching methods, and teacher education. His current research interests centre mainly on sustainable assessment practices that assist undergraduates to develop lifelong learning behaviours.

Cenk Budayan received a PhD in civil engineering from Middle East Technical University. He is an associate professor in the Civil Engineering Department, Construction Management Division in the Civil Engineering Faculty of Yildiz Technical University. His current interests include project management in construction projects, time management in construction projects, engineering economy, public–private partnership projects, building information modelling, and data analysis in construction projects. He is the author/co-author of 25 articles published in leading academic journals and 15 national/international conference papers.

Gökhan Demirdöğen has been working as research assistant in Construction Management in the Civil Engineering Department, Yildiz Technical University since 2013. Currently, he is studying for his PhD at Yildiz Technical University, Istanbul. Previously, he completed his MSc degree in Structural Engineering at Yildiz Technical University in 2015. His research interest areas are project management, facility management, energy management, BIM, innovation, Big Data analytics, and technology transfer.

Chiranjib Dey has over 20 years of professional experience in the construction industry on a wide variety of building projects as a cost manager, project manager, and architect. He did a Bachelor's degree in Architecture (1998) at the Academy of Architecture, India, and a Master of Construction Management (2017) from The University of Melbourne. Having trained and worked in multiple roles in a variety of countries, Chiranjib has gained a considerable knowledge base of different types of industry as well as country practices. Chiranjib's experience at all levels of delivery provides an appreciated balance and understanding of the issues to be addressed throughout the project life cycle of building construction.

Carrie Sturts Dossick, P.E. is a professor of Construction Management and the Associate Dean of Research in the College of Built Environments, University of Washington. Dr Dossick also holds an adjunct professor appointment in the Department of Civil and Environmental Engineering and is currently the vice-chair of the National BIM Standard – US Planning Committee. Dr Dossick co-directs the Communication, Technology, and Organizational Practices lab in the Center for Education and Research in Construction (CERC). Dr Dossick has over two decades of research and teaching experience focused on emerging collaboration methods and technologies such as building information modelling (BIM).

Faris Elghaish is a teaching fellow in Construction Project Management at the University of Portsmouth, UK. His responsibilities include the delivery of modules relating to construction management and building information modelling (BIM). He has also been intensively participating in the development of a new online course for a Master's of Project Management in Construction that will be managed in a collaboration between the University of Portsmouth (UoP) and the Cambridge Education Group (CEG). He has carried out extensive research in the field of BIM, cost optimization/automation, and decentralized systems (i.e. smart contracts).

Teo Ai Lin Evelyn is presently associate professor of Building in the School of Design and Environment, National University of Singapore, where she has previously served as programme director of the Bachelor of Science in Project and Facilities Management and director of External Affairs. She currently serves as director for the NUS Centre of Excellence in BIM Integration and 5G Advanced BIM Lab. She received the Excellent Educator Award in 2011, 2012, and 2013 (this award is presented to one who is innovative and who has demonstrated the abilities to enhance students' learning in a creative way through competition) from the dean of Engineering College at Kyung Hee University, Korea.

Zeynep Işık has been working as a faculty member of Construction Management in the Civil Engineering Department, Yildiz Technical University since 2009. She completed a doctoral degree in civil engineering at the Middle East Technical University, Ankara, in 2009. Her research areas are construction management, project management, strategic management, sustainable structures, performance management, contract administration, risk management, urban regeneration, cost management, and BIM.

Mohsen Kalantari is an associate professor of geomatics engineering at the University of Melbourne. His research and teaching activities cover topics in building information modelling. Assoc. Prof. Kalantari has designed the first BIM course for engineering at the University of Melbourne. He has also developed the first wholly online graduate-level BIM subject in Australia.

Bimal Kumar is a professor of Built Environment Digital Futures in the Department of Architecture and Built Environment at Northumbria University in the UK. He is a Fellow of the Institution of Civil Engineers (FICE) and a Chartered Engineer (CEng). He has been involved in built environment digitalization research and teaching since the mid-1980s and has led several projects funded by the government, research councils, and the private sector.

Elisa Lumantarna is a lecturer and academic specialist at the University of Melbourne, with expertise in advanced structural design and modelling of structures under earthquake loading. Dr Lumantarna has 15 years of teaching and research experience in the field of structural and earthquake engineering. She has been a chief investigator in collaborative projects in Australia and Indonesia, including the Bushfire and Natural Hazard Cooperative Research Centre and the Australia–Indonesia Centre projects, aimed at improving the earthquake resilience of building structures.

Igor Martek is currently an academic at Deakin University, Australia. He earned his PhD in 'Enterprise Strategies in International Construction' from the University of Melbourne. He has an MBA from the Australian Graduate School of Management, NSW, and an MA in International Relations from the Australian National University, Canberra. He has worked extensively in industry in evaluating, generating, and managing large capital projects in various locations around the world. He has worked in Europe, including Eastern Europe, the Maghreb, the Levant, China, Korea, and was managing director, Far East, of a British consultancy firm based in Tokyo, for ten years. His research interests include the procurement and facilitation of capital projects as an instrument of national competitive strategy, and the competitive behaviours of international construction firms.

Sandra Matarneh is currently an assistant professor at the Civil Engineering Department, Al Ahliyya University, Jordan. She has been a practitioner in the construction industry for almost 20 years, holding various positions in managing large-scale international corporates. She is a member of the Institute of Workplace and Facilities Management (IWFM) and BIMe Initiative Community. She holds a PhD in Building Information Modelling (BIM) from the University of Portsmouth, UK, Her main areas of research revolve around BIM, information management, and construction management. Currently, she is a member of the Jordanian Engineering Association and a member of Autodesk Design Academy. She is also an active reviewer for many top-ranked academic journals.

Sas Mihindu has well developed skills, knowledge, and experience gained through working for academia, research, and business enterprises in Europe and Australasia over many decades and is the founding member of the United Kingdom's first BIM Research & Special Interest Group. His current research interests include human-centric technology infrastructures for future workplaces, strategies of distributed knowledge and intelligence management for the establishment of future cybersecurity infrastructures, virtual environments facilitating, collaboration amongst cybersecurity professionals, community collaboration, virtual reality and virtual technology infrastructures to influence future AI systems. He has been leading and delivering numerous European Commission projects with Information Society Technologies Programmes via the University of Salford, Greater Manchester.

Nicholas Nisbet has been involved with the development of BIM since 1976. He is now an independent consultant supporting government and commercial interests in construction process and product improvement through information management. He supports standards development in buildingSMART, ISO, and BSI, including the authoring of BS1192, BS1192-4 and PAS1192-6. Having helped develop the Singapore ePlanCheck and applications for the US International Code Council and the US Army ERDC, he is now active in automated code compliance within buildingSMART and the UK Digital Compliance initiatives. He can be contacted through nn@aec3.com.

Oluwole Alfred Olatunji is an associate professor of Construction Management. His research interests are in broad areas of construction, including digital construction methods (and the legal nexus thereof), applied psychology, cost studies, decision analysis, major and mega projects, and forensic analysis of procurement. Associate Professor Olatunji is well-published. He provides editorial leadership for four academic journals.

Mehran Oraee is a lecturer in Building Information Modelling (BIM) within the Faculty of Architecture, Building and Planning at the University of Melbourne, Australia. Dr. Oraee has a PhD in BIM and Construction Management and has broad research and industry expertise in the built environment and construction sector. His research is mainly focused on BIM, digital engineering, construction technologies, and construction management, and he has published quality research papers in reputable international journals and conferences in the construction context. Dr. Oraee is the coordinator of the BIM Specialisation at the University of Melbourne, which provides the students with technical skills and management knowledge in the BIM area. Also, he has over ten years' experience in the construction industry in Australia and overseas, working in various positions as a builder, project supervisor, and coordinator.

Eleni Papadonikolaki ARB, MAPM, FHEA, is an associate professor in Digital Innovation and Management at University College London (UCL) and a management consultant. She has a PhD in Design and Construction Management from TU Delft, the Netherlands; an MSc degree (cum laude) in Digital Technologies, also from TU Delft; and a Dipl.-Eng. in Architectural Engineering (cum laude) from the NTUA, Greece. Eleni is the author of over 60 peer-reviewed publications on digital transformation in construction. She has secured and delivered collaborative research projects of total worth circa £10M as principal and co-investigator funded by the European and UK research councils. She is the director of the MSc in Digital Engineering Management at UCL where she develops the new generation of project leaders for digital transformation.

Ergo Pikas defended his joint Doctor of Science dissertation in 2019, titled *Causality and Interpretation: Integrating the Technical and Social Aspects of Design* at the Aalto University in Finland and at the Tallinn University of Technology in Estonia. Ergo is currently working as a postdoctoral researcher at the Aalto University in Finland. During his doctoral studies, he spent ten months as a Fulbright Visiting Scholar/Researcher at the University of California, Berkeley. He is currently focused on building design and building design management, construction management, and construction digitalization research topics.

Benny Raphael is a professor in the Department of Civil Engineering at IIT Madras. He did his BTECH and M.S. in Civil Engineering at IIT Madras and obtained his PhD in Civil Engineering from the University of Strathclyde, Glasgow. He worked at Infosys Technologies Ltd. Bangalore, EPFL Switzerland and NUS Singapore, before joining IIT Madras. He co-authored the books *Fundamentals of Computer Aided Engineering* (John Wiley, 2003) and *Engineering Informatics* (John Wiley, 2013) which are used as textbooks in many universities. He won the distinguished R&D award from the Minister for National Development (Government of Singapore) in 2011 for innovations in developing the Zero Energy Building at the BCA Academy, Singapore. He has published about 50 papers in peer reviewed international journals and about 50 papers in the proceedings of international conferences. He has won the best paper award of the American Society of Civil Engineers (ASCE) *Journal of Computing in Civil Engineering* in the years 2008 and 2013. His research interests are in the areas of optimization, computer aided engineering, BIM, automation and control, and energy efficient buildings.

Juan S. Rojas-Quintero is an Industrial Engineer and Civil Engineer and holds an MSc from the University of Los Andes, Colombia. He has experience in financial structuring and management of real estate projects, technical designs coordination, BIM implementation and management, along with risk management in construction companies, and lean construction. His main area of interest is the analysis of strategic decision making and improvement of management processes through integrating innovative tools and practices from civil and industrial engineering.

Rafael Sacks is a professor of Construction Engineering and Management at the Technion - Israel Institute of Technology. He researches building information modelling (BIM) and lean construction. His work has included the development of BIM-enabled lean production control systems and semantic enrichment of BIM models. He is the lead author of the *BIM Handbook*, currently in its third edition, and of *Building Lean, Building BIM: Changing Construction the Tidhar Way*. He has received numerous awards, including the Thomas Fitch Rowland prize of the American Society for Civil Engineering's Construction Institute. Prof. Sacks is the author or co-author of 87 papers in international peer-reviewed academic journals, 105 peer-reviewed conference papers, and numerous research reports.

Mark Swallow is a senior lecturer in construction project management at Sheffield Hallam University. He has over 11 years of experience in construction management and over 6 years of experience teaching in higher education. His teaching and research within the built environment includes building information modelling, health and safety, digital technology, and project management. Within these areas, he has published journals and conference papers with a focus on 4D planning, immersive technologies, and health and safety management.

Reza Taban is a PhD candidate in BIM education at the University of Melbourne. He has experience in teaching and learning as well as architecture and construction industry practice. His practice includes BIM implementation in building projects and teaching and learning in higher education.

Saeed Talebi is a senior lecturer at the Department of Built Environment, Birmingham City University, UK. He has a PhD in the Built Environment and is a Fellow of the Higher Education Academy (FHEA). Bringing practical experience from working on several projects, Saeed's research is focused on helping teams in managing the interfaces between theory and practice. He has been engaged as either principal investigator or co-investigator in more than £200K worth of funded research projects. A recent project, funded by Network Rail, used emerging surveying methods and Artificial Intelligence to automate the inspection of infrastructure projects. He is a board member of the Community of Practitioners-Lean Construction Institute, North West Chapter, and is an active member of the International Group for Lean Construction.

Allen Wan, trained in the UK and Canada, is a chartered HSEQ professional with over 25 years of experience, including in chemical and commercial testing, environmental management and auditing, occupational safety and health compliance, and teaching in Hong Kong. Dr Wan's current role is an HSEQ manager for a specialized contractor in Hong Kong and Macau, with research interests in building information modelling for site safety, life-cycle assessment, and green building.

Song Wu is an internationally renowned leading researcher in digital construction with more than 20 years of experience in the UK higher education sector. He has worked in many leadership roles including programme leader, departmental lead, director of teaching and learning, and associate dean of teaching and learning. He is also a Fellow of the Higher Education Academy, a Fellow of the Chartered Institute of Building, and a Chartered Member of the Chartered Management Institute.

Wei Wu, Ph.D., LEED AP, GGP, CM-BIM, A.M. ASCE, is an associate professor and current department chair of the Department of Construction Management in Lyles College of Engineering at California State University, Fresno. He received his Bachelor of Engineering in Built Environment and Equipment Engineering from Hunan University in China in 2004, Master of Science in Environmental Change and Management from the University of Oxford in the UK in 2005, and Doctor of Philosophy in Design, Construction and Planning from the University of Florida in 2010. Dr Wu teaches construction graphics, design-build project delivery, building information modelling (BIM) for construction management, and construction management capstone. He was the recipient of the 2018 Teaching Excellence Award of the Associated Schools of Construction (Region 7), and the 2019 Provost Award in Innovation of Fresno State. Dr Wu's research focuses on BIM, green building, sustainable construction, cyberlearning and educational technology, construction workforce development, and engineering education. He has published more than 50 articles and conference proceedings in these areas. Dr Wu's research has been funded by regional and federal agencies, including a National Science Foundation (NSF) grant on investigating Mixed Reality (MR) for career-specific competency cultivation among construction management and engineering students.

Sambo Zulu is a reader in Quantity Surveying and Construction Management at Leeds Beckett University. He has over 20 years of teaching experience in higher education. His research interests span many areas, including sustainable infrastructure, organizational and management issues, strategic project management, information and communications technology, building information modelling, leadership, and built environment education. He has published journal and conference papers related to these areas, including aspects linked to the impact of technological advances on organizational and management practices in the construction industry.

Introduction

Building information modelling (BIM) is at the forefront of digitalization in the architecture, engineering, and construction (AEC) industry. The BIM market is projected to grow at a compound annual growth rate (CAGR) of 14.5%, from USD 4.5 billion to USD 8.8 billion, during the forecast period of 2020–2025. BIM's market growth has resulted in new work practices, with the roles and responsibilities related to BIM gradually becoming established professional positions within the AEC industry context. Leaving the question of BIM benefits aside, this development presents the industry with challenges on multiple fronts. Chief among them, in seeking to increase BIM adoption, is the need for companies to recruit employees with the requisite skills to work on BIM-enabled projects, hence the ever-present and always vexing demand for "BIM-savvy professionals". In the short term, construction companies can resort to outsourcing BIM tasks to fulfil their immediate needs. From a long-term, strategic perspective, the most viable solution to the lack of professionals with BIM capabilities is to provide an ongoing pipeline of BIM-ready graduates, most of whom would be from universities. Industry sources recommend BIM education as a foundational activity, a critical need for both industry and academia, and a priority due to the apparent skill shortage. Therefore, BIM training is placed among the top three priority areas of investment by the AEC industry.

Despite significant advancements in the development of BIM education, what universities and training institutions offer remains restricted. Evidence shows that existing curricula, if any, remain in a fledgling state. The available programmes vary significantly in quality and content across countries, universities, and disciplines. The variety in standards for BIM pedagogical delivery and assessment methods across institutions could lead to different perceptions among graduates in terms of their learning and behaviours when in actual BIM practice. It is questionable whether intended learning outcomes are aligned with the needs of the industry. Thus, even the best BIM programmes might fail to provide BIM-ready graduates. Given this scenario, BIM practitioners, educators, and service users, as a community, need to address the disparity in BIM education by converging to present consistent content. Moreover, despite the importance of BIM education in universities, educators – particularly in developing countries – too often take on this responsibility with little or no resources and with scarcely any guidance available to avoid things going wrong in their programmes.

With the above scenarios, producing a book devoted to BIM training and education is deemed timely and imperative. It is an essential step towards converging disparate efforts and assisting educators and BIM trainers to incorporate BIM into their curricula across all universities and training institutions. We set out to write a book to guide the practical and theoretical considerations of effective BIM education and training. Thus, we have focused on the key pedagogical foundations related to the development and delivery of BIM education programmes. These are addressed in our range of chapters for educators that clarify expectations of BIM education programmes in the current market and discuss various key dimensions of designing BIM programmes. Throughout these chapters, we highlight examples of good practice inspired by real-world cases.

Maintaining consistency while embracing diversity in what is offered to students at various universities and training institutions has been a strong motivation for writing this book. Our chapters for students focus on the main topics related to BIM in the current market, ranging from the elementary and basic threshold concepts of BIM to advanced discussions on areas hitherto overlooked in the BIM education domain. These chapters do not claim to be exhaustive, but are an essential starting point for students and educators. They provide the type of knowledge that students need to become BIM-ready for their future careers. We recognize that BIM educators face many competing demands in running their BIM programmes. Hence, the chapters are not tightly interdependent; they can be read out of sequence, focusing on special topics and programmes, or in their entirety for a more holistic BIM education.

Throughout the book, we refer to *students* and *universities*. Of course, many training institutions that are not universities offer BIM training in the AEC industry, and many studying BIM are trainees and not students. This book is also designed to appeal to other training institutions and to trainees.

We wish to thank all chapter contributors for their help, advice, and criticism. We are grateful that they obligingly revised their chapters several times to minimize the overlap between various chapters and with existing books and publications in the market.

Section 1

For students and trainees

Section 1–1

Foundations and threshold concepts

1 Foundational concepts for BIM

Rafael Sacks and Ergo Pikas

Introduction

BIM (building information modelling) is many things to many people, but the one thing around which there is consensus is that BIM represents a paradigm change for the construction industry. Since civilizations began building things that required the cooperation of more than one individual, people have struggled to conceptualize the structures they were building and to communicate their ideas with other people. Design and construction, which are both individual and collaborative activities, depend entirely on people forming mental models of structures and communicating those models with their collaborators. This began with simple descriptions in language and rudimentary sketches in the sand, and developed into drawings on various forms of papyrus, parchment, stone, and paper. Some of the earliest drawings can be found in Vitruvius's *De architectura* (On Architecture), a ten-volume handbook for Roman architects.

Communicating mental models of building structures with anything more than speech essentially began with the consolidation of Euclidean geometry (Booth, 1996), some 2,300 years ago, and did not improve much until the invention of *descriptive geometry* in the 1760s. In 1799, Gaspard Monge published a book defining a precise method for describing physical objects with three dimensions using arrays of 2D drawings (Monge, 1799). The method, with parallel projection and fixed orthogonal views, was used by generations of architects and engineers and remains deeply rooted in the industry today.

In the late twentieth century, descriptive geometry underpinned the development of computer-aided design and drafting (CAD) software for the construction industry. The only difference was that now, instead of using ink on vellum or mylar films, architects and engineers could draw the same 2D views using a computer. While CAD had numerous advantages, such as automating the menial tasks of drawing repetitive parts, automatic accurate dimensioning, and electronic storage and file transfer, it is no different in concept from drawing with pen and paper.

BIM, then, is the next step in the technological evolution of conceptualization and communication of mental models of buildings and other structures. It is quite different in concept and in practice from its predecessor, descriptive geometry and its 2D drawings. It is a radical change in *concept* because the primary depiction is multi-dimensional rather than 2D and because models are composed of objects that represent the conceptual parts of buildings, rather than compositions of abstract symbols that require interpretation according to convention. It is a radical change in *practice* because it offers the industry the first practical means to build and test prototypes before constructing buildings. Figure 1.1 provides an overview of the process in which we model a new building digitally (model the product), simulate its

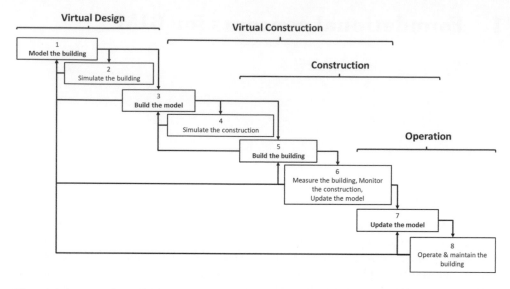

Figure 1.1 An overview of a comprehensive BIM process, encompassing virtual design, virtual construction, physical construction, and operation of a building using BIM.

functions (to test its performance), digitally simulate construction (model the process), build the building and monitor and control the construction process, update the model, and subsequently operate and maintain the building, all using BIM. We explain these ideas in the next section of the chapter, "What is BIM?"

Students who learn the foundational concepts of BIM at the start of their professional training will enjoy an advantage denied their predecessors – the freedom to accept BIM as the standard way of doing things, rather than struggling with it as a disruptive innovation in their professional lives. They will be not be restricted by the need to represent a 3D world in 2D (and indeed, they may explore design and communication with more dimensions, such as time, cost, etc.). This is the subject of the second section, titled "How do we work with BIM?"

The third section outlines some of the practical benefits of BIM from the perspectives of various stakeholders in building and construction processes. Finally, the chapter provides a glossary of the key terms that all students and practitioners must become familiar with; jargon remains an inescapable hallmark of any true profession.[1]

After studying this chapter, you should be able to:

- explain what BIM is, covering both process and technology aspects;
- list the principles for working with digital prototypes;
- explain the importance of simulation for evaluating building performance before construction;
- comprehend BIM jargon.

What is BIM?

Building information modelling, or BIM, is a paradigm shift in designing, engineering, constructing, and managing buildings and other infrastructure (Sacks et al., 2018a). It is fundamentally different to the paradigm that has dominated the architecture/engineering/construction (AEC) industry since the Renaissance in three specific ways. BIM platforms:

a)   use multi-dimensional (3D, 4D, and so on) representations of buildings;
b)   model buildings as compositions of digital objects that are faithful to the form, function, and behaviour of the physical building elements they represent;
c)   serve as digital prototypes that enable simulation and analysis of the functional performance of buildings.

With the new capabilities of BIM technology, new design and construction processes have emerged to exploit these capabilities to the full. Thus, BIM encompasses both computer technologies and business processes. Four key characteristics distinguish BIM technologies and BIM processes:

1)   They offer *object-oriented* representation of form, function, and behaviour.
2)   They have the ability to build and test *digital prototypes* of buildings.
3)   They allow for the *integration* of the work of all the people involved in a construction project.
4)   They provide an *information environment* and is an enabler of "construction tech".

Object-oriented representation

In much the same way that buildings themselves are composed of physical elements, such as walls, doors, windows, beams, and columns, BIM models are composed of digital objects. However, the objects in BIM models can also represent abstract concepts that have no material manifestation, such as rooms or systems. Architects, engineers, and construction managers think of buildings in terms of functional systems as well as in terms of physical components, and it is these conceptualizations that give BIM models an additional layer of useful information. They include aggregations of physical elements and volumes bounded by physical elements. All of these, whether physical or abstract, are objects. All BIM modelling software tools use object-oriented programming. This means that each object has associated software modules – called *methods* – that implement the function, the form, and the behaviour of the objects.

Function refers to the design purpose intended for the objects, that is, what the object(s) are expected to do. For example, a door provides access from one space to another; a concrete slab is designed to carry live loads imposed on a floor to the beams, walls, or columns that support it; walls enclose spaces, which are designed to fulfil specific functions as rooms – bedrooms, kitchens, bathrooms, etc. Functions can also be modelled as relationships between objects, such as "object A supports object B", or "object C encloses object D". In general, an object's function is represented explicitly by its class within the BIM system (e.g., a "Wall" object) and/or by alphanumeric values assigned to its object's properties (its attributes).

Form refers to the geometry of physical objects, and it is always represented using solid geometry. There are two ways to model solid geometry – with boundary representation (B-rep) or constructive solid geometry (CSG). In B-rep, sets of faces define volumes with closed manifolds. In CSG, basic parametric volumetric shapes are combined using Boolean operations. CSG solids are parametric and thus easier to edit than B-rep solids. One can generate B-rep forms from CSG formulations automatically, but deriving a CSG from a B-rep is difficult. As such, CSG forms are preferred.

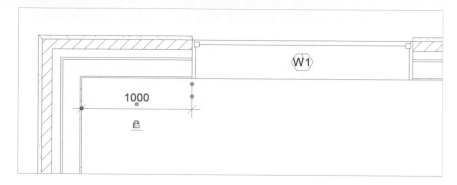

Figure 1.2 A dimensional constraint applied to fix the distance of a window from the corner of a room. The closed padlock symbol below the dimension indicates that if the left-hand wall were moved to the left or right, the window would move accordingly to maintain the 1,000 mm distance.

Behaviour refers to the way in which objects are created and adapt parametrically to their context. For example, a window can only be placed in a wall; if the wall is made thicker, the window frame must adapt automatically; if the wall is deleted, the window must "delete itself" too. Behaviour is implemented using parametric constraints and relationships. An equivalence relationship of the window frame's thickness parameter to the wall's width parameter allows a software function of the window object to reset the window's geometry when the wall's geometry is changed. These relationships can include constraints that enforce minimum distances between objects, parallel or perpendicular orientations, coincidence of points, etc. Figure 1.2 illustrates an example of a dimensional constraint.

Object-oriented modelling also gives BIM software objects properties of *encapsulation*, *data abstraction*, *polymorphism*, *instantiation*, and *inheritance*. While these terms have specific meanings for software programmers, they also lead to behaviours in BIM systems that users of BIM software should be aware of. *Encapsulation* means that the software functions which implement an object's behaviour (its methods) are intrinsic to that object's class and the object's properties and the details of its methods are hidden from calling functions. Thus, the same external command in a BIM system may cause different behaviours in different objects. *Polymorphism* means that objects may take on different forms in different situations; the same object – a door, say – will be shown in quite different ways in different engineering views. *Instantiation* means that each object in a BIM model belongs to a class of objects, and that the behaviour and properties are specified for the class. *Inheritance* means that all classes inherit both properties and methods (behaviours) from a parent class, in a hierarchy. A result of this is that object classes can be specialized. A generic door may have properties for opening width and height, and these are common for all doors. A more specialized class of door, say a double swing door, will have additional properties for the length of each panel of the door and for their opening directions.

Digital prototypes

The importance of digital prototypes for the construction industry cannot be overstated. Until the advent of digital models of buildings, the only way in which architects and engineers could test the function of their designs was to build physical models of the systems. The high cost of preparing models for testing and the fact that most tests involved testing to destruction meant that this was only done for the most unique architectural and structural designs. The result is that many building details, designed and implemented for the first time in each new building, do not function as intended. Mathematical models of structural behaviour offer opportunities to predict behaviour before construction, but they are approximations, and for any but the smallest structures, they are impractical without computers. As such, digital models of buildings are essential for engineering design, with methods such as Finite Element Analysis and Computational Flow Dynamics designed specifically for computer simulation. Yet until the advent of BIM, models for procurement, construction, and many physical analyses (acoustics, thermal behaviour, etc.) were expensive and thus uncommon.

In general, the digital models of buildings that are needed for a wide range of analyses and simulations can be compiled from BIM models. This means that the cost of testing prototypes is a fraction of what it would cost to build and test physical prototypes. Even one-off building designs can be tested quickly for aesthetics, lighting, acoustics, circulation, handicapped access, fire egress, thermal insulation, air flow, static and dynamic structural performance, and other aspects.

The key benefit of digital prototyping is better design of the construction product and the construction process. One of the simplest analyses available is *clash-checking*, in which the physical components of a building are compared using Boolean intersection operations to identify and locate any physical interferences. Given that physical objects cannot occupy the same space in a building, clashes must be resolved in advance of construction if the losses of productivity that result when such situations crop up during building are to be avoided (the management processes that coordinate the sequences of design and analysis to implement this correctly are discussed in the next section).

Yet clash-checking can only solve the simplest of design failures, ensuring physical fit between different disciplinary design models. To improve the quality of design such that buildings perform their intended functions well throughout their service life requires diligent and repeated application of the simulations and analyses described above. Ensuring that these are done at the right time – before design is finalized – is not a trivial matter, requiring careful and intelligent management of design processes.

Finally, a word of caution – designers' willingness to perform functional analyses is constrained by their cost, a large part of which consists of preparing the model data for the analysis software. In many cases, this process still requires significant effort on the part of engineers. For example, a model for simulation of the energy performance of a building requires significant remodelling of the spaces or volumes of a building, of its thermal envelope, etc. The BIM software industry, including both established vendors and myriad start-up companies, invest significant resources to automate these information handovers. Researchers too continue to try to solve the problem of information transfer (the technical term for the issue is *interoperability*).

Increasingly, there is a realization that achieving seamless automated preparation of information for multiple analyses and simulations will require artificially intelligent software modules that implement a process called *semantic enrichment* (Belsky et al., 2016). Semantic enrichment is a process in which specialized software can supplement a BIM model with the information – objects, properties, and relationships – needed for a given use. For example, an intelligent tool may deduce from the geometry of a beam and a column in a reinforced concrete model that the connection should be analyzed as a fixed structural connection.

Integration and collaboration

In addition to facilitating analyses of various kinds, the prototypical buildings that BIM models represent serve another important function in the design and construction process, that of *boundary objects* that support collaboration among the people involved in a project. Boundary objects are representations of information that are understood by people from different communities, enabling them to engage in meaningful conversations to develop a subject despite having different perspectives (Star and Griesemer, 1989).

In the context of the AEC industry, people from a wide range of quite different professional disciplines need to collaborate closely to design, construct, use, and operate buildings and other facilities. They each understand the building in different ways, even using different language to refer to the same building elements (for example, an architect will refer to *floors* and *walls*; a structural engineer will be concerned with the *slabs* that support the floors, and any walls that do not carry loads will be considered *partitions* rather than structural walls). Until the advent of BIM, sets of drawings were essential for teams of professionals to engage with their designs as communities of practice (Wenger, 1999). Yet drawings were not equally accessible to all. Clients and other stakeholders who were not trained in architecture or engineering were, to a large extent, excluded from the discourse among the design professionals, and were unable to express their needs or opinions until the physical construction provided a more material boundary object, one they could understand.

BIM models provide a common platform for collaboration that is far superior to sets of drawings. Visual aids make the models accessible to people without professional training. BIM software supports the generation of photo-realistic renderings of buildings and other facilities in natural environments, and many of the analysis and simulation tools provide visual depictions of their results in easily interpreted figures and animations. Examples include contour maps showing thermal performance, daylight and shadow simulations illustrated in rendered animations, and fire egress animations produced from discrete event simulations of occupant circulation. Virtual reality tools, including both augmented reality and immersive tools, are perhaps the ultimate technological expression of the demand for easily understood boundary objects in the built environment (Wenger, 1999).

Yet even in their "raw" form as design models in BIM modelling tools, models support much closer integration of design and construction teams than was possible with drawings alone. As a result, new forms of project organization have been developed to exploit the opportunity for collaboration. The term *virtual design and construction* (VDC) was coined to describe a process in which architects, engineers, and construction professionals work together to compile highly detailed prototypes of the facilities they are building. VDC is the process of collaborative prototyping of products and processes in construction. When executed well, teams using VDC can significantly reduce costs and schedule durations, increase productivity, and reduce waste when compared with traditional methods. *Integrated project delivery* (IPD) is a method for construction procurement in which owners, designers, and contractors form alliances for projects in which they try to align their commercial interests in such a way that creates a supportive business context or very close collaboration in achieving project goals. BIM is an essential tool for IPD projects, because models reduce the risk that can arise in construction projects from the uncertainty in design and the lack of common understanding of the scope of a project (Whyte, 2002).

Information environment

BIM models form the backbone of the information generated and stored in construction projects. Information may be stored with the objects in models, as geometry or properties of the objects, or stored independently as data that is associated with the objects in models by reference. References commonly use either a *globally unique ID* (GUID) or other fields to associate data records with one or more BIM objects. In general terms, we refer to the broad set of software and data in a project as the *BIM environment* of the project.

A BIM environment may be implemented as a set of BIM tools and platforms with the information itself stored in files in a *common data environment* (CDE) (ISO, 2018), or the environment may be provided through a *BIM Server*, which is an online cloud-based service that delivers both BIM objects and software tools to manipulate them. Cloud services offer multiple advantages in terms of data storage and backup, but their main advantage is in facilitating the work of teams. In theory, designers could collaborate on models concurrently, because their work could be coordinated closely. To achieve such a workbench requires solving the technical challenges posed by simultaneous access for multiple users to edit objects, with the obvious conflicts that arise where designers may override the work of others. This requires not only procedures for locking and releasing objects, but also formal work procedures for managing the sequence of design interventions and users' permissions. However, collaboration among building designers is not simply a matter of tools; it is a complex socio-technical process in which designers use discipline-specific tools and procedures that cannot be done concurrently, as will be explained in the next section, "How do we work with BIM?"

As the information source for construction projects, BIM environments are key to supporting multiple digital functions that go beyond traditional design. These include automated checking of designs for conformance to building codes and regulations; automated issue of construction permits; support for off-site fabrication, design for manufacturing, and modular construction; big data applications for decision-making and setting policy in the built environment; applications of artificial intelligence (AI) for design, such as generative design and design optimization; increased automation and robotics on construction sites; and new models of planning and control that use *digital twins* (a digital twin is a compilation of information, collected automatically through a variety of technologies on the construction site, that represents the current state of construction. Digital twins are associated with BIM models of projects and their objects are directly comparable) (Sacks et al., 2020).

How do we work with BIM?

Since the start of the industrial revolution, technology has been applied to increasingly automate and aid material processing activities. For example, nowadays, it would be unimaginable to build roads without mechanized earthmoving equipment. The second half of the twentieth century brought significant advances in computing and communication technologies, which increasingly automate information processing activities (Björk, 1999).

Many problems in construction are caused by poor design or construction management processes, that is, by poorly performed information processing activities. Yet as customers demand buildings with increasingly refined functional performance and with more complex geometrical forms than before, which require broader processes and larger organizations, so have information processing activities become essential in delivering construction projects. The need to manage and work with the resulting scope of information has made BIM indispensable in the delivery of major construction projects.

BIM is used throughout the entire life cycle of built assets, including the delivery and operations stages. BIM is not simply a technology add-on to existing project delivery processes, but an entirely new way of working. BIM has been described and explained from many different perspectives, but generally, the term has three different meanings:

1. BIM is a *building information modelling technology*, used to produce, communicate, and analyze building information.
2. BIM processes result in *building models* – digital prototypes of assets.
3. BIM also refers to the *management of building information* throughout the life cycle of a facility.

General context for implementing BIM

Over time, academic and practical institutions have developed hundreds of documents (guidelines, standards, templates, etc.) to govern the use of BIM. As the large number of documents suggests, these developments were highly fragmented and were inspired by many different sources: policies, purposes, needs, and state of the art of processes and technologies. As an outcome, this has not helped to reduce or eliminate the misunderstanding of BIM in the construction industry but has rather exacerbated it.

Fortunately, the construction industry has taken a major step forward within the last five years. The International Organization for Standardization (ISO) has developed a series of BIM standards, which have been or are being adapted and adopted worldwide (for example, by the European Committee for Standardization). The purpose of BIM standards is to establish a common language for the construction industry to describe its assets, services, and processes. Standardization is a prerequisite for developing and implementing information and communication technologies, including BIM in the construction industry. In the context of this chapter, the ISO 19650 – "Organization and digitization of information about buildings and civil engineering works, including building information modelling (BIM)" – is the most relevant. ISO 19650 is situated within a set of broader standards, including standards for asset management (ISO 55000), project management (ISO 21500), classification of building terms (ISO 12006-2), and the Industry Foundation Classes (IFC) (ISO 16739).

ISO 19560 defines the life-cycle concepts and principles for asset and project information management using building information modelling. The assumption is that when building information is properly produced and managed during the delivery and operation phases of assets, the processes deliver benefits for all stakeholders. Specifically, Part 1 of the standard describes general concepts and principles for the information management of facilities; Part 2 specifies requirements for the management of information during the production and delivery; Part 3 specifies information management during the operations stage of built assets; Part 4 defines requirements for exchanging information; and Part 5 lays out principles and requirements for security-minded information management.

Asset and project information management is a whole life-cycle activity and should be initiated by the client. The standard distinguishes between information management and information production and delivery processes. The four main processes in information management, shown in Figure 1.3, include (1) specification of client's information requirements and tendering; (2) detailed planning of information delivery, and mobilization of resources; (3) production and delivery of asset and project information; and (4) review of information deliverables during and at the end of project delivery. These four information management processes are carried out in different stages of project delivery: (1) before and during the tender, (2) during the detailed planning and mobilization of resources, (3) during the production

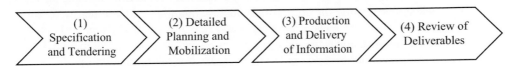

Figure 1.3 Strategic asset and project information management processes (adapted from ISO 19650).

and delivery of asset and project information, and (4) during and after the creation and delivery of asset and project information.

However, note that the up-front activities, including the specification and tender and detailed planning and mobilization, are revisited during all stages of the production and delivery of asset and project information. For example, new observations may lead to changes in the initial specification of information requirements and plans for producing and delivering the asset and project information. Also, this is especially important when new project team members are involved in the production and delivery of asset and project information.

Specifying information requirements and tendering

Whenever a new project is initiated, and before appointing lead designers, engineers, and general contractors, clients should assign individuals in their organizations or from external organizations to take responsibility for information management. This reflects the clients' commitment, a paramount prerequisite to the successful delivery of construction assets and projects and associated information.

Specifying information requirements consists of multiple steps. At first, based on the client's business case, needs, and processes, the organizational information requirements (OIR) should be established. OIR is the prerequisite to developing the asset information requirements (AIR) and project information requirements (PIR). In the asset information requirements, the information production and delivery obligations concerning the needs of the operations stage of an asset are established. In the project information requirements, the information delivery milestones, standards, references for information sources and resources, and production methods and procedures are established. The following aspects need to be kept in mind when creating project information requirements: the project's scope, its purpose, the project's plan of work, the procurement method, key decision points and decisions, and the required qualifications and competencies. These decisions, made by the client relating to the project standard, have a strategic impact on the actual asset and on project information production and delivery.

One such decision concerns interoperability – the ability of BIM applications to exchange BIM data and operate on it. There are four ways to assure interoperability and enable collaboration in the production and delivery of information: (1) use one software vendor's products (proprietary approach); (2) use applications that can exchange files directly (one-to-one approach); (3) use applications that support open industry-wide standards, such as the Industry Foundation Classes (open standards approach); or (4) use model servers for data exchange (common data environment–based approach).

Another critical decision relates to the federation strategy and the breakdown structure for project information containers. In the context of BIM, the traditional construction project standards, specifying the breakdown of information containers and delivery, are often not applicable. Furthermore, clients should define requirements on the level of information (also known as the *level of development, level of detail, level of definition,* etc.) of BIM elements for each information deliverable, stating the appropriate quality, quantity, and granularity of information. A convenient way for a client to oversee and manage the production and delivery of asset and project information is to establish a common project data environment (CDE).

All these decisions should be made before the tender. For the preparation of the tender invitation, clients should also establish requirements on the tender's exchange of information, references for information sources and resources, and response and evaluation criteria. In the tender response and evaluation criteria, requirements on pre-contract BIM execution plans, competencies, and capabilities of individuals and organizations responsible for information management tasks and the production and delivery of project information, mobilization plan, and risk assessment are established. In the tender response, suppliers (e.g., architects, engineers, and contractors) need to state why, what, how, who, where, and when they will meet the client's expectations and requirements.

Detailed planning for information delivery and mobilization of resources

After the appointment of a design and construction team, comprehensive planning for the management, production, communication, and delivery of project and asset information should be carried out by the selected organizations. The results of the information production and delivery planning are compiled in the post-contract BIM execution plans (BEP). If the pre-contract BIM execution plans state the intentions, then the post-contract BIM execution plans define how the team plans to deliver the asset and project information. Typically, the following aspects are covered in the BIM execution plan: purposes of the BIM project, BIM uses, information deliverables, responsible parties and users for BIM deliverables, information exchange requirements, level of information requirements, main BIM workflows, and project infrastructure (software and hardware).

BIM execution plans are prepared based on a master information delivery plan (MIDP) (see Figure 1.4). Master information delivery plans are developed based on task information delivery plans (TIDP). Each execution team (an architectural practice, for example) prepares a TIPD to plan their production and delivery of asset and project information: which BIM models will be delivered when, how they will be delivered, and what information they will contain.

Furthermore, before project work commences, teams must confirm the availability of resources and carry out proper onboarding and training. For example, project teams should be educated on the project scope and requirements and trained to use the CDE. Project information technologies need to be mobilized; software licences must be acquired, and applications installed. Also, before the actual implementation of asset and project delivery plans, the project's information production methods and procedures should be tested, refined if necessary, and communicated to all parties.

Figure 1.4 Asset and project information delivery planning documents and methods (adapted from ISO 19650).

Producing and delivering project information

The production and delivery of asset and project information continue from the early stages of design until the building is commissioned, and after. This is where the actual value of implementing BIM is captured and delivered by the appointed project partners. BIM-based workflows have engendered a discussion in the construction industry about the need to shift the paradigm regarding the timing and scope of information delivery. In particular, people recognize that the move from 2D to BIM should be used to generate higher quality and more granular information at earlier stages of project delivery. This helps at the earlier stages when the potential impact of decisions on project investment needs and costs is greatest, while the cost of making changes is the lowest. This shift was cogently expressed by Patrick MacCleamy, former chairman of HOK Architects, and depicted in the "MacCleamy Curve" (MacCleamy, 2004).

The range of BIM uses is rapidly expanding and extends from fully automated design to construction progress monitoring. Almost all uses require collaboration and sharing of information, and to make this work, all participants in a project must comply with the project and asset information requirements and with the BIM execution plan. Among other things, these preclude producing either more or less information than required. Over-production is one of the seven wastes defined in Lean Thinking (Jones and Womack, 1996), and in BIM it often occurs when modellers select and apply objects with more detail than the design intent required (the level of detail required at each stage is commonly specified in the Level of Development specification in the BEP).

Common data environments support project managers and project task teams to collaborate and coordinate their work. CDEs are different to traditional project information repositories. In a CDE, project information is managed at the building object level rather than at a file level, which allows management and synchronization of parallel development of BIM disciplinary models.

Information review and quality assurance

To assure the usefulness of BIM information, the content of BIM models must meet the syntactic and semantic requirements of any downstream applications. Thus, teams must ensure that the objects in BIM models have the right format and contain the right level of information. All parties involved in information management, production, and delivery are responsible for review and quality assurance.

Quality assurance is a continuous process, not a one-time event. As suggested in the systems theory model of the project life cycle – the "V-model" (Forsberg et al., 2005) – there are four levels of information verification and validation: unit, sub-system, system, and acceptance levels. Before submitting information for review and approval by lead project partners, all individuals should check their BIM elements and models (sub-systems) against the established requirements. In turn, leading partners are responsible for the review and authorization of information models across all sub-disciplines and systems. Finally, BIM models should be reviewed and checked by clients or their representatives. Thus, quality assurance and checking should be carried out by all stakeholders, at all levels of resolution, and throughout all stages of the asset and project information production and delivery.

As BIM uses expand, quality requirements and volumes of building information also increase, and the amount of work required for information quality assurance grows too. For this reason, and for the sake of consistency, there is mounting interest in automated model checking software, and several applications are available. Rule-based verification of BIM objects and models can be used to check for the presence, proper naming, and level of information of BIM objects; for conformance of BIM objects and models to building codes; and for conformance of BIM objects and models to asset and project information requirements.

What value does working with BIM provide?

Why do building owners, designers, builders, and managers invest the effort in education, equipment, and software to work with BIM? BIM provides many advantages over alternative ways of working. Here we list only the most salient and obvious benefits. More detailed explanations can be found in Section 1.7 of the BIM Handbook Sacks et al., 2018).

For *owners and operators* of buildings and other facilities, the major benefits in the delivery phase stem from BIM enabling them to understand the design much better and much earlier in the process than would be the case with traditional methods. When used correctly, BIM's digital prototypes allow mistakes to be made virtually. Designs can be refined and focused to meet the owners' needs with greater confidence, avoiding both over- and under-design. Owners can guide the design and construction teams in meeting the owner's goals better, providing feedback sooner and at a far lower cost than would otherwise be the case.

BIM improves the construction process as a whole, removing much of the waste that is so common in traditional processes – wastes of material, time, space, information, rework, over-production, etc. Over time, this leads to a win–win situation in which suppliers can lower prices and schedule durations while making more money faster – thus reducing cost and schedule while increasing quality for owners.

During the operations phase, information can be drawn from BIM models to populate operations and maintenance management systems, and BIM models will be the basis for designing any future renovations. More significantly, however, good use of the digital prototypes during design should lead to buildings with better energy performance, thus increasing sustainability and reducing operating costs.

For *design professionals*, productivity is the most tangible benefit they experience. BIM reduces the amount of time needed to design, coordinate, communicate, and document designs. BIM platforms automatically generate all the different views of a model that are needed, they ensure internal consistency, renderings and other visualizations are easy to produce, quantities and other information can be extracted easily, and so on. During construction, better quality design reduces the number of queries from the site. Because designers produce information, their employees' time is the most significant cost in their operations. Reducing work hours is a direct boon to their profitability.

Designers also benefit from greater versatility in terms of geometry and materials, as BIM supports the use of parametric design tools and automated design optimizations. Buildings with creative geometries become easier to design, to detail, and to construct, and there has been a proliferation of novel, interesting architecture since BIM became available.

BIM environments support close-knit collaboration among designers of various professions, both among themselves and with construction engineers and product suppliers, which ultimately reduces schedule durations and costs and leads to better quality designs. Designers can spend more time designing and less time on the mundane ancillary tasks that the paradigm of 2D drawings demanded of them.

For *construction contractors and builders*, the ability to prototype construction processes themselves, to the most fine-grained level of detail, before building on site is a major change to their way of producing buildings. The VDC process described earlier enables them to avoid much of the wastes that inevitably come from poorly coordinated design and production control. Better production flow leads directly to reduced material wastes, reduced waste of workers' time, fewer claims, and more. Much detail on these benefits, including quantitative results, can be found in "Building Lean, Building BIM", a study of four construction companies who adopted lean construction and BIM in tandem (Sacks et al., 2018b).

BIM glossary

4D model　A time-dependent view of a BIM model. Objects in a 4D BIM model are associated with the duration and types of activities in construction plans. The type of activity determines what happens to the objects represented in the 4D BIM model before, during, or after the start and completion of an activity. For example, scaffolding in a 4D BIM model is a temporary activity. Thus, elements should occur in the 4D model for a limited time. 4D models can be animated, and construction production workflows can be simulated to identify potential space–time conflicts between different kinds of works.

5D model　A cost-dependent view of a BIM model. The measurable properties of BIM objects in a 5D model are associated with construction budget line items. The use cases for 5D models include conceptual and progressive cost estimation to forecasting the costs and monitoring of actual costs.

As-built BIM model　An as-built BIM model is prepared during the construction stage, reflecting the objects in the building as they were installed or built. It can be used to detect discrepancies between the design model and the constructed facility.

Asset information model (AIM)　A model used by the owner to record the spaces and assets in the facility, also known as the *facility maintenance (FM) BIM model*. It has many use cases, extending from space planning and management to ticket (task) management. Typically, the AIM does not need to include all the information from the design intent model nor the as-built model. It should contain only enough equipment and systems (connectivity) data to enable FM operations.

Asset information requirements (AIR)　Specification of information requirements established by the client concerning the operation of a facility, as required by ISO 19650.

BIM Collaboration Format (BCF)　BCF is an XML file format for exchanging and collaborative resolution of clashes and other issues. It is developed and maintained by buildingSMART. The main benefit of the BCF format is that clashes and issues can be shared across different software applications together with the properties of the model camera views.

BIM environment　A set of BIM platforms and libraries that are interfaced to support multiple information and process pipelines that encompass the various BIM tools, platforms, servers, and libraries within the project workflows of an organization. A BIM environment is supported by a set of policies and practices that facilitate the management of BIM project data.

BIM execution plan (BEP)　A plan that explains how the information management aspects of a project will be carried out. It is often a contractual agreement among the project team that defines how a BIM project will be executed, how the BIM models will be updated over the life of the project, and what information is to be delivered in the various exchanges. BEP is developed in connection with the asset information requirements (AIR) and project information requirements (PIR) established by the client.

BIM guide　A compilation of best practices in a document to guide individuals, teams, or organizations in BIM topics. Guides typically support BIM users in making decisions regarding BIM implementation in their projects to achieve the target goal. Synonyms include *BIM manual*, *BIM handbook*, *BIM protocol*, *BIM guidelines*, and *BIM project specifications*.

BIM maturity model   A tool for benchmarking the level of BIM implementation of a project, an organization, or a region. BIM maturity represents the level of capability in BIM implementation.

BIM model checking or model quality checking   Quality assurance for BIM models and for their contents. Rule-based verification of BIM objects and models can be used, for example, to check the conformance of BIM objects to the level of information requirements.

BIM model integration tools   Software applications that permit users to merge multiple BIM models to form a federated model and to perform different kinds of checks and analyses on the contents. Some tools also provide useful functions for construction management tasks, such as functions for quantity take-off or 4D planning.

Building objects   Digital representations of parts of buildings with associated properties and property values. Objects are faithful to the form, function, and behaviour of the building elements they represent. They may be physical or abstract (such as spaces or systems). Building objects can be aggregated into more complex objects, such as "assemblies" and "systems", which are sets of elements that exist in some relation (spatial, temporal, functional, or any other relationship) to each other.

BIM platform   A BIM model authoring application that generates information for multiple uses and incorporates various tools directly or through interfaces with varying levels of integration. BIM platforms automatically generate all the different views of a model that are needed; they ensure internal consistency; they can generate renderings and other visualizations; quantities and other information can be extracted; and so on.

BIM process   A process for managing, producing, analyzing, and communicating asset and project information using BIM applications for design, analysis, fabrication detailing, cost estimation, scheduling, or other uses.

BIM standard   BIM standards establish a common language, a detailed description (prescriptive or performance-based) of assets, services, and processes, functioning as references to be measured against. Standardization is a prerequisite for coherent and predictable development and implementation of BIM in the construction industry.

BIM tool   A software application for BIM. Also known as a *BIM application*, a BIM system is used for building information modelling. BIM tools can be further qualified to denote specific application areas (e.g., architecture, structural engineering, construction management) or functions (e.g., model authoring, simulation, and analysis).

Boolean operations   A set of operations, named after George Boole, used in computer graphics to operate on geometric shapes and solids. Boolean operations, which include *union*, *subtract*, and *intersect*, are fundamental in advanced 3D modelling.

B-rep (boundary representation)   A method used in solid modelling and computer-aided design to represent 3D geometry. A solid element is described as a collection of connected surface elements creating a closed manifold envelope.

Building data model or building product data model   A data schema suitable for representing a building, including information about building parts, systems, users, energy loads, or processes. A building data model may be used for XML-based web data exchange, or to develop a database schema for a repository. IFC is the most well-known example of a building data model.

Building information modelling (BIM)   As a verb or an adjective, this phrase refers to tools, processes, and technologies used to produce, communicate, and analyze building information. Building information modelling processes result in building models, digital prototypes of assets.

Building model or building object model A digital representation of a particular building. The digital building model is made of digital building objects.

Clash–checking A process in which models are coordinated, with the aim of identifying clashes between the objects of different design and engineering disciplines; checking that objects in BIM models do not occupy the same space (hard clash) or are too near to each other (soft clash).

Common data environment (CDE) A common, shared source of information for any given project or asset, for collecting, managing, and disseminating information through a managed process. Common data environments support project managers and project task teams to collaborate and coordinate their work.

Construction BIM model The development of construction BIM models is led by the general contractor and subcontractors or prepared by them. Typically, construction BIM models are modelled before construction operations begin on site and in sufficient detail to permit pre-construction activities: planning, procurement, fabrication/logistics, installation, monitoring, etc.

Construction operations building information exchange (COBie) In the National BIM Standard, United States Version 3, COBie denotes the format for exchanging information on building assets such as equipment, products, materials, and spaces.

CSG (constructive solid geometry*)* In constructive solid geometry, complex solids are created through Boolean operations applied to basic shapes or solids. Complex objects are made by combining primitive ones. This is a core capability of parametric modelling.

Design BIM model or design intent model The BIM model(s) generated by designers (architects and engineers) to represent and to express a building's function, behaviour, and form for evaluation in building visualization and performance applications (energy simulation and analysis, conceptual cost estimation, etc.).

Exchange information requirements (EIR) A defined set of information units that need to be exchanged to support a particular business requirement at a specific phase of the process(s) or stage(s) (Source: ISO 29481-1).

Federated model A composite information model made of separate information containers. The independent information containers used during the federation typically come from different task teams (Source: ISO 19650-1).

IFC (Industry Foundation Classes) An international neutral and public standard specification (schema) for representing building information. It uses ISO-STEP technology and libraries. IFC is a non-proprietary "BIM file format" developed and maintained by buildingSMART. All major BIM applications support the import and export of IFC files (Source: ISO 16739).

Interoperability BIM interoperability is the ability of BIM applications to exchange BIM data and operate on it. There are four approaches to interoperability: (1) use one software vendor's products (proprietary approach); (2) use applications that can exchange files directly (one-to-one approach); (3) use applications that support open industry-wide standards, such as the Industry Foundation Classes (open standards approach); or (4) use model servers for data exchange (common data environment based approach).

LOD (level of definition, level of detail, level of development) A set of definitions in a BIM execution plan that specifies the detail, the intent, and the timing of delivery of information required in BIM models, which guide the production and delivery of model information during design and construction processes. The required level of 3D modelling is determined based on BIM goals and model uses.

Master information delivery plan (MIDP)   An MIDP describes how individual project teams execute and deliver information to meet the conditions established in a BIM execution plan. The post-contract BIM execution plan consists of master information delivery plans, which, in turn, are developed based on task information delivery plans (TIDP).

Model View Definition (MVD)   A subset of the IFC schema, defined to satisfy one or many exchange requirements of the AEC industry. For example, MVDs have been established for design coordination, energy analysis, building operations, and other similar exchanges. The method is used by buildingSMART and is defined in the Information Delivery Manual, IDM (Source: ISO 29481).

Object class or object family   Object classes or object families are used in the parametric modelling of buildings. They represent the information structures that define object instances in BIM platforms and tools. Examples include object classes for columns, beams, footings, and so forth, in structural engineering BIM tools. BIM software objects have properties of encapsulation, data abstraction, polymorphism, instantiation, and inheritance, which define how instances of a class are structured, how they are edited, and how they behave when their context changes.

Object–based parametric modelling   Object-based parametric modelling technology is used in BIM applications to define individual objects whose shape and other properties can be controlled through pre-defined object parameters. The same principles apply to assemblies of objects, systems, etc., allowing the models to be controlled parametrically.

Organizational information requirements (OIR)   Specification of information requirements at the level of an organization's business information systems. The contents depend on an organization's business case, workflows, and processes. An OIR is a prerequisite to developing asset information requirements (AIR) and project information requirements (PIR).

Project information requirements (PIR)   The information delivery milestones, standards, references for information sources and resources, and production methods and procedures for a construction project. The following are considered when creating project information requirements: the project's scope, purpose, plan of work, procurement method, key decision points and decisions, and required qualifications and competencies.

Semantic enrichment   Semantic enrichment is a process in which specialized software can supplement a BIM model with the information – objects, properties, and relationships – needed for a given use. For example, an intelligent tool may deduce from the geometry of a beam and a column in a reinforced concrete model that the connection should be analyzed as a fixed structural connection.

Solid modelling   Solid modelling is a method for the digital representation of three-dimensional solids. Solid modelling uses include the creation, visualization, animation, exchange, analysis, and annotation of digital models of physical objects. It forms the foundation of BIM modelling technologies. B-rep, constructive solid geometry, and feature-based modelling are all forms of solid modelling.

Task information delivery plan (TIDP)   A schedule of information delivery dates for a specific task team. TIPDs are prepared by project teams to plan their production and delivery of asset and project information – which BIM models will be delivered when, how they will be delivered, and what information they will contain.

Virtual design and construction (VDC)   VDC is the practice of virtual prototyping of construction processes, used by designers and builders to test both the product and the construction process virtually and thoroughly before the execution of work on site.

Questions

1. List three key characteristics of BIM software and explain how each characteristic contributes to make working with BIM more effective and efficient than working with 2D drawings.
2. BIM allows architects, engineers, and contractors to build digital prototypes. What are digital prototypes of buildings, how are they used, and in what ways do they improve construction?
3. List and explain three generally recognized meanings of BIM.
4. What are the four main processes and methods for managing the production and delivery of asset and project information?
5. Why is information review and quality assurance of building models important, and who is responsible for it?
6. What is VDC? Which stakeholders in construction projects benefit from VDC? In your opinion, which stakeholders benefit most from VDC?
7. What are the key benefits of BIM for designers?

Note

1   Paraphrasing the original quote by Kingman Brewster, Jr.

References

Belsky, M., Sacks, R., Brilakis, I., 2016. Semantic Enrichment for Building Information Modelling. *Computer-Aided Civil and Infrastructure Engineering* 31, 261–274.

Björk, B.-C., 1999. Information Technology in Construction – Domain Definition and Research Issues. *International Journal of Computer Integrated Design and Construction* 1, 1–16.

Booth, D.W., 1996. Mathematics as a Design Tool: The Case of Architecture Reconsidered. *Design Issues* 12, 77–87.

Forsberg, K., Mooz, H., Cotterman, H., 2005. *Visualizing Project Management: Models and Frameworks for Mastering Complex Systems*, Third ed. Wiley, Hoboken, NJ.

International Organisation for Standardization (ISO), 2018. ISO 19650-1, *Organization and Digitization of Information about Buildings and Civil Engineering Works, Including Building Information Modelling (BIM) – Information Management Using Building Information Modelling – Part 1: Concepts and Principles*. International Organisation for Standardization, Geneva, Switzerland.

Jones, D.T., Womack, J.P., 1996. *Lean Thinking: Banish Waste and Create Wealth in Your Corporation*. Productivity Press, New York.

MacLeamy, P., 2004. WP-1202, *Collaboration, Integrated Information and the Project Lifecycle in Building Design, Construction and Operation*. The Construction Users Roundtable, Cincinnati, OH

Monge, G., 1799. *Géométrie descriptive: Leçons données aux Écoles Normales*, l'an 3 de la Republique. Imprimeur du Corps législatif et de l'Institut national, Paris, France.

Sacks, R., Brilakis, I., Pikas, E., Girolami, M., 2020. Digital Twin Construction. *Data-Centric Engineering* 1, 1–26. DOI: 10.1017/dce.2020.16.

Sacks, R., Eastman, C., Lee, G., Teicholz, P., 2018a. *BIM Handbook: A Guide to Building Information Modeling for Owners, Designers, Engineers, Contractors, and Facility Managers*, Third ed. Wiley, Hoboken, NJ.

Sacks, R., Korb, S., Barak, R., 2018b. *Building Lean, Building BIM: Improving Construction the Tidhar Way*. Routledge, New York.

Star, S.L., Griesemer, J.R., 1989. Institutional Ecology, Translations' and Boundary Objects: Amateurs and Professionals in Berkeley's Museum of Vertebrate Zoology, 1907–1939. *Social Studies of Science* 19, 387–420.

Wenger, E., 1999. *Communities of Practice: Learning, Meaning, and Identity*. Cambridge University Press, Cambridge, UK.

Whyte, J., 2002. *Virtual Reality and the Built Environment*, First. ed. Architectural Press, Oxford, UK.

2 BIM technologies, tools, and skills

Hamid Abdirad and Carrie Sturts Dossick

Introduction

Building information modelling (BIM) is a common phrase that swirls around the architecture, engineering, and construction (AEC) industry. At the core of BIM implementation processes are the technology features that enable or constrain information development and exchange in projects (Holzer, 2016). Understanding these features, however, is challenging for industry newcomers because of the sprawling definitions, developing technologies, and heterogeneous processes that carry the label *BIM*. With the variations in roles and responsibilities in the industry and the surge of BIM software applications (hereinafter also referred to as *BIM tools*), each practitioner may experience working with a different facet of BIM technologies. Thus, understanding technology features that signify similarities and differences between BIM tools is a highly desired learning outcome for BIM teaching and learning (Abdirad & Dossick, 2016) and the focus of this chapter.

Many technology features that now drive BIM tools were in fact envisioned decades ago. Examples include the automated visualization of three-dimensional (3D) models based on building specifications and using pre-defined shapes and data templates for building elements to model repetitive components efficiently (Eastman et al., 1974; Engelbart, 1962). Since the 1980s, software developers have materialized, commercialized, and advanced these features through a multitude of proprietary tools under different labels such as computer-aided design (CAD), building product modelling (BPM), and building information modelling (van Nederveen & Tolman, 1992). In fact, the proliferation of the term *BIM* in the industry occurred in the 2000s after the trend of investment in desktop computers in the 1990s. With this history of incremental and fragmented advances, BIM software developers have not evenly incorporated technological features into BIM tools, nor have AEC firms implemented BIM tools with a similar point of view, pace, and rigour. Accordingly, this chapter gives an overview of noteworthy technological features that help educators and learners dissect BIM tools and their technology features. These features include geometric representation, parametric modelling, object-based modelling, and reality capture. This chapter will clarify that the ever-expanding list of BIM tools in the market support or limit these technological features in different ways. Consequently, the current practice in teaching, learning, and implementing BIM deems it necessary to assemble custom *BIM toolkits* (Holzer, 2016), whereby BIM tools with complementary features are coupled together to support developing coherent and coordinated building information models.

With the uptake of BIM technologies and proliferation of BIM tools in the industry, there is a growing demand for upskilling professionals and training industry newcomers with skill sets that address BIM technology implementation. This chapter concludes by discussing some core skills and competencies for BIM technology implementation, including technology integration and management, content development and workflow standardization, software customization and programming, and tool application skills. This chapter serves as a general guide to structuring introductory training programmes, courses, or self-learning paths for BIM implementation.

BIM technologies

One of the key takeaways from this chapter for educators and learners is that there is no binary divide or single overriding technical feature that determines whether a software application supports BIM. BIM training must clarify that BIM technologies encompass a multitude of features that fall along a continuum; each tool may support or constrain these features based on the modelling goals and computational performance. Thus, many tools currently used on computers or mobile devices may fall under the large umbrella of BIM despite their different technology features. The following sections overview key features that, in varying degrees, have driven BIM tools over time.

Geometric representation

BIM tools have inherited computational geometric representation methods from developments in CAD. These methods, and by extension the BIM tools that adopt them, can vary significantly in their underlying geometric definition for 2D and 3D shapes, their precision, and the manipulability of geometries. Therefore, it is necessary that educators and learners understand whether geometric representation methods in a tool can satisfy heterogeneous modelling needs for visualization (e.g., 2D and 3D representation), static analysis and simulation (e.g., checking collision between geometries), dynamic analysis (e.g., deformation of geometries), and fabrication of geometries (Shapiro, 2002).

Geometric types, primitives, and arbitrary shapes

Geometric primitives, in the context of CAD and BIM, are pre-defined types and shapes of geometries that a tool can process. *Geometry types* (or topological entities) may include points (or vertices and nodes), curves (or edges), surfaces (or faces), and solids. In each tool. *primitive geometric shapes* (or forms) include only a subset of common analytic shapes and topologies (e.g., straight line and open conic curves; rectangle and closed conic curves to bound surfaces; cube, cylinder, or pyramid for solid forms). The key consideration here is that BIM tools can vary in their support of different geometry types and processing geometric primitives (Johnson, 2016). For example, depending on the representation methods and user access, a BIM tool may or may not support breaking down a higher-level geometry like a solid cube to retrieve and work with lower-level geometries like its surfaces, boundary edges, and vertices.

Geometric primitives may be used as a basis for creating custom and more complex geometric compositions often referred to as *arbitrary shapes* (e.g., drawing five lines to create an irregular pentagon). In more advanced geometric representation methods, arbitrary shapes can be created via *free-form modelling*, especially when the limitations in primitives cannot effectively control, smoothen, and approximate complex curves, surfaces, and solids. The following sections briefly overview geometric representation methods that address modelling primitive and arbitrary curves, surfaces, and solids. It is recommended that educators and learners identify the capabilities and limitations of these representation methods as they explore the functionality of BIM tools.

Curve and wireframe modelling

While straight line, analytic polygons, and conic curves (e.g., arcs, circles, ellipses) provide the majority of primitives for curve modelling, the need for smooth free-form curves led to the development of parametric curves. In parametric curves, instead of defining a curve through points on it, control points near the curve define its path and curvature for smooth representation. The variations in parametric curve representation methods include Bezier, B-spline (Basis splines), and NURBS (Non-Uniform Rational B-splines). NURBS is the most generalized representation that can represent both primitive and complex free-form curves with high flexibility, closest precision, and an optimum number of segments and control points (Figure 2.1). With these representation methods, free-form curves can be accurately documented and replicated given their underlying parameters and control points. However, BIM tools often vary in their adoption of parametric curves and, therefore, in their ability to exchange, visualize, approximate, and manipulate free-form curves (e.g., a NURBS curve created in a tool may be converted to polyline when imported into another tool; Figure 2.1).

Surface modelling

Surface modelling often involves boundary representation (B-Rep) schemes that define surface limits either through specific topologies (e.g., primitive surfaces like a rectangle) or through lower-level geometry types such as custom polygons, curves, and vertices. In its basic applications (level 1; Figure 2.2), B-Rep supports modelling surfaces solely through planar faces bounded by polygons (a polygon face, a polysurface of connected faces, or a tessellated surface as in triangular or quadrangular meshes). Intermediate B-Rep applications (level 2; Figure 2.2) provide further support to model planar or non-planar surfaces through analytic shapes and topologies that are often defined as primitive shapes or surfaces in modelling tools (e.g., circles, cones). Advanced B-Rep applications (level 3; Figure 2.2) support modelling both analytic and free-form surfaces through parametric curves with high flexibility and

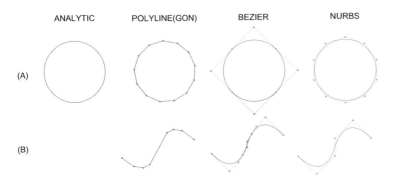

ANALYTIC   POLYLINE(GON)   BEZIER   NURBS

(A)

(B)

Figure 2.1 Different curve representation methods for (a) primitive circle and (b) arbitrary free-form curve (emphasis on curve segments, vertices, and nearby control points).

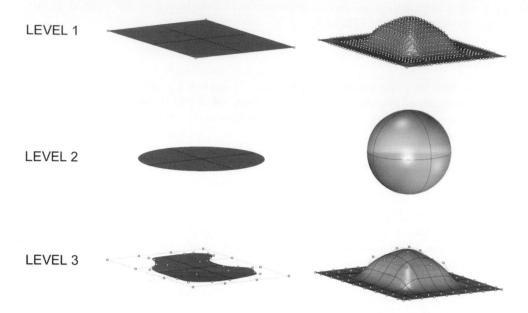

LEVEL 1

LEVEL 2

LEVEL 3

Figure 2.2 Different levels of complexity in B-Rep surface modelling; Level 1: Polygon B-Rep; Level 2: Analytical B-Rep; and Level 3: NURBS B-Rep.

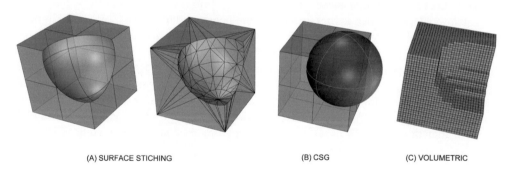

(A) SURFACE STICHING   (B) CSG   (C) VOLUMETRIC

Figure 2.3 Different paradigms for modelling and representing solid forms; (A) B-Rep polysurface and triangular mesh; (B) Boolean difference for cube and sphere primitives; (C) volumetric voxel approximation.

precision (Helpenstein, 2012). In sculptured NURBS surfaces, parametric curves are generated in two directions to create seamless patches formed by nearby control points.

Solid modelling

Solid modelling involves the geometric representation of 3D closed forms. Solid representation schemes generally follow one or more of the three fundamental paradigms (Rossignac, 2007). The first and most common paradigm is surface stitching to defines the boundaries of a solid (as in B-Rep schemes; Figure 2.3-A). In this paradigm, low-level topologies such as the vertices, edges, or faces of a solid can be processed. The second paradigm is constructive solid geometry (CSG; Figure 2.3-B), in which baseline solid primitives (e.g., cubes, spheres) are instantiated and combined in procedural operations to build arbitrary forms. These operations include but are not limited to Boolean operations that process the union, intersection, and difference of solids. In CGS, low-level topologies of a solid (e.g., vertices, edges, or faces) are not necessarily retrievable unless the CGS output is further processed with B-Rep. The third paradigm is volumetric geometry development (Figure 2.3-C), in which 3D cells (e.g., cubes or other polyhedrons) or thickened planar slices are stacked to define forms and approximate complex geometries. Since a volumetric representation offers discrete granular geometries, they may not process faces, edges, and vertices (Johnson, 2016; Rossignac, 2007). The key consideration for BIM educators and learners is that these solid representation methods can significantly vary in their flexibility to model complex forms, precision (e.g., for volume and area calculation), and level of access to lower-level topologies. Therefore, in implementing different BIM tools, the capabilities and limitations of geometric representation methods should be weighed against the intended modelling needs (e.g., visualization, analysis, or fabrication). For further reading on solid representation schemes, the interested reader is referred to Rossignac (2007).

Parametric modelling

In the 1960s, early modelling tools began to support generating 2D and 3D geometries based on a sequence of operations. The major limitation of these applications was that, in each sequence, the resulting geometries became fixed and the history of original geometries, operations, and variables was lost. This was a significant issue for the productivity of design and keeping the knowledge relevant to the changes applied to complex geometries. In the 1970s, parametric modelling tools were developed to address this limitation. With parametric modelling, geometries are instantiated through a series of parametrized steps. In this approach, tools not only provide the resulting geometries but also record the input parameters and constraints that end-users define for each step; in more advanced scenarios, they allow users to simply modify a subset of existing parameters to redefine the resulting geometries in real-time. In other words, geometric shapes and operations can be defined as a function of input parameters and constraints to instantiate or modify geometries (Eastman, 1999; Hoffmann & Joan-Arinyo, 2002).

This section illustrates such a parametric modelling feature with a series of parametrized steps for creating a box and a cylinder (Figure 2.4). In this example, two operations generate two primitive solid geometries based on input parameters. For creating the cylinder, the input parameters include coordinates for the centre point of the base circle, radius, and height. For creating the box, the input parameters include coordinates for the centroid of the box, length, width, and height. After generating the primitive solids in Figure 2.4, a set of Boolean operations are applied to them to extract the geometric difference of the solids from one another, the geometric union of the solids, and the geometric intersection of the solids.

In Figure 2.5, two of the recorded parameters for the cylinder are updated to move the cylinder along the Y axis and change its height; the parametric software automatically regenerates the outputs of Boolean operations. In a non-parametric modelling software, rather than updating just two parameters, all the parameters and operations must be recreated by the user to generate the outputs shown in Figure 2.5. More advanced parametric modelling platforms not only allow these operations, their parameters, and their values to be recorded

Creating Primitive Geometries:

1) Operation Cylinder:
>*Parameters:*
>>*Base Point: (0,0,2)*
>>*Radius: 2*
>>*Height: 8*

2) Operation Box:
>*Parameters:*
>>*Centroid: (0,0,0)*
>>*Length: 6*
>>*Width: 6*
>>*Height: 6*

Boolean Operations on Primitives:

Alternative 1)
Operation Boolean Difference:
>*Parameters:*
>>*First Solid: Output of Operation Box*
>>*Second Solid: Output of Operation Cylinder*

Alternative 2)
Operation Boolean Difference:
>*Parameters:*
>>*First Solid: Output of Operation Cylinder*
>>*Second Solid: Output of Operation Box*

Alternative 3)
Operation Boolean Union:
>*Parameters:*
>>*First Solid: Output of Operation Box*
>>*Second Solid: Output of Operation Cylinder*

Alternative 4)
Operation Boolean Intersection:
>*Parameters:*
>>*First Solid: Output of Operation Box*
>>*Second Solid: Output of Operation Cylinder*

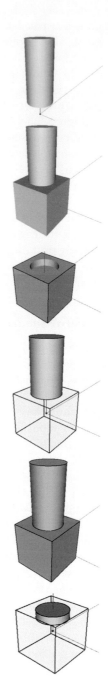

Figure 2.4 Example of the use of parametric modelling for sequential geometric operations.

Updating Primitive Geometries:
1) Operation Cylinder:
  Parameters:
      Base Point*: (0,-3,0)
      Radius: 2
      Height*: Radius + 1

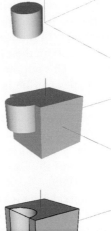

2) Operation Box:
  Parameters:
      Centroid: (0,0,0)
      Length: 6
      Width: 6
      Height: 6

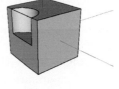

Boolean Operations on Primitives:
Alternative 1)
Operation Boolean Difference:
  Parameters:
      First Solid: Output of Operation Box
      Second Solid: Output of Operation Cylinder

Alternative 2)
Operation Boolean Difference:
  Parameters:
      First Solid: Output of Operation Cylinder
      Second Solid: Output of Operation Box

Alternative 3)
Operation Boolean Union:
  Parameters:
      First Solid: Output of Operation Box
      Second Solid: Output of Operation Cylinder

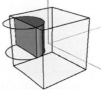

Alternative 4)
Operation Boolean Intersection:
  Parameters:
      First Solid: Output of Operation Box
      Second Solid: Output of Operation Cylinder

* = Updated Parameters

Figure 2.5 Updating the parameters for primitive geometries and regeneration of Boolean operations.

and exposed, but they can also show relational behaviours and support using expressions. For example, in Figure 2.5, the expression for the height parameter of the cylinder calls the value of the radius parameter.

In parametric modelling, different templates of parameters can be defined to drive geometries such that specific instances of a geometry can be created or modified based on a template. In the foregoing example, the box and the cylinder were instantiated and modified based on their pre-defined parametric templates (steps 1 and 2 in Figure 2.4 and Figure 2.5). This approach in parametric modelling is often called *variant modelling*, where similar shapes can be instantiated or modified based on a fixed set of parameters. It is important to note that proprietary software may define primitive geometries with different parametric templates (e.g., requiring eight points for a box; or requiring four points for a base rectangle and a value for height to create a box). Furthermore, adjusting parametric values may be supported graphically through picks and clicks on screen (e.g., clicking on a spot to insert the base point) or by typing in pre-defined fields that prompt user input (Eastman, 1999).

Variant modelling is a precursor to more advanced parametric modelling processes such as constraint-based modelling and feature-based modelling, wherein custom parameters or attributes can be used to generate and control non-template geometries (Hoffmann & Joan-Arinyo, 2002).

When geometric needs for a modelling task are not supported by variants of pre-defined templates, the parametric software needs to process a custom set of parameters and constraints (or rules) as defined by a user. For this reason, parametric modelling tasks that involve custom geometry creation are also referred to as *constraint modelling*. These constraints define parametric values of a geometry, its parts (e.g., curve, surface, solid), or its relationship with its context or other geometries. Some types of constraints are listed as follows (Hoffmann & Joan-Arinyo, 2002; Sacks et al., 2018):

- geometric dimensional relationships (e.g., distance and angle);
- geometric descriptive relationships (e.g., concentricity, perpendicularity, parallelism);
- equational operands and relationships (e.g., math expressions as in Figure 2.5);
- conditional operators (e.g., if/then to change a geometry or its parts based on test results or conditions);
- semantic relationships (e.g., below, on top); and
- topological relationships (e.g., incidence or connectivity).

Parametric constraints are often created, solved, and applied onto a geometry (or related geometries) in sequential order (a series of steps in a behind-the-scene directed graph) to limit the number of alternative solutions that could be generated when multiple unordered constraints are defined (e.g., in Figure 2.5, the change to the cylinder was applied before the Boolean operations).

Such advances in constraint modelling paved the way for incorporating technical design, engineering, and manufacturing information into parametric design tools as attributes that constrain or specify geometric and non-geometric characteristics of modelled components. This advanced application of parametric design is often referred to as *feature modelling* (Hoffmann & Joan-Arinyo, 2002), wherein users associate geometries with knowledge useful for reasoning about their behaviour and characteristics. The geometric features can build on conventional parametric constraints and include features for form (the nominal shape of a geometry), tolerances (possible deviations from a nominal shape), and assembly (grouping, positioning, and mating relations). Non-geometric features address technical, functional, and visual characteristics of modelled elements (e.g., material, performance, usage), and users may utilize them as conditions that drive and change geometric characteristics.

In the following sections, this chapter overviews the development of parametric feature modelling in the BIM ecosystem under the label *object-based parametric modelling*, whereby technical and geometric features of conventional building elements are incorporated to facilitate creating and exchanging building information models. In AEC practice, the need for parametric modelling (i.e., driving and modifying geometries through variant, constraint, and feature modelling) in projects often decrease during the transition from design to construction (Figure 2.6). In the early design and engineering stages, parametric modelling provides practitioners with the flexibility needed for changing design parameters, high-level coordination and adjustments, and generating alternatives (e.g., form-finding, comparing design options, optimizing multi-criteria design and engineering configurations). The need for making substantial changes to a project configuration, and therefore, the need for parametric modelling features, gradually diminishes when the design is fully documented for construction. Accordingly, the current ecosystem of BIM encompasses a variety of software applications that address parametric modelling in different ways for different design and construction purposes. For example, some software applications that are commonly used for design and construction documentation have built-in or add-on features to support parametric modelling, especially for geometric operations. Other BIM software applications do not have parametric modelling functionalities. Some of these applications support importing data or geometries from other applications without having the flexibility or features to change them. Therefore, each software application may consider a different position along the continuum of parametric modelling.

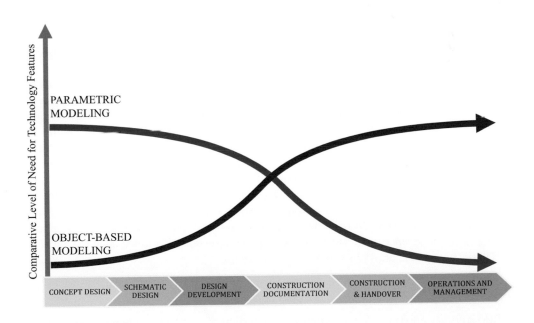

Figure 2.6 The changing need for parametric and object-based BIM features at different project stages.

Object-based modelling

Object-oriented programming: Classes, attributes, and methods

Object-based modelling in BIM is a feature built upon object-oriented programming (OOP) techniques in software development. OOP enables software developers to think about the nature and structure of data rather than just focusing on isolated data objects. With this approach, software developers create master templates, called *classes*, to categorize data as objects and define their behaviours, characteristics, and relations to each other. These objects can be simple "things" like numbers or letters as well as more complex "things" like databases, files, folders, and disk drives on computers. Based on pre-defined templates, many similar data objects of each class can be instantiated, stored, and modified (Ko & Steinfeld, 2018).

For example, for a geometric modelling software, one can define a high-level class named *Geometry* with some sub-classes like *Point, Line, Polyline, Surface*, and *Solid*. Rethinking the example shown in Figure 2.4 in terms of such an object-oriented software, the operations *Cylinder* and *Box* instantiated two specific objects of the class *Solid*. Therefore, the box and the cylinder created by these operations both follow the template of *Solid* as well as the higher-level template of *Geometry* (by inheritance). This means that *Point, Line, Polyline, Surface*, and *Solid* classes are similar in how they inherit the characteristics and behaviours of their parent class *Geometry*; however, they are not entirely similar and need their own specific characteristics and behaviours in their class definitions. For example, objects of *Solid* can have *Volume* as a characteristic unique to solids because *Point, Line, Polyline*, and *Surface* objects cannot have volume. In OOP, this kind of characteristics is called *attributes*. For the *Geometry* class, one can define a common attribute named *colour*. All sub-classes of *Geometry* inherit this attribute; when invoked on a specific geometric object, this attribute can report or change its colour.

In addition to attributes, each class can have specific behavioural operations called *methods*. For example, in the *Geometry* class, one may define the method *RotateAroundZ* that, given a value for angle as the input, rotates any geometry instance around the Z axis in the model. Each subclass can have its specific methods as well; e.g., *CalculateDistanceFromOrigin* for *Point, CalculateCenterPoint* for *Line, CalculateNormalVector* for *Surface*, and *CalculateCentroid* for *Solid*. In sum, classes, as master templates in OOP, determine what behaviours, what characteristics, and what relations and inheritance structures make up objects in software.

OOP is a critical feature for the performance of software products and productivity in software development and maintenance. For example, in Figure 2.4, the pre-defined template for creating a solid box (think of it as a constructor method for class *Solid*) requires only four input parameters, while behind the scenes, the template generates the eight points, 12 lines, and six surfaces needed to create a solid box. Therefore, once defined as a class, a template can be re-used many times to reduce the number of operations and parameters needed for creating similar objects. Consequently, end-users may not be exposed to some of these underlying objects, attributes, or methods that form the backbone of objects in software applications. These examples clarify that parametric modelling, as explained in the previous section, is best achieved by OOP. In fact, advances in OOP and parametric modelling have gone hand-in-hand since the development of interactive geometric modelling applications in the 1960s (Eastman et al., 1974; Ko & Steinfeld, 2018).

Object-based BIM: Categories, templates, types, and instances

In BIM software applications, the fundamental concepts of OOP have influenced not only the underlying structure of behind-the-scene data objects but also the very tangible digital objects with which users interact through their human–computer interface. Object-based BIM relies upon templates for the classes, attributes, and behaviours of heterogeneous objects or concepts that are meaningful to the AEC industry. These objects or concepts include, but are not limited to, building elements (e.g., walls, floors, columns, ducts, air diffusers, pipes); drafting and drawing elements (e.g., sections, plans, grids, dimensions, symbols); construction materials (e.g., concrete, structural steel, finishes); or spatial concepts (e.g., rooms, spaces, building stories, site; Abdirad & Lin, 2015). Although the need to classify these concepts and identify their attributes is not new, object-based BIM sheds new light on both the urgency and limitations of ontology studies on AEC objects and concepts. In fact, since the 1960s, the industry has developed different classification systems to standardize information development and exchange for different purposes like cost estimation and specification development (e.g., classification systems like MasterFormat, UniFormat, OmniClass in the United States; UniClass in the United Kingdom; etc.). However, BIM software applications do not necessarily conform to the industry classification systems in how they define templates for the classes, attributes, and behaviour of AEC objects. Many software applications have their custom classification systems, and therefore, proprietary object-based BIM contents and data structures.

From an end-user's perspective, in object-based BIM, the nomenclature for describing the classes, attributes, and behaviour of building objects is different from conventional nomenclature in OOP. The high-level classes of objects define the functional role elements in buildings; they are often pre-defined in BIM software applications and called *categories*. For example, *Walls, Floors, Roofs, Columns, Windows, Beams,* and *Ducts* are some categories that provide high-level definitions for objects, their attributes, and their behaviour. Under each category, object templates differentiate between a variety of assemblies or products that can be used in each category. These object templates (often referred to as *Families* or simply *Objects*) incorporate geometries (2D or 3D) as well as non-geometric information of the assemblies. For example, under the category *Structural Columns*, different object templates can be created based on the underlying shape and material of assemblies (e.g., *Concrete Rectangular Column, Concrete Round Column, Wide Flange Steel Column*). Where applicable, each object template may have sub-templates called *Types* to accommodate a set of sizes or logical alternatives that can be pre-defined for each assembly. For example, wide flange steel columns can have different sizes in terms of width, depth, and thickness (e.g., W12×40, W14×30) that could be incorporated into one object template. Once created, such object templates and types can be stored in BIM object libraries and re-used to allow for the creation of any number of instances in a project or across different projects (Eastman et al., 2011).

Object instances are specific objects that are created in project models. Although these object instances inherit the attributes and behaviours of categories, object templates, and types, they need to additionally have non-template instance-level attributes to support having multiple instances of a building component (or an assembly) in a project and adjusting their specificities per project conditions. For example, instances of W12×40 columns may be rotated, placed at different grids and building stories, and assigned unique mark numbers for project documentation.

The attributes that are inherently the same for all instances of an object template or type are often called *Type Properties* or *Type Parameters* (e.g., all instances of W12×40 columns have the same value for *FlangeThickness*). The attributes that allow for variability across instances are often called *Instance Properties* or *Instance Parameters* (e.g., *GridLocation* for column instances). Both type and instance properties may be used to embed geometric or non-geometric attributes into objects. Therefore, it is important to understand the intended use of project models to determine what properties should be embedded into them. For example, for walls, one can use *area (or volume)* and *cost per unit* for cost estimation, *material* for design documentation and visualization, *u-value* for energy analysis, *fire-rating* for code compliance checking, and *task ID*, *start date*, *end date* for scheduling. In fact, the ability to embed this sort of data into 3D models facilitated the gradual transition from a focus on 3D modelling to the exploration of information management and exchange across disciplines (Holzer, 2016). BIM educators and learners can refer to Level of Development (LoD) Specification (BIMForum, 2019) to review an in-depth effort to standardize and articulate object-based content and attributes for distinct categories of BIM objects.

The need for object-based modelling in the sequence of projects gradually increases (Figure 2.6). In the early concept design stages, the type, geometry, and attributes of building assemblies or equipment are rarely known. This information is incrementally generated, analysed, revised, and finally fixed and documented in project submittals and models during the subsequent design, construction, hand-over, and facility operation stages. Accordingly, the current BIM ecosystem encompasses a variety of software applications that address object-based modelling in different ways. For example, some geometric modelling software lacks object-based features that differentiate between building objects and their characteristics. Some software applications that are commonly used for design and construction documentation have built-in features to support object-based modelling with add-on libraries of templates for building objects. Some applications only support importing object-based models or data from other applications without having the flexibility or features to change them. Therefore, in the ecosystem of BIM software, each application may offer distinct features along the continuum of object-based modelling.

Object-based parametric modelling

Object-based parametric modelling – originally developed in the 1970s and commercialized in the construction industry in the 1980s – incorporates both parametric modelling and object-based modelling features into software applications (Sacks et al., 2018). To this end, software applications not only support developing object templates based on assembly categories, types, and properties but also offer the built-in flexibility to change and update objects parametrically and automatically without redefining all of their geometric or non-geometric characteristics from scratch. It is only with the recent advances in the computational power and affordability of computers and software applications that object-based parametric modelling features have become mainstream.

With object-based parametric modelling, properties that control geometric or non-geometric behaviours of objects are not necessarily fixed or frozen. Object templates can have properties that support the needed flexibility to change object attributes, refer or relate to other objects in the model, and follow a set of rules to automatically adjust geometries and non-geometric characteristics of objects based on user preferences or changes in related objects. With such flexibilities, when developing a model that has thousands of building elements, users do not need to manually adjust every geometric or non-geometric aspect of objects (Sacks et al., 2018). For example, a wall instance can host opening, door, or window instances. The object-based parametric relationships between the host and hosted elements rule that wall geometries should be cut when these elements are placed into walls, and these elements should move with their host wall when its location changes. As another example, for modelling a floor slab supported by columns, instead of using a specific numeric value for the *Top Elevation* of the slab and columns, the user can assign a specific building level to control the *Top Elevation* of these elements. With this approach, even if changes in design move the level higher or lower (e.g., due to the changing floor-to-floor height in a high-rise building), the slab and columns would automatically follow the assigned level regardless of its specific numeric value for elevation.

Despite the demanding computational requirements, this kind of object-based parametric modelling feature has had significant implications for the management, productivity, and accuracy of model production for design and construction documentation. Accordingly, the value of object-based parametric modelling is highest during design development, when building assemblies, materials, equipment, and their specifications are gradually established. However, some flexibility is important in the later stages of design and early construction as change orders increase the need for the flexibility of models to efficiently incorporate project changes. The following sections illustrate features of object-based parametric modelling software in more detail. First, the project view concept and drawing generation capabilities are explained to illustrate how object-based parametric modelling technology synchronizes information across orthogonal and 3D views and automates the creation of drawing sets. Next, data tables and schedule views are discussed to show how object-based parametric modelling provides software users the ability to acquire, filter, and sort data embedded into models.

Project views and drawing generation capabilities

The advanced features in object-based parametric modelling, which facilitate developing data-rich 3D models, can be extended to support 2D drawing generation as well. This is especially important because the industry's information-exchange norms and conventional contractual requirements still dictate the use of 2D drawings. Furthermore, without object-based parametric modelling, generating 2D drawings is highly cumbersome and error-prone because objects must be manually drafted and coordinated in each and every *representational view* on drawings (e.g., plans, sections, and elevations) (Sacks et al., 2018).

With object-based parametric modelling, the object templates and types can include settings that control how objects should be represented in 2D plans, sections, and elevations (e.g., with scale-aware annotations or symbols); this is an essential feature because, for some objects, extracting 2D views directly from 3D geometries does not satisfy technical symbolic representation needs. For example, Figure 2.7 shows that each of the plan and elevation views of a door object template includes view-based 2D lines for symbolic representation of the swing direction and hinged side (which are not represented in the 3D view).

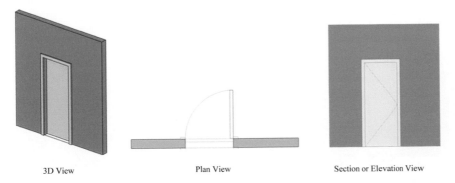

3D View   Plan View   Section or Elevation View

Figure 2.7 An object template for a door controls how the door should be represented in 3D and 2D views.

With this capability, in project models, each object is instantiated only once to facilitate using one model as a single source of truth for all representational needs. The tools support the extraction of plans, sections, elevations, and enlarged detail views from the main model to represent objects based on their pre-defined template drawing standards. This also enables users to directly insert or edit 3D model objects from these orthogonal views. These views are "live" in the sense that they automatically regenerate graphics to represent the latest state of the model, i.e., a change in one view is propagated and reflected across all views in the project. Also, template view settings can be created to make the graphical representation of objects consistent across similar views. These settings can control and filter the visibility of objects categories and object templates, and even override the pre-defined settings nested into object templates (e.g., colour, line style, transparency). For example, the plan views shown in Figure 2.8 represent the same floor level from a structural model of a tower; different template settings for each view control what object categories should be visible and how the visible objects should be represented. This clarifies that users can model 3D model objects only once and then the tools support the representation of these objects in different ways without the need to adjust settings for every single view and every single object. Furthermore, the 2D drafting elements (e.g., annotations, tags, lines) placed on these views can be *View-Specific*, i.e., visible only on the views that they are placed on.

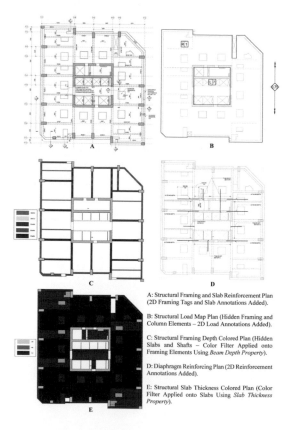

A: Structural Framing and Slab Reinforcement Plan (2D Framing Tags and Slab Annotations Added).

B: Structural Load Map Plan (Hidden Framing and Column Elements – 2D Load Annotations Added).

C: Structural Framing Depth Colored Plan (Hidden Slabs and Shafts – Color Filter Applied onto Framing Elements Using *Beam Depth Property*).

D: Diaphragm Reinforcing Plan (2D Reinforcement Annotations Added).

E: Structural Slab Thickness Colored Plan (Color Filter Applied onto Slabs Using *Slab Thickness Property*).

Figure 2.8 Five view settings that control graphical representation of building objects; the same tower floor shown on all plans (image courtesy of Magnusson Klemencic Associates).

Data tables and schedule views

Object-based parametric modelling facilitates the extraction of data directly from modelled objects into data tables. These tables can be considered non-graphical project views that can tabulate, filter, group, summarize, or itemize data from 2D and 3D modelled objects based on their categories, object templates, types, and relevant data stored as object properties or attributes. This feature enables users to create an unlimited number of schedules for drawing sets (e.g., air handling unit schedule for mechanical drawings, beam schedule for structural drawings) or for heterogeneous BIM-based data-mining workflows (e.g., quantity take-off and cost estimation, model quality control). These data tables are live and interactive as they report real-time data from models and update as model objects change. In addition, their cells and fields can be used to directly change the properties of model objects as the connection between the data table and the model objects is bi-directional.

Reality capture

One of the BIM implementation challenges is to connect and align digital models with the realities of construction job sites. This alignment is important for documenting (a) the existing buildings and sites, and (b) the progress and output of construction processes to model as-built conditions. To carry out these tasks, the industry has turned to reality-capture technologies as more efficient and precise alternatives to conventional manual processes. Two key technologies are increasingly being used in practice to capture visual and geometric characteristics of the physical world: photogrammetry and laser scanning (Son et al., 2015).

Photogrammetry

The origin of photogrammetry dates back to the 1800s, when photography was invented and used to better understand the theories of perspective and mathematics of geometry. Early examples include the application of stereoscopic techniques to imitate humans' perception of depth and distance by pairing two slightly offset photos of a scene with two eyes (viewpoints in the photos are separated by the width of the eye). The widespread use of photogrammetry for industrial sectors did not occur until the mid-1900s, when photogrammetry became the science of measuring distances in photos by using the mathematics of perspective. Today, photogrammetry is used in both long-distance and close-up photography to facilitate measuring and modelling objects of different scales (e.g., large terrains or building interiors or equipment). The main idea is that tools calculate the coordinates of points in three dimensions if any point represented in two or more photos is taken from different positions (photos partially overlap in the objects they represent). Advances in computer vision facilitate the recognition of points across a set of photos using their features (e.g., shadows, textures, colours, focal points). Accordingly, depending on the number and quality of photos taken from a scene, photogrammetry tools create a 3D model of the objects represented in the scene. This is often a surface model that can be imported into BIM tools.

Until very recently, photogrammetry for the construction industry was expensive, and the use of photogrammetry to capture buildings was also focused on historical preservation. With the increase in computing power and the prevalence of smartphone technologies, the use of photogrammetry has seen a resurgence in recent years. With an ageing infrastructure and an increased interest in reusing existing buildings, the industry has explored the use of photogrammetry to create model geometries for existing buildings (Armesto et al., 2009).

Furthermore, with the introduction of drones, photogrammetry has increased in popularity for site surveying and connecting BIM data to reality capture with aerial photography. In recent years, BIM software developers have acquired tools for photogrammetry and have built integration workflows of these tools with BIM tools. These types of tools make photogrammetry technology accessible to BIM users in design and construction processes. Design and construction firms encourage individuals with BIM expertise to learn techniques for photo capture that support photogrammetry analysis as well as techniques for using the resulting photogrammetry in integration with BIM.

Laser scanning

Laser scanning is another type of 3D scanning technology used to capture the shape and appearance of objects. This technology works on the principle of analysing the way the laser reflects off surfaces to create a collection of points known as *point clouds* (Son et al., 2015).

This idea of using reflected light to measure and capture the surface shapes of objects can be traced back to the 1960s. With advances in this area, in the 1980s, scientists began to explore laser technology for scanning applications. Today, the most common laser scanning technique in the construction industry is light detection and ranging (LIDAR), i.e., time of flight technology where a scanner (mounted like a camera) sends out a series of laser beams and records the time it takes each beam to reflect back from a physical point to the scanner in order to measure distance. Current LIDAR scanners also include photography capabilities to map photographs onto the point clouds to create a coloured set of points for visualization. One of the limitations of laser scanning is that the point cloud only captures the objects that are within the direct line of sight of the scanner. If building components are behind one another, only the components in front will be seen in the point cloud. This requires the technician to take multiple scans of a space from different locations to get the full 3D point clouds of objects. These multiple point clouds are then merged together in software to create the larger point cloud of the scan, which represents a whole scene.

Point clouds may be imported directly into BIM tools and used as reference data. Converting point clouds to surface/solid models or reducing the point cloud density is important for their usability. In advanced BIM tools, point clouds are used for object recognition, wherein surface/solid models are generated from point clouds and further processed to extract semantic object data (e.g., recognizing objects like walls and identifying their materials).

Reality-capture applications

Both photogrammetry and laser scanning technologies allow BIM users to combine new models with points, surfaces, images, or solid geometries from reality-capture technologies. The need for reality capture may arise at any stage during a project from design and construction to hand-over and renovation work (Figure 2.9). The application of these technologies in the construction market is, therefore, diverse and can address the analysis of new construction as well as the documentation of existing conditions. Therefore, BIM educators and trainees from different AEC disciplines can often connect reality-capture applications to their core knowledge areas. These applications include but are not limited to

- existing conditions survey (e.g., interiors or exteriors – hard to access tunnels and spaces);
- as-built survey modelling and construction visualization;
- verifying site conditions;

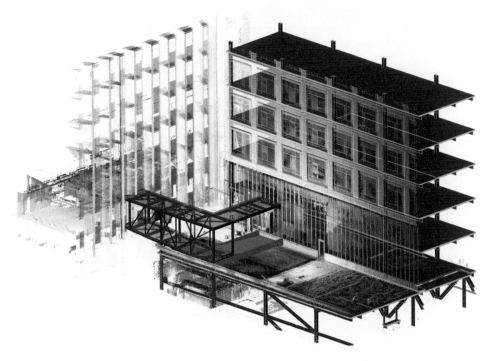

Figure 2.9 Overlaying a point cloud of existing buildings onto a structural BIM for modelling existing conditions in a tenant improvement project (image courtesy of Magnusson Klemencic Associates).

- pre-design engineering for facility upgrades or expansions;
- clash detection (e.g., new building components against existing components);
- construction sequencing, scheduling, and simulations;
- design vs. construction discrepancy analysis;
- construction analysis (e.g., measurements, deformation, and clearance analysis);
- collecting geospatial data and survey mapping.

BIM tools and toolkits

Centralized BIM vs. distributed BIM

The early thought leadership around BIM envisioned having software applications that supported creating and maintaining a centralized model during a project life cycle to satisfy all visualization, documentation, and analysis needs for all disciplines. However, this idea has lost its momentum over time because BIM is used for a variety of purposes, often called *model uses* (also known as *BIM uses*; see Table 2.1), and each model use may require a different set of software functionalities, geometric representation, parametric behaviours, and object-based properties for building elements (Succar et al., 2016).

For example, for structural analysis, building information often includes wireframe and surface geometry of structural elements as well as data for live and dead loads; structural member characteristics (e.g., sizing, material, strength); gravity; seismic and wind design

Table 2.1 A sample list of model uses

Sample model uses (or BIM uses)	
Capturing and representing	
2D documentation	Laser scanning
3D detailing	Photogrammetry
As-constructed representation	Surveying
Generative design	Visual communication
Planning and designing	
Conceptualization	Lift planning
Construction planning	Operations planning
Demolition planning	Selection and specification
Design authoring	Space programming
Disaster planning	Urban planning
Lean process analysis	Value analysis
Simulating and quantifying	
Accessibility analysis	Quantity take-off
Acoustic analysis	Reflectivity analysis
Augmented reality simulation	Risk and hazard assessment
Clash detection	Safety analysis
Code checking & validation	Security analysis
Constructability analysis	Site analysis
Construction operation analysis	Solar analysis
Cost estimation	Spatial analysis
Egress and ingress	Structural analysis
Energy utilization	Sustainability analysis
Finite element analysis	Thermal analysis
Fire and smoke simulation	Virtual reality simulation
Lighting analysis	Life cycle assessment
Wind studies	
Constructing and fabricating	
3D printing	Construction logistics
Module prefabrication	Construction waste management
Casework prefabrication	Mechanical assemblies prefabrication
Concrete pre-casting	Sheet metal forming
Operating and maintaining	
Asset maintenance	Building inspection
Asset procurement	Hand-over and commissioning
Asset tracking	Relocation management
Space management	
Monitoring and controlling	
Building automation	Performance monitoring
Field BIM	Real-time utilization
Structural health monitoring	

Source: Adapted from Succar et al., 2016

criteria; and simulation results. For environmental analysis (e.g., lighting and energy simulation), building information often includes surface modelling of structural and non-structural building envelope or interior elements in addition to data for location, weather, and the physical characteristics of materials used in building elements (e.g., thermal and visual properties). A model created for construction documentation may not be readily used for structural and environmental analysis, or vice versa. Therefore, for creating a centralized model that supports all model uses, each building object needs to store a set of substantially different geometric objects (e.g., wireframe, surface, solid, and drafting symbols) along with a massive dataset for object properties needed during a project life cycle. Such a model can make the conventional industry processes particularly challenging because the digital model itself becomes significantly sizeable and thus impractical to maintain and manage. Furthermore, working on the same model, project parties across disciplines may not be able to take ownership of building elements to create or change them for their internal workflows or flexibly test alternative design and construction ideas (Mackey & Sadeghipour Roudsari, 2018). Therefore, the idea of centralized BIM has shifted toward *consolidated* or *federated* BIM, whereby practitioners merge their distributed models only at certain milestones when their design is fixed and ready for coordination and exchange with other project parties.

The foregoing challenges for the creation, utilization, and maintenance of centralized models by different project parties in addition to the competitive market for BIM software development have led the industry to use many speciality *BIM tools* that often have distributed functionalities. Each tool may implement BIM technology features (e.g., geometric capabilities, parametric modelling, and object-based modelling) in a different proprietary manner and focus the technological features on specific model uses (Mackey & Sadeghipour Roudsari, 2018). In many practices, these tools are still disconnected, and the models created in them are used in isolation from other project information. In other words, for different model uses, practitioners create often partial building models in disconnected and distributed tools with manual data re-entry and translation of information from dispersed documents and markups. While there are functional, organizational, and contractual reasons for creating and using parallel information models, this distributed toolset is also problematic and a source of reduced efficiency in BIM implementation. The challenge of exchanging model data from one tool (or one format) to another is known as *interoperability* (Allen et al., 2005).

To mitigate the issue of redundant modelling efforts across BIM tools, the industry's leading organizations and software developers have attempted to promote solutions for BIM interoperability in order to facilitate data exchange among tools (Allen et al., 2005). However, being prone to data loss and data translation issues, full BIM interoperability across BIM tools is still developing. For this reason, the current workarounds often involve exporting and importing information to and from intermediary open formats (or a limited number of proprietary formats) to mobilize at least a subset of geometric or non-geometric data across tools. As a way of connecting distributed BIM tools and models, this data mobilization process often provides project teams with reference information on the basis of which they can develop and update their own models for specific uses.

BIM toolkit and tool ecologies

BIM implementation today often involves a variety of BIM tools utilized at different project stages to support different model uses. The need to connect these distributed tools (through interoperability or data mobilization) has led to the transition of current practice from implementing a singular tool to implementing processes that connect multiple tools in a *toolkit*. To assemble a toolkit, practitioners need to have an understanding of not only how and when each speciality BIM tool in a toolkit can be used, but also how the BIM tools collectively work together for continuous, coherent, and coordinated information development and exchange in projects (Mackey & Sadeghipour Roudsari, 2018).

With the growing expectations for collaborative BIM implementation among stakeholders in projects, assembling a BIM toolkit requires considering also how project information should flow into different tools used by different parties to facilitate multi-party information exchange (Holzer, 2016). Aligning the implementation of a BIM toolkit with the scope and requirements of a project requires establishing a *BIM tool ecology* to address project specificities. BIM tool ecologies clarify (a) what tools will be used in a project for specific model uses or other information processing purposes, (b) where and in what format models or information will be stored, and (c) how and in what sequence information will be exchanged among tools (Figure 2.10).

BIM tool ecologies may also include other generic tools that are not specific to the AEC industry but can be coupled with speciality BIM tools to facilitate data processing, information extraction and exchange, and file management (e.g., common productivity tools and image viewers). Therefore, BIM tool ecologies across two different projects, even in the same business unit and firm, can have drastic differences. Although this tool ecology gives each project party a general idea and road map for implementing BIM tools in a project, specific details like timelines and specific software versions can be determined in a BIM execution plan and further refined over the course of the project.

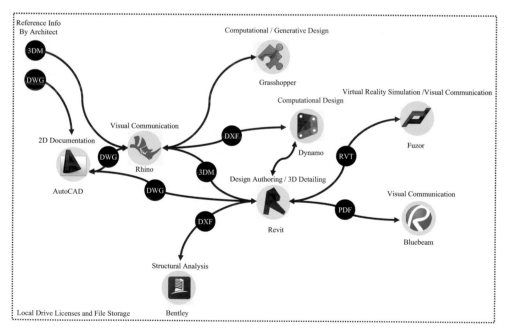

Figure 2.10 BIM tool ecology for schematic structural design of a museum (image courtesy of Magnusson Klemencic Associates).

BIM tools and related skills

This section overviews four tracks of technology-related BIM competencies and skills that educators and learners can include in their training programmes. These tracks include (a) technology integration and management skills; (b) BIM content development and workflow standardization skills; (c) software customization, programming, and computational skills; and (d) tool application skills. Many BIM courses and training programmes only address tool application skills for implementing individual tools in an educational setting. While helpful as a bare minimum, focusing solely on individual tools often cannot help trainees realize the diversity of BIM technologies and tools and the advanced technical and managerial skills required to efficiently implement BIM technologies in project and organizational settings.

Technology integration and management skills

With the gradual transition of BIM from using standalone local tools toward assembling a toolkit, BIM technology management nowadays requires an understanding of not only how a BIM toolkit contributes to design, engineering, and construction processes, but also how information technology (IT) infrastructure must support this toolkit. Therefore, the most fundamental skills in implementing BIM technologies are to assemble a BIM toolkit, identify the IT infrastructure required for its implementation, and plan and coordinate with IT management to set up the infrastructure and deploy the toolkit. As the BIM technologies discussed in this chapter may be available in a number of commercial tools, developing these skills necessitates a feasibility analysis of implementing BIM tools in terms of investment costs (e.g., subscriptions and training), market share, maturity of technologies, interoperability and model exchange capabilities, and technical support (Abdirad, 2017).

While BIM technology management has a focus on design, engineering, and construction processes for project delivery, IT management has a broader focus on a variety of hardware, software, local networks and servers, and cloud services as well as remote connectivity infrastructures that support firm-wide, office-wide, and job-site IT needs. Accordingly, the organizational responsibilities for IT management and BIM technology management may merge into one role in smaller business units or be allocated into different roles in larger business units. In AEC firms, the roles or titles that often address BIM toolkit management and its interface with IT management include *design technology management, BIM management,* or *virtual design and construction* (VDC) *management.* The interested reader can refer to Holzer (2016) for further details on the BIM and IT interface as an integral part of smooth BIM technology management.

BIM content development and workflow standardization skills

To reinforce productivity and consistency in the quality of workflows and deliverables, each firm needs to routinize its information development, organization, and representation across projects (Abdirad et al., 2020). This attempt is aligned with the basic premise of object-based and parametric features in BIM technologies, i.e., object templates can be created, customized, and re-used to drive geometries, graphics, and data in project models, drawings, and documentation. Therefore, the uptake of BIM now demands BIM content development and content management as indispensable skills in BIM implementation (Weygant, 2011).

The content development and management responsibilities address object templates for building components, object templates used for the symbolic or iconic presentation of drafting and drawing elements, heterogeneous settings in tools to drive the visualization and representation of objects and symbols (e.g., line weight, line style, colours, transparency, hatch patterns, etc.), pre-drafted standard details (also known as *typical details*) of project assembly regularly used across projects, attributes or properties that must be embedded into models, and reference design, engineering, and construction data used for BIM-based documentation or analysis (e.g., assembly cost per unit). Content development and management requires not only an understanding of how proprietary tools develop, use, or implement contents, but also a continuous dialogue with project delivery staff (e.g., project architects, engineers, field staff) to identify content needs and review existing contents. Therefore, it is important that content libraries are regularly maintained in accordance with changes in internal workflows, advances in industry practice, and frequent upgrades made to BIM tools. Furthermore, the standard contents must be readily accessible for project delivery and be properly documented, shared, and used as a part of internal standard operating procedures in each firm.

Software customization, programming, and computational skills

Even though out of the box (OOTB) BIM tools and interoperability solutions are continuously advancing, their efficient implementation is still constrained by pre-defined commands, operations, and data structures built into the tools. These OOTB functionalities do not necessarily align with the existing or planned information modelling, production, and exchange workflows in each firm, and therefore, they may impose alternative workflows on BIM processes. Furthermore, many built-in commands and operations in BIM tools (e.g., placing objects, renaming objects, modifying templates) can become repetitive and tedious manual tasks when used in large projects (e.g., adding a custom property to thousands of object templates in a content library). Although the software development market has become prolific in offering tools and software add-ons that provide workflow automation capabilities, many AEC firms have invested in their internal software customization, programming, and computational skills to tackle such issues. These skills facilitate developing interoperability and data mobilization solutions across BIM tools, transferring BIM files or data into the broader realm of IT (e.g., data analytics and databases, online repositories, cloud and web services), and extending (or customizing) the core functionality of individual BIM tools using their application programming interface (API).

APIs provide software developers with access to behind-the-scenes classes, methods, and attributes of objects and backbone structures of BIM data. Depending on the supported OOP languages (commonly used languages include python, C#, VB), developers can script codes to create new functions or operations in BIM tools and expose them to the end-user as new commands. Due to the learning curve associated with learning software programming languages, some BIM tools offer a built-in or add-on visual programming interface (VPI; also known as *visual scripting interface*) that can run as a complementary interface on top of BIM tools. In a VPI, instead of text-based programming, a subset of commonly used API functionalities are built into a series of individual nodes that can connect to one another in a customizable directed graph to automate a sequence of tasks on objects (e.g., placing new objects instances, moving existing object instances, assigning different templates or types on object instances, modifying properties).

Creating custom solutions using API or VPI requires an understanding of how proprietary object-oriented classes, methods, and attributes are built into each tool (software developers often publish VPI and API documentation for this purpose). This also requires computational thinking skills (Wing, 2006) that enable users to formulate problems, recognize patterns among objects of interest, identify logical steps (or algorithms) to solve the problem and represent a solution in a format processable by API or VPI. There is a growing demand for skills that incorporate computational thinking and programming skills not only to build on technical workflows in firms but also to tackle unique problems that come to the fore in specific projects. For example, computational design skills (automating the parametric and/or object-based modelling tasks), and in most recent advances, generative design skills (automating the exploration of design, manufacturing, and fabrication alternatives based on performance criteria) are increasingly sought after in the industry to optimize process and products considering project specificities.

Tool application skills

With the advancing technology features and growing number of tools that users incorporate into BIM toolkits, firms need to continuously train their staff to build on or develop their tool application skills. This skill-building effort can be time-consuming and costly, and therefore, is known to be the most significant barrier to BIM adoption in the industry (Ku & Taiebat, 2011).

The availability of tool application skills in a firm can affect the choice of tools in tool ecologies for projects, and tool ecologies planned for a project, in turn, can impose training and skill-building efforts on organizations and staff (Holzer, 2016). These skills involve implementing BIM technology features in integration with discipline knowledge relevant to each model use. From the technology standpoint, it is important to understand annotative, geometric, parametric, and object-based features in tools. From the core discipline standpoint, it is essential to know how information is developed (e.g., design, engineering, and analysis tasks), documented and communicated (e.g., graphic or non-graphic info), and exchanged (e.g., project documentation and media) among project parties. This integration of technology features and industry practice through model uses in BIM is a departure from the earlier CAD era, when skills for drafting, 3D modelling, and graphic representation were less integrated with the core design, engineering, and construction workflows. Therefore, applying tool application skills in projects requires an understanding of how proprietary developments in each tool address technology features in the context of industry practice.

Basic tool application skills involve navigating a project model in a BIM tool; identifying the built-in geometric, parametric, and object-based capabilities of the tool; and understanding how its OOTB user interface, settings, menus, commands, and internal database support or constrain applying the intended model uses. In more advanced tool application skills, the emphasis is on the actual implementation of model uses in projects. These skills include retaining the knowledge of workflows, picks, and clicks in tools to efficiently plan and implement model uses; customizing object templates and BIM contents (e.g. assembly details) to address project specificities; and mastering the interoperability and data mobilization capabilities of the tool for integration within a tool ecology to support distributed BIM implementation (Sacks & Pikas, 2013).

Conclusion

This chapter provided an overview of BIM technologies, tools, and skills that are fundamental to BIM education and training in the AEC industry. The first part of this chapter discussed the technology features that signify similarities and differences between BIM tools. These features included geometric representation, parametric modelling, object-based modelling, and reality capture. This chapter further clarified that commercial BIM tools may support or limit these technological features in different ways, providing different BIM implementation experiences and outcomes for different practitioners and learners. The second part of this chapter presented the notion of distributed BIM in custom toolkits, through which a variety of BIM tools with complementary technology features are coupled together to facilitate BIM implementation per organizational, project, and training needs. The third part of this chapter overviewed four skill development tracks relevant to BIM technologies and tools. These tracks outline skill sets that span a variety of technology management, development, and application competencies for BIM practitioners, educators, and learners in the industry. BIM educators and trainees can use this chapter as a reference for developing an all-around introductory BIM training programme that removes disciplinary bias toward BIM while covering fundamental technology features and skills that are integral to BIM implementation in the industry.

Acknowledgements

The authors wish to thank M. Reza Hosseini (School of Architecture & Built Environment, Deakin University), Brian Johnson (Department of Architecture, University of Washington), and Andy Fry and Kevin Carroll (Magnusson Klemencic Associates) who generously supported this work by offering their constructive comments and information from sample projects.

Questions

1. Why are there different interpretations of BIM technologies and applications in practice?
2. Explain how geometry representation methods may constrain BIM implementation.
3. How does parametric modelling (and its variations) build on the traditional geometric representation technologies?
4. Illustrate examples of how two *BIM* uses (from Table 2.1) may demand different technological features or capabilities for *geometric representation*, *parametric modelling*, and *object-based modelling*.
5. What are the advantages (or disadvantages) of *distributed* BIM implementation in comparison to *centralized* BIM implementation?
6. How can unique technological BIM requirements in projects influence BIM implementation in terms of management and staffing?
7. Explain how BIM technological advances (or constraints) have influenced the emergence of new speciality skill paths in the industry.

References

Abdirad, H. (2017). Metric-based BIM implementation assessment: A review of research and practice. *Architectural Engineering & Design Management, 13*(1), 52–78. doi:10.1080/17452007.2016.1183474

Abdirad, H., & Dossick, C. (2016). BIM curriculum design in architecture, engineering, and construction education: A systematic review. *Journal of Information Technology in Construction (ITcon), 21*(17), 250–271.

Abdirad, H., Dossick, C., Johnson, B., & Migliaccio, G. (2020). Disrupted information exchange routines in construction projects: Perception and response patterns. *Building Research & Information*. doi:10.1080/09613218.2020.1750939

Abdirad, H., & Lin, K. -Y. (2015). Advancing in object-based landscape information modelling: Challenges and future needs. *Journal of Computing in Civil Engineering 2015* (pp. 548–555).

Allen, R. K., Becerik, B., Pollalis, S. N., & Schwegler, B. R. (2005). Promise and barriers to technology enabled and open project team collaboration. *Journal of Professional Issues in Engineering Education & Practice, 131*(4), 301–311.

Armesto, J., Lubowiecka, I., Ordóñez, C., & Rial, F. I. (2009). FEM modelling of structures based on close range digital photogrammetry. *Automation in Construction, 18*(5), 559–569. doi: https://doi.org/10.1016/j.autcon.2008.11.006

BIMForum. (2019). *2019 Level of Development Specification*. Washington, DC: buildingSMART International.

Eastman, C. (1999). *Building Product Models: Computer Environments, Supporting Design and Construction*: Boca Raton, FL: CRC Press.

Eastman, C., Fisher, D., Lafue, G., Lividini, J., Stoker, D., & Yessios, C. (1974). *An Outline of the Building Description System*. Pittsburg: Carnegie Mellon University.

Eastman, C., Teicholz, P., Sacks, R., & Liston, K. (2011). *BIM Handbook: A Guide to Building Information Modelling for Owners, Managers, Designers, Engineers and Contractors (2nd Edition)*. Hoboken, NJ: Wiley.

Engelbart, D. C. (1962). Augmenting human intellect: A conceptual framework. *SRI Summary Report AFOSR-3223*. Washington, DC: Air Force Office of Scientific Research.

Helpenstein, H. J. (2012). *CAD Geometry Data Exchange Using STEP: Realisation of Interface Processors*. Berlin: Springer.

Hoffmann, C. M., & Joan-Arinyo, R. (2002). Parametric modelling. In G. Farin, J. Hoschek, & M.-S. Kim (Eds.), *Handbook of Computer Aided Geometric Design* (pp. 519–541). Amsterdam: Elsevier Science.

Holzer, D. (2016). *The BIM Manager's Handbook: Guidance for Professionals in Architecture, Engineering, and Construction*. Hoboken, NJ: Wiley.

Johnson, B. R. (2016). *Design Computing: An Overview of an Emergent Field*. New York: Routledge.

Ko, J., & Steinfeld, K. (2018). *Geometric Computation: Foundations for Design*. Oxfordshire, UK: Taylor & Francis.

Ku, K., & Taiebat, M. (2011). BIM experiences and expectations: The constructors' perspective. *International Journal of Construction Education & Research, 7*(3), 175–197. doi:10.1080/15578771.2010.544155

Mackey, C., & Sadeghipour Roudsari, M. (2018). The tool(s) versus the toolkit. In K. De Rycke, C. Gengnagel, O. Baverel, J. Burry, C. Mueller, M. M. Nguyen, P. Rahm, & M. R. Thomsen (Eds.), *Humanizing Digital Reality: Design Modelling Symposium Paris 2017* (pp. 93–101). Singapore: Springer.

Rossignac, J. R. (2007). Solid and physical modelling. In *Wiley Encyclopedia of Electrical and Electronics Engineering*. Hoboken: John Wiley & Sons, Inc. doi:10.1002/047134608X.W7526.pub2

Sacks, R., & Pikas, E. (2013). Building information modelling education for construction engineering and management. I: Industry requirements, state of the art, and gap analysis. *Journal of Construction Engineering & Management, 139*(11), 04013016. doi:10.1061/(ASCE)CO.1943-7862.0000759

Sacks, R., Eastman, C., Lee, G., & Teicholz, P. (2018). *BIM Handbook: A Guide to Building Information Modelling for Owners, Designers, Engineers, Contractors, and Facility Managers*. Hoboken, NJ: Wiley.

Shapiro, V. (2002). Solid Modelling. In G. Farin, J. Hoschek, & M.-S. Kim (Eds.), *Handbook of Computer Aided Geometric Design* (pp. 519–541). Amsterdam: Elsevier Science.

Son, H., Bosché, F., & Kim, C. (2015). As-built data acquisition and its use in production monitoring and automated layout of civil infrastructure: A survey. *Advanced Engineering Informatics, 29*(2), 172–183.

Succar, B., Saleeb, N., & Sher, W. (2016). Model uses: Foundations for modular requirements clarification language. Singhaputtangkul, Natee, ed. In: *40th Australasian Universities Building Education Association Conference (AUBEA 2016)*, 06–08 July 2016, Cairns, 1–12.

van Nederveen, G. A., & Tolman, F. P. (1992). Modelling multiple views on buildings. *Automation in Construction, 1*(3), 215–224. doi:https://doi.org/10.1016/0926-5805(92)90014-B

Weygant, R. S. (2011). *BIM Content Development: Standards, Strategies, and Best Practices*. Hoboken, NJ: Wiley.

Wing, J. M. (2006). Computational thinking. *Communications of the ACM, 49*(3), 33–35.

3 Understanding BIM to translate it into action

Cenk Budayan and Yusuf Arayici

Introduction

Due to new technologies and the high expectations of society, the construction industry has to evolve and improve construction processes to increase efficiency and quality, which are scarce in the construction industry compared to the other industries and are not adequate for society's needs. Therefore, the construction industry should start rethinking the ways and methods that are used to build (Egan, 1998). Newly graduated engineers are crucial in realizing these changes since they are more open to changes and innovations. Therefore, civil engineering students should be prepared for these changes in their education. However, unfortunately, in most universities, curricula are not revised according to the newly demanded skills and competences. Therefore, newly graduated engineers must overcome the problems that arise as they deal with these changes on their own.

In recent years, building information modelling (BIM) has become popular in different countries and BIM implementation is becoming mandatory in construction projects. However, the maturity level of BIM implementation has not still been at the desired level since there are many important challenges in applying BIM effectively, such as resistance to change; a lack of understanding of the potential and value of BIM; a lack of personnel with BIM skills; a lack of understanding of BIM technologies, tools, and applications; a lack of collaboration; a lack of understanding of the responsibilities of different stakeholders; and low demand for BIM implementation (Arayici et al., 2011, Khosrowshahi and Arayici, 2012). As these challenges demonstrate, it can be said that the industry does not fully comprehend the concept of BIM.

The industry lacks highly skilled cross-trained staff with both construction and informational technology (IT) skills (Cook, 2004). Newly graduated students can be crucial resources for overcoming these challenges since they have the required skills and they can change the industry paradigm as the future leaders of the construction community (Ahn et al., 2013). Besides, although most professionals know about BIM software, they struggle to implement it effectively. In other words, the utilization of the real benefits of BIM is limited in the industry. Therefore, students and professionals need to understand the different aspects of BIM in order to translate BIM into action. Hence, the goal of this chapter is to make BIM understandable for students and professionals by considering not only the technological aspects of BIM but also other important aspects required to translate BIM into action.

Understanding BIM

Although BIM is popular and widely known in the construction industry, there are some challenges to understanding BIM explicitly. Even the acronym *BIM* has different meanings, for instance, *building information model* and *building information modelling.* Therefore, the first step in implementing BIM is to understand BIM clearly. To understand BIM, we need to define it comprehensively. In the literature, there is no consensus on the definition of BIM. Some researchers focus on the output of BIM as the representation of the physical and functional characteristics of a facility digitally (National BIM Standard, 2020, Van Nederveen et al., 2010). However, BIM is not only a product but also a socio-technical system that leads to broad changes in all stages of a project, from inception onward (Sacks et al., 2018). Therefore, some researchers mention the changes which BIM makes to the process and the working forces in the construction market in their definitions (HM Government, 2012, Arayici, 2015).

Secondly, although the popularity level of BIM is high in the construction industry, the adoption level is low in many markets, such as the United Kingdom, New Zealand, and Turkey (Ghaffarianhoseini et al., 2017). Therefore, BIM should be translated into action. To achieve a high level of adoption, we need to understand how the organizations function and behave in their daily operations and in the long term; in other words, understanding organizations is critical. For that purpose, institutional theories can be used. These theories can help to explain why BIM adoption may produce a variety of performance outcomes and how they may be influenced by various individuals and organizations that collaborate. One of the most important institutional theories is proposed by Scott (1998). Based on Scott's theory, three institutional pillars, namely regulative, normative, and cultural-cognitive, shape organizations' behaviours. The regulative pillar involves regulation, monitoring, and sanctioning activities. From a BIM perspective, the processes conducted in organizations can be considered as this pillar. This can be seen in the form of standards, environments, and delivery systems. The second pillar is about the culture, the workplace environment, and the ambience of the workplace (Walker and Llyod-Walker, 2019). Therefore, it comprises the behaviours, values, and actions of the workers. The last pillar is related to skills, knowledge, dexterity, resilience, and reflective capacity (Walker and Lloyd-Walker, 2019). Similarly, Khosrowshahi and Arayici (2012) stated three important pillars, namely system methodology (people-oriented), information engineering (data-driven approach), and process innovation (process-oriented approach), in BIM implementation. Consequently, by considering the BIM definitions and Scott's institutional theory, three pillars of BIM – technology, process, and people – can be considered for better understanding and adoption levels of BIM in the construction market.

Technology

There have been enormous technological advances in recent years that have also led to the development of many different BIM tools and technologies to facilitate the various tasks and activities performed throughout the life cycle of a building. However, the main functionalities and key competencies of BIM technologies are not well comprehended by practitioners, and therefore they are not fully exploited for the right tasks and activities (Arayici, 2015). All terminologies and technologies related to BIM should be understood by all parties so that they speak a common language between them (Antwi-Afari et al., 2018).

With technological development, the representation of buildings has evolved over time. Traditionally, the geometry of buildings has been presented in two-dimensional CAD drawings. Although there are tools for two-dimensional CAD drawings, the information embedded in these drawings is disseminated through paper-based drawings. Errors and omissions in these paper documents can lead to cost overruns and delays due to lack of communication, which in turn may lead to conflicts between various parties in a project team. Besides, 2D drawings-based communications can lead to an increase in time and cost at the design stage due to the use of unintegrated delivery systems. Consequently, efforts are being made to transfer models from two dimensions to three dimensions to address these problems.

Geometric modelling

The purpose of geometric modelling is to simulate building components in a CAD platform and represent the geometry of a building. This model is a powerful tool for providing information about quantities and for producing 2D drawings. It is effective for transferring information between parties since it provides different representations of buildings. This model can also be extended into BIM because it contains rich data about the building; indeed, the first step in developing a BIM model is the development of a geometric model.

Solid modelling is the first generation of practical 3D modelling, and two approaches can be followed to perform solid modelling, namely *explicit modelling* and implicit modelling. Explicit modelling, also known as *boundary representation*, represent objects in terms of closed, oriented sets of bounded surfaces. In this method, the points, edges, and surfaces are used to create objects, and it allows that shapes to be swept or rotated or a defined face manipulated into a third dimension to form a solid. On the other hand, implicit modelling, also known as *constructive solid geometry*, employs a sequence of construction steps to describe a volumetric body. In this method, the solid primitives typically include a rectangular prism, a cylinder, a cone, a wedge, and a sphere, which are added or subtracted from each other (Shih, 2020).

There are debates about which one of these two methods is better. However, it is agreed that the combination of these two methods provides the best performance. Implicit modelling is used for editing, and explicit modelling is used for visualizing, measuring, clash detection, and other non-editing issues.

Parametric modelling

Parametric modelling is the creation of a digital model based on a series of pre-defined dependencies and constraints. All design and revision procedures are automated due to the usage of parametric features. Parameters are used to demonstrate the properties of each object. Parameters can be geometric dimensions such as height, width, and position. This model can also contain some non-geometric properties and features such as price, spatial relationship, manufacture, geographic information, vendor, materials, code requirements, and any other parameter related to object use (Jiang, 2011). These dimensions can vary at any time during the design process. However, in parametric modelling, there are also dependencies to show the relationships between parameters. Therefore, a variance in the dimension of an object will affect the dimensions of the other objects related to this object.

Parametric modelling provides a flexible model that can be changed easily and quickly according to newly emerging demands and requirements. Therefore, it provides the ability to compare design alternatives to determine the best design. Due to the dependencies, any change to an object within the model will be reflected automatically in the rest of the views of the project. On the other hand, in 3D modelling, any modification and addition must be updated and checked in all views, which in turn, lead to extra efforts. For instance, a deleted door in one view must be deleted in all views. Moreover, existing design data can be reused in different designs since the objects are designed in terms of building systems and elements rather than as geometrical primitives. The resulting model is a "data-rich", "object-oriented", "intelligent and parametric digital representation of the facility", which can be used for decision making and improving the processes in a project (Gardezi et al., 2013).

Interoperability

A single shared model is not recommended because of the lack of accountability, which can lead to problems in terms of legal aspects. Besides, there is no superior BIM tool that includes all the required functions across the project lifecycle. Yet, BIM is conducted by using various BIM applications throughout the life cycle (Arayici, 2015). All stakeholders use different applications for different purposes. For instance, an architect uses applications for architectural design, while a structural engineer uses other applications for structural design and analysis, and so on. Therefore, these BIM applications should be interconnected with each other to facilitate better performance and to improve interoperability between BIM applications. The following standards are proposed:

Industry Foundation Classes (IFC)

There are different efforts to standardize the processes conducted at different stages to achieve interoperability between these BIM applications. One of the widely known efforts, namely IFC, has been proposed by the buildingSMART alliance. IFC is a schema, not a format, and it represents an extensible set of consistent data used for exchange between different BIM applications.

IFC is an object-oriented building data model representing the project's identity, semantics, characteristics or attributes, relationships between objects, abstract concepts, processes, and people (buildingSMART, 2020). Thus, the model developed by any party involved can be used by other parties to perform their businesses without any interruption. IFC can also be used as a means of archiving project information, whether incrementally throughout the life cycle of a building or as an "as-built" collection of information for reuse in future projects or the future operation of the building.

IFC provides a solution to the problem of accountability since each party remains the author and owner of their own model in the IFC-based workflow. Although IFC is a publicly and internationally recognized standard in the market, it is weak in terms of its implementation (Sacks et al., 2018). It has been accused of losing information while importing and exporting BIM models and it also leads to the development of models that are no longer parametric and require punctual remodelling activities if newly imported in proprietary languages (Daniotti et al., 2020).

Construction Operations Building Information Exchange (COBie)

COBie is another effort developed by the US Army Corps of Engineers to transfer information throughout the design and construction stages to the client organization for operating and maintenance of the facility and to provide more general facility management information; it is also used as the National Building Information Modelling Standard (NBIMS) in the United States and as the British code of practice (Lea et al., 2015).

In traditional methods, data is generally provided by a bulk of paper at the handover stage. Therefore, extracting useful information from all the data is a tiresome activity that in turn leads to inefficiency and loss of information during operation and maintenance. In

COBie, data is captured and stored incrementally in a digital format throughout the design, construction, and commissioning stages.

In COBie, data is transferred in a hierarchical format by using spreadsheets to decrease the complexity of the systems. Spreadsheets are used to provide building information. However, COBie does not provide any data about geometric modelling directly. This information is available as an IFC model view or a subset of IFC (Daniotti et al., 2020).

Although COBie is critical for operation and maintenance, there are problems in its implementation. Unfortunately, designers and contractors do not provide fully completed spreadsheets due to misunderstanding on the part of end-users and insufficient software implementation (Schwabe et al., 2015).

XML-based schemas

XML (extensible mark-up language) is used for storing and transporting a wide variety of data on the web. It can be used to define a data model and to carry the actual information. With the use of XML, different XML-based schemas are created for different purposes in BIM applications, such as

- gbXML, which transfers building information between BIM authoring programmes and energy simulation tools;
- aecXML, which carries descriptions and specifications of buildings and their components (Sacks et al., 2018);
- ifcXML, a subset of the IFC schema;
- BIMXML, which carries building data in a simplified spatial building model through web services.

However, all these schemas define the entities, attributes, and rules differently. Therefore, these are not compatible with each other (Sacks et al., 2018).

Cloud-based BIM technology

Cloud and mobile technologies have the potential to increase the effectiveness of the standalone BIM technologies since cloud and mobile-based BIM technologies can enable project teams to cooperate and collaborate due to the continuous flow of the most updated information through effective communication platforms (Wong et al., 2014). Due to these benefits of cloud and mobile technologies, the use of these technologies has grown in the construction industry in recent years (Abanda et al., 2015). Therefore, different applications have been developed for BIM implementation in mobile operating systems such as iOS, Android, and other mobile platforms.

Advanced analysis can be performed via a cloud computing platform, which can lead to effective, attractive, and cheaper information technology applications (Alreshidi et al., 2016). In complex BIM implementations, in particular, advanced computer systems are required, which is considered one of the important barriers to BIM adoption due to their high cost (Azhar, 2011). Alreshidi et al. (2016) stated the benefits of cloud technologies in BIM implementation as accessibility, scalability, reliability, advanced interoperability for BIM applications, security, automatic backup in real time, cost, and green credentials. Consequently, many researchers have considered cloud computing as an important part of BIM technologies (Fathi et al., 2012, Abanda et al., 2018).

There are different cloud BIM applications, such as Autodesk BIM 360, Cadd Force, Onuma System, and BIMcloud. Chong et al. (2014) tabulated the features of six cloud BIM applications, namely Autodesk BIM 360, Cadd Force, BIM9, BIMServer, BIMx, and Onuma System. However, these systems are still evolving, and new features are introduced with the new releases of these systems. The following list provides a quick overview of the most widely used cloud systems in the construction industry.

Autodesk BIM360 provides a unified cloud platform for all project members from the beginning of a project through construction. The core products of BIM 360 are BIM 360 Docs, BIM 360 Design, BIM 360 Coordinate, and BIM 360 Build. All these products cover different phases of the project. BIM 360 Docs is used to manage and properly control 3D drawings and other documents. BIM 360 Design provides a collaborative design environment that allows multiple parties to work on the same design simultaneously. BIM 360 Coordinate is used to coordinate the project stakeholders by merging the published models, providing an immediate view of identified clashes, and filtering the clashes by model, discipline, or user. Finally, BIM 360 Build is used for project management, cost management, and quality and safety management.

BIMcloud is another cloud BIM software and developed by Graphisoft to provide an integrated BIM workflow for project team members. However, it requires conventional design software for developing a BIM model. ARCHICAD and BIMx can be directly connected to BIMcloud. However, BIM projects developed by using different BIM software can be aggregated and coordinated in BIMcloud via IFC connection and the MEP Modeler functionality of ARCHICAD.

Internet of things (IoT)

IoT can be defined as "things having identities and virtual personalities operating in smart spaces using intelligent interfaces to connect and communicate within social, environmental, and user contexts" (Srivastava, 2006). In other words, with the IoT, smart devices and their services communicate with each other to make smart decisions. Therefore, IoT can be used by different sectors to develop a smart environment in which information and communication technologies are used for making aware, interactive, and efficient decisions about city administration, education, healthcare, public safety, real estate, transportation, and utilities (Bélissent, 2010). IoT devices and sensors are continuously being installed globally, and the number of connected devices is estimated to have reached 24 billion by 2020 (Gubbi et al., 2013). Therefore, IoT technologies provide real-time valuable information that can be also used for construction projects throughout their life cycle (Främling et al., 2014). In other words, IoT has the potential to change the traditional methods for capturing knowledge in the construction industry. Hence, large amounts amount of accurate information can be captured via IoT technologies.

IoT is also very valuable for BIM implementation, and the potential of the integration of IoT and BIM has been widely recognized in the literature, since this can eliminate the limitations of each (Tang et al., 2019). IoT technologies provide real-time and recordable high amounts of information about all activities performed in the construction and operation stages. Thus, the most up-to-date information can be embedded into BIM models to provide real-time feedback for making accurate decisions. By reviewing 97 papers published about BIM and IoT, Tang et al. (2019) have determined four domains where BIM and IoT can be integrated:

- construction operation and monitoring;
- health and safety;

- construction logistics and management;
- facility management.

They also stated 12 subdomains. Due to this potential, various efforts are being made to integrate BIM and IoT. For instance, Dave et al. (2018) proposed a framework for integrating BIM and IoT through open standards for energy usage, occupancy, and user comfort.

Process

Traditionally, the construction industry is fragmented in nature and has been criticized due to the low levels of collaboration and coordination between the parties involved, even though all the processes conducted throughout the life cycle of construction are interconnected with each other. In other words, there should be a continuous connection and intense information exchange among the parties. Therefore, changes should be made to all processes in the construction industry to achieve high efficiency and productivity.

BIM can be crucial for increasing levels of information sharing and collaboration. However, BIM implementation leads to new challenges in the building process. For instance, when traditional project delivery systems, such as the Design-Bid-Build delivery system, are used, stakeholders will probably use different BIM applications, and the integration of these models can add complexity and introduce potential errors to the project. Furthermore, the integration of these models can be impossible in some situations. However, to use BIM effectively, these models must be integrated to enable a collaborative environment.

Stakeholders should be willing to share their knowledge with other stakeholders, and to achieve this, an environment of trust must exist. Unfortunately, the various parties in the construction industry prioritize their own goals and compete with each other to achieve them throughout the project (Lee et al., 2018). There is a resistance to change in the construction industry. Therefore, construction professionals slow down the rate of progress in the adoption of sophisticated IT systems. Consequently, the culture of construction companies should change and the processes in their organizations should change radically.

The key BIM standards

There are different standards for BIM implementation. Some of them are at the national level, for instance, UNI 11337 for Italy, BIM manual for Norway, and PAS 1192-2:2013 for the United Kingdom. Some of them are at the international level such as BS EN ISO 19650.

PAS 1192-2:2013 was developed by the British Standards Institution for supporting the BIM strategy of the British Government, which mandates that public projects should have reached maturity Level 2. The aim of PAS 1192-2 is to provide a specific guideline for managing information created and captured throughout projects delivered using BIM. Therefore, it provides a consistent and structured environment for the efficient and accurate exchange of information during a project (British Standards Institution, 2013). The application of PAS 1192-2:2013 requires three documents:

1. A Master Information Delivery Plan, which consists of plans for how information is created and delivered throughout the project life cycle;
2. A BIM execution plan, which is prepared by the suppliers to explain how the information modelling aspects of a project will be carried out and define the required level of development for project deliverables;

3. An Employer's Information Requirements document, which represents the information required by the owner and facility management for the development and operation of the project.

The British Standards Institution has also released a series of standards for BIM implementation at the national level (UK BIM Framework, 2019):

- PAS 1192-3:2014 is developed to provide a guideline for information management at the operational phase of the assets via BIM.
- BS 1192-4:2014 is developed to provide a guideline for the collaborative production of information. The details of COBie are provided in this standard.
- PAS 1192-5:2015 is developed to provide a guideline for security-minded building information modelling, digital built environments, and smart asset management. However, this standard is not specific to BIM.
- PAS 1192-6:2018 is developed to provide a guideline for collaborative sharing and the use of structured Health and Safety information via the use of BIM.
- BS 8536-1:2015 is developed for the brief of design and construction and as a code of practice for the facilities management of building infrastructures.
- BS 8536-2:2015 is developed for the brief of design and construction and as a code of practice for asset management of linear and geographical infrastructure.

PAS 1192-2: 2013 has been terminated and replaced with an international standard BS EN 19650, which is developed based on BS1192:2007 + A2:2016 (principles) and PAS1192 part 2. It consists of two parts called BS EN ISO 19650-1 and BS EN ISO 19650-2, according to UK reference codes. Moreover, to assist the UK construction industry with some necessary process adjustments, BS EN ISO 19650-1 includes a National Foreword explaining the terminology and is published alongside PD 19650-0:209, which provides transitional guidance.

BS EN ISO 19650-1 gives information about the basic concepts and principles and provides recommendations on the management of building information. BS EN ISO 19650-2 describes the delivery phase of the asset and provides recommendations throughout the whole life cycle of the asset.

BS EN ISO 19650 provides a project information delivery cycle. This cycle is also divided into two components, namely the project information model (PIM) and the asset information model (AIM). PIM is developed for the design and construction phase of a project, although the information provided by the PIM can be also used in the operation stage. Therefore, the relevant information should be transferred from the PIM to the AIM at the end of the delivery stage.

An AIM is a model that provides the required data and information for the operation of an asset. It includes three deliverables: documents, non-graphical information, and graphical information. An AIM can be composed of the original design intent, 3D models, and the desired information for the asset. At the start of the delivery phase, relevant information should be transferred from the AIM to the PIM. Moreover, all information management activities in AIM and PIM are performed within a common data environment, detailed information about which is given in the next section.

Common data environment (CDE)

Throughout the life cycle of a building, a lot of information at different levels of detail is created by different parties using different software and technologies. Therefore, data exchange between these parties is a challenging process. However, it is crucial for the effective implementation of BIM. Therefore, an information management system is required for structuring, combining, distributing, managing, and archiving digital information. To address this, the concept of a common data environment (CDE) is proposed in the UK design regulation BSI 1192-1:2007 and recalled in PAS 1192-2:2013.

CDE represents a common digital project space that different project stakeholders can access to collect, manage, evaluate, and share data with clear status definitions and a robust workflow description for sharing and approval processes (Preidel et al., 2015). Therefore, all project participants should use this environment to store all domain-specific partial models and documents and to retrieve the required input data. Thus, a centralized platform for data storage is established that in turn facilitates collaboration between stakeholders and eliminates the risk of data redundancy and duplication. Moreover, it enhances the reusability of information in support of all life-cycle stages. Therefore, it reduces the time and cost required to reproduce the available information.

ISO 19650-1 defines CDE by combining the CDE workflow and the solution; therefore the CDE should be considered as a technology or solution even if the CDE workflow is defined in detail, as shown in Figure 3.1, which shows four scopes of action: *work in progress*, *share*, *publish*, and *archive*. The reliability and maturity of the data are checked at the end of each stage, and the CDE provides a platform for the exchange of reliable and accurate data.

The industry misperceives that CDE is a technology solution only, and that there is only one single solution. However, the industry can require a range of technologies to develop CDE. Therefore, the connection between these systems should be considered. Instead of

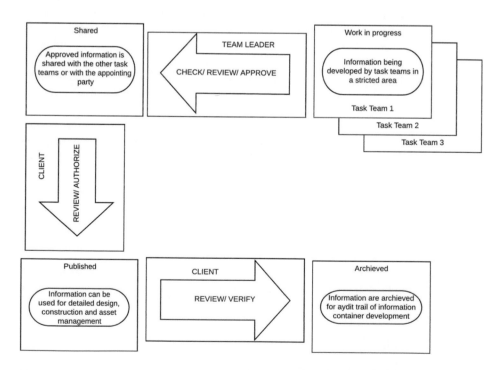

Figure 3.1 CDE concept demonstrated in ISO 19650-1 (UKBimAlliance (2019)).

focusing on the number of systems, the required functionality should be established with the support of single and multiple systems (UKBimAlliance, 2019).

Level of development

Since one of the properties of any project is progressive elaboration, projects should be elaborated continuously throughout their life cycle. However, the level of elaboration is an important question that the parties should decide on, since any increase in elaboration would also require a dramatic increase in resources for planning and modelling. Therefore, the level of elaboration should be established while constructing the building. In BIM, what is called the LOD scale is used as a measure of the quantity and quality of information embedded into the model.

Initially, this scale was defined by the term *level of detail*. However, this term led to misconception as it emphasized the geometry of the objects. Therefore the *term level of development* was invented to eliminate the direct connection with geometry. The LOD shows how much information is provided in the model. More precisely, LOD describes the steps through which a BIM model can logically progress from the lowest level of conceptual approximation to the highest level of representational precision (Brewer et al., 2012).

LOD was first proposed by AIA (the American Institute of Architects) as comprising five levels to indicate the extent to which a BIM model is developed. However, the AIA has expanded the number of LOD levels by collaborating with American BIMforum to propose six LOD levels. Finally, eight levels of development are proposed by INNOvance (Daniotti et al., 2020). They are represented as LOD000, LOD100, LOD200, LOD300, LOD350, LOD400, LOD500, and LOD550. These levels require coordination amongst all stakeholders involved in a project to identify who will be responsible for the development of each component and to what extent the BIM model will be detailed (Gaudin, 2013). There are also different LOD standards in the market; for instance, UNI 11337 introduced a different LOD referring to objects and measured in letters from A to G.

In PAS 1192-2:2013, LOD means *level of definition*, which is defined by combining *level of detail* and *level of information*. The level of detail provides information about the graphical content of models. On the other hand, the level of information is about non-graphical or alphanumerical attributes. With BS EN ISO 19650, the term *level of information need* was introduced to synthesize different standards. Level of information need shows the right level of information required to satisfy the purposes of each information exchange. It should be established by the appointing party for each information container.

Integrated project delivery

Integrated project delivery (IPD) is "a project delivery approach that integrates people, systems, business structures, and practices into a process that collaboratively harnesses the talents and insights of all participants to optimize project results, increase value to the owner, reduce waste, and maximize efficiency through all phases of design, fabrication, and construction" (The American Institute of Architects, 2007). The aim of IPD is to integrate all parties, which are traditionally functional adversaries, into a single process by adopting "best for project" behaviour. In other words, parties should engage in "best for project" behaviour rather than "best for stakeholder" thinking.

BIM is a crucial tool that can facilitate integration between project parties. Although an IPD project may be delivered without using BIM and vice-versa, the real benefits will be seen only when BIM methodologies are applied to IPD processes (NAFSA et al., 2010). Moreover, BIM and IPD yield an eight-in-ten chance of completing a project on schedule and within budget compared to other project delivery systems; this is a remarkable improvement (Deutsch, 2011), since construction activities can be highly knowledge-intensive due to the complexity of the projects, regulatory requirements, stakeholders' needs, and time and budget concerns (Glick and Guggemos, 2009). A project manager generally spends 75%–90% of their time distributing or obtaining project information (Scanlin, 1998). The American Institute of Architects (2007) also considers IPD to be the best project delivery system for the most comprehensive use of BIM.

Although BIM plays important roles in the application of IPD processes, BIM not only manages information in the IPD processes which end with the closing of the project following construction; BIM can also extend the life of the project and can be used in subsequent projects.

People

The other important dimension of BIM implementation is people, and people's BIM capabilities and understanding are crucial for successful BIM implementation. BIM capabilities are not limited to technical capabilities but also include non-technical skills such as good interpersonal relationships, effective team working, and collaborative skills (Ku and Taiebat, 2011). Professionals should change their perspectives through their businesses and administrative processes conducted in their organizations. In other words, there should be a fundamental transition from traditional views and opinions towards a BIM-based production-line approach, since the introduction of BIM has meant significant changes in administrative structure, including the emergence of new roles and the need for new key capabilities and skills (Khosrowshahi and Arayici, 2012).

BIM roles

With the implementation of BIM in the construction industry, roles and responsibilities have emerged for the creation, management, and coordination of building information models. The most important roles are defined as the BIM manager, BIM coordinator, and BIM modeller (Borrmann et al., 2018). There are other roles, such as BIM analyst, BIM modelling specialist, BIM consultant, and BIM researcher (Barison and Santos, 2010). Moreover, in the UK market and ISO standards, the terms *information manager*, *interface manager*, and *information originator* are also used. UNI 11337 proposes four BIM professionals: BIM manager, BIM coordinator, CDE manager, and BIM specialist.

BIM manager

The role of BIM manager has emerged with the introduction of BIM in the construction industry, since BIM implementation is not only a technology but also represents a radical change in how business is done. These changes should be managed carefully to avoid the conflicts and frictions in the organization by achieving coordination and support throughout the life cycle of a project. However, due to the fragmented structure of the construction industry, all parties have different structures and processes, which in turn may lead to data discontinuities. Therefore, traditional roles cannot be valid for coordination between organizations due to projects' scope and technical complexity. This leads to the emergence of a new, independent role, namely the BIM manager (Holzer, 2016).

The BIM manager is considered one of the important components in ensuring the success of a project, since the BIM manager acts as an integrator of the other components, namely people, process, technology, and policy (hochtief-vicon, 2020). Therefore, the main responsibility of the BIM manager is to manage BIM implementation and/or maintenance while integrating all the components of BIM. BIM managers may set design templates, coordinate the integration of entity models, and coordinate access to the BIM model (Barison and Santos, 2010).

Skills and capabilities

Successful implementation of BIM requires individuals who have in-depth BIM knowledge and competent BIM skills. Therefore, the required skills should be identified carefully to improve the applicability of BIM in the construction industry. Various efforts are being made to determine present and future requirements. For instance, ACIF and APCC (2017) presents a framework that covers the skill sets and competencies required of all the roles and responsibilities in BIM projects. Based on this framework, the expected level of knowledge and skills required are graded into five levels. Similarly, BIM Academic Forum UK develops a different framework for embedding BIM within the taught curriculum. They develop BIM teaching impact matrix, which presents the required knowledge and understanding, practical skills, and transferable skills at different programme levels.

Davies et al. (2015) group the required skills for BIM implementation in three categories, namely soft skills, technical skills, standards, and frameworks, via interviews. Soft skills are related to personal and interpersonal management skills. The most important soft skills are determined to be communication, negotiation and conflict management, authority and leadership, and attitude. Barison and Santos (2011) analyzed 22 job adverts and 24 studies from the literature to extract the competencies demanded in the market. They categorized these competencies under five categories: aptitude, qualifications, skill/ability, knowledge, and attitude. Similarly, Uhm et al. (2017) analyzed 242 online job posts to determine the roles available and the required competencies for these roles. They assign 43 competencies for eight BIM jobs determined.

Bosch-Sijtsema et al. (2019) determine the competency expectations of BIM actors in terms of BIM actor characteristics, BIM actor education and experience, and task expectations. The common points of all these studies are a combination of soft skills and technical knowledge. In other words, professionals should have technical and non-technical skills. However, these skills should be arranged according to the role, since each BIM role requires different competencies.

Education and training

Education and training are crucial since dramatic changes have taken place in the process and technology within the construction industry in recent years. BIM implementation in particular demands new skills from people. However, these new skills are not limited to technical skills, such as learning a BIM tool. People should learn to work in a BIM environment; in other words, they should focus not only on technology but also on the processes in the BIM project environment.

There are different certification programmes for teaching BIM. For instance, the methods proposed in the literature for learning BIM mainly cover four courses, namely i) an introduction to BIM, ii) BIM technology, iii) BIM project execution planning, and iv) BIM adoption, implementation, and return on investment. Similarly, there are other BIM courses provided by industrial bodies for which a BIM certificate is awarded, consisting of modules such as BIM initiation, BIM process, BIM collaboration and integration, and BIM technology.

The British Standards Institution, for example, provides four type certifications: the BSI Kitemark for Design and Construction, the BSI Kitemark for BIM Asset Management, the BSI Kitemark for BIM Objects, and the BSI Kitemark for BIM Level 2. There are also many other training and certification programmes offered by different public and private organizations from different countries. The common point of these training and certification programmes is that they do not focus only on BIM technology but also consider the integrated structure of BIM implementation and evaluate different aspects of BIM.

At the university level, there are many different programmes and courses for BIM. Sacks et al. (2018) categorized BIM courses into three models.

- In the first model, an integrated environment for design and engineering is proposed for undergraduate students, graduate students, and advanced BIM classes. In this model, changes in the curriculum are limited to BIM.
- The second model can be considered an intermediate approach. There are changes in the curriculum for introducing new BIM classes related to the BIM tools and basic BIM concepts.
- In the third model, the existing programme is developed into a BIM-oriented programme.

There are also different studies about BIM curricula models. For instance, Ghosh et al. (2015) proposed a model for developing a BIM curriculum based on vertical integration. Similarly, Anderson et al. (2019) present findings in curriculum design for BIM-enabled projects performed by students from different universities around the world and they concluded that BIM curriculum should be developed by including the managerial challenges of BIM, such as teamwork and communication. Finally, they emphasized the importance of preparing a BIM execution plan, and they stated that students faced difficulties in the preparation of a BIM execution plan.

Conclusion

BIM has been popular and used widely in the construction industry in recent years, and in most countries, new legislation has made the use of BIM mandatory in public projects. Therefore, understanding BIM is crucial for success in the modern global construction industry. As we can understand from its name, BIM is related to information, which is valuable for performing every process throughout the life cycle of a construction project, since good decisions can be made only on the basis of the accurate information available.

Traditionally, managing information has generally been ignored in the construction industry. Moreover, due to the fragmented structure of the construction industry, construction projects must reinvent the wheel many times throughout the project. With the introduction of BIM, all the traditional approaches used in construction projects must be changed. In other words, radical changes have taken place in all construction companies that are using BIM. Therefore, implementing BIM depends on three important pillars, namely people, process, and technology, and these three pillars and the relationships between them should be deemed strong if BIM is to be implemented appropriately.

The first pillar, namely technology, comprises the methods and tools used in BIM. With improvements in technology each year, new tools, devices, and software for BIM implementation are introduced. Therefore, people in the industry have dealt with new features and approaches in BIM implementation. In particular, there have been tremendous efforts to increase the interoperability between different software and technologies. Mobility has become an important capability, and construction companies expect that professionals can work at mobile workstations, which shows that cloud BIM has become critical in BIM implementation. Similarly, new systems are proposed and developed for the integration of IoT and BIM, which leads to reliable decisions based on real-time data.

The process pillar comprises the rules and regulations. The processes performed in construction have changed dramatically with the introduction of BIM. Therefore, to manage these new processes, different standards have been proposed by different organizations in different countries. In particular, the standard BS EN 19650 developed by ISO proposes a comprehensive framework for BIM implementation. This standard should be understood and implemented to apply BIM systematically. Other important changes in construction processes compared to traditional processes have occurred in relation to the shareholders involved in construction projects. The environment has become more integrated. In other words, all construction projects demand the involvement of all parties from the beginning of a project to the end of the project. Therefore, integrated project delivery is also crucial in BIM implementation.

The last pillar is related to people. BIM cannot be applied without appropriately skilled people. New roles and responsibilities have emerged, and new capabilities and skills to fulfil these responsibilities should be developed via training and education. Therefore, the other aspect of this pillar is the training and education given to people in construction organizations.

BIM is not only about technology; it is also about process and people. Process is to do with streamlining design construction and FM processes into an integrated building life cycle within which people or stakeholders have clear roles and responsibilities for which they use parametric technologies for authoring, analysis, and simulation purposes. Consequently, technology, process, and people are considered the core pillars of BIM, since it is impossible to implement BIM successfully if these are not tackled coherently. In other words, it is not only the technology pillar that needs to be addressed in implementing BIM but also the process and people pillars too. Hence, all BIM implementation framework developments should offer technological, operational, and behavioural insights.

References

Abanda, F., Mzyece, D., Oti, A. & Manjia, M. 2018. A study of the potential of cloud/mobile BIM for the management of construction projects. *Applied System Innovation*, 1, 9, https://doi.org/10.3390/asi1020009.

Abanda, F. H., Vidalakis, C., Oti, A. H. & Tah, J. H. 2015. A critical analysis of building information modelling systems used in construction projects. *Advances in Engineering Software*, 90, 183–201, https://doi.org/10.1016/j.advengsoft.2015.08.009.

ACIF and APCC (2017). *BIM Knowledge and Skills Framework* [Online]. Available: https://www.acif.com.au/documents/item/799/ [Accessed 12 April 2021].

Ahn, Y. H., Cho, C.-S. & Lee, N. 2013. Building information modeling: Systematic course development for undergraduate construction students. *Journal of Professional Issues in Engineering Education and Practice*, 139, 290–300, https://doi.org/10.1061/(ASCE)EI.1943-5541.0000164.

Alreshidi, E., Mourshed, M. & Rezgui, Y. 2016. Cloud-based BIM governance platform requirements and specifications: Software engineering approach using BPMN and UML. *Journal of Computing in Civil Engineering*, 30, 04015063, https://doi.org/10.1061/(ASCE)CP.1943-5487.0000539.

Anderson, A., Dossick, C. S. & Osburn, L. 2019. Curriculum to prepare AEC students for BIM-enabled globally distributed projects. *International Journal of Construction Education and Research*, 16, 1–20, https://doi.org/10.1080/15578771.2019.1654569.

Antwi-Afari, M. F., Li, H., Pärn, E. A. & Edwards, D. J. 2018. Critical success factors for implementing building information modelling (BIM): A longitudinal review. *Automation in Construction*, 91, 100–110, https://doi.org/10.1016/j.autcon.2018.03.010.

Arayici, Y. 2015. *Building Information Modelling*, London, Bookbon Publisher.

Arayici, Y., Coates, P., Koskela, L., Kagioglou, M., Usher, C. & O'Reilly, K. 2011. Technology adoption in the BIM implementation for lean architectural practice. *Automation in Construction*, 20, 189–195, https://doi.org/10.1016/j.autcon.2010.09.016.

Azhar, S. 2011. Building information modeling (BIM): Trends, benefits, risks, and challenges for the AEC industry. *Leadership and Management in Engineering*, 11, 241–253, https://doi.org/10.1061/(ASCE)LM.1943-5630.0000127.

Barison, M. B. & Santos, E. T. 2010. An overview of BIM specialists. *In*: Proceedings of the International Conference on Computing in Civil and Building Engineering, ICCCBE2010. 141, Nottingham, UK.

Barison, M. B. & Santos, E. T. 2011. The competencies of BIM specialists: A comparative analysis of the literature review and job ad descriptions. *In*: International Workshop on Computing in Civil Engineering 2011, June 19–22, Miami, FL. 594–602.

Bélissent, J. 2010. *Getting Clever about Smart Cities: New Opportunities Require New Business Models*. Cambridge, MA, Forrester, 193, 244.77.

Borrmann, A., König, M., Koch, C. & Beetz, J. 2018. Building Information Modeling: Why? What? How?. *In*: *Building Information Modeling*, New York, Springer.

Bosch-Sijtsema, P. M., Gluch, P. & Sezer, A. A. 2019. Professional development of the BIM actor role. *Automation in Construction*, 97, 44–51, https://doi.org/10.1016/j.autcon.2018.10.024.

Brewer, G., Gajendran, T. & Le Goff, R. 2012. *Building Information Modelling (BIM): An Introduction and International Perspectives*. Newcastle: University of Newcastle Australia.

British Standards Institution. 2013. PAS1192-2: 2013, *Specification for Information Management for the Capital/Delivery Phase of Construction Projects Using Building Information Modelling*, London, UK: British Standard Institution.

BuildingSMART. 2020. *Industry Foundation Classes (IFC) - An Introduction* [Online]. Available: https://technical.buildingsmart.org/standards/ifc/ [Accessed 13 March 2020].

Chong, H.-Y., Wong, J. S. & Wang, X. 2014. An explanatory case study on cloud computing applications in the built environment. *Automation in Construction*, 44, 152–162, https://doi.org/10.1016/j.autcon.2014.04.010.

Cook, C. 2004. Scaling the building information mountain. *AEC Magazine*, 17.

Daniotti, B., Pavan, A., Spagnolo, S. L., Caffi, V., Pasini, D. & Mirarchi, C. 2020. *BIM-Based Collaborative Building Process Management*, Cham, Switzerland, Springer.

Dave, B., Buda, A., Nurminen, A. & Främling, K. 2018. A framework for integrating BIM and IoT through open standards. *Automation in Construction*, 95, 35–45, https://doi.org/10.1016/j.autcon.2018.07.022.

Davies, K., Mcmeel, D. & Wilkinson, S. 2015. Soft skill requirements in a BIM project team. *In*: J. Beetz, L. V. B., T. Hartmann, & R. Amor (ed.) 32nd CIB W78 Information Technology for Construction Conference. Eindhoven, Netherlands.

Deutsch, R. 2011. *BIM and Integrated Design: Strategies for Architectural Practice*, Hoboken, NJ, Wiley.

Egan, J. 1998. Rethinking construction–the Egan report. *In*: *Department of Trade and Industry*, London, Construction Task Force.

Fathi, M. S., Abedi, M., Rambat, S., Rawai, S. & Zakiyudin, M. Z. 2012. Context-aware cloud computing for construction collaboration. *Journal of Cloud Computing*, 2012, https://doi.org/10.5171/2012.644927

Främling, K., Kubler, S. & Buda, A. 2014. Universal messaging standards for the IoT from a lifecycle management perspective. *IEEE Internet of Things Journal*, 1, 319–327, https://doi.org/10.1109/JIOT.2014.2332005.

Gardezi, S. S. S., Shafiq, N. & Khamidi, M. F. B. 2013. Prospects of building information modelling (BIM) in the Malaysian construction industry as a conflict resolution tool. *Journal of Energy Technologies and Policy*, 3, 346–350,

Gaudin, B. 2013. *Impacts of Building Information Modelling (BIM) on Project Management in the French Construction Industry*. MSc thesis, University of Birmingham.

Ghaffarianhoseini, A., Tookey, J., Ghaffarianhoseini, A., Naismith, N., Azhar, S., Efimova, O. & Raahemifar, K. 2017. Building Information Modelling (BIM) uptake: Clear benefits, understanding its implementation, risks and challenges. *Renewable and Sustainable Energy Reviews*, 75, 1046–1053, https://doi.org/10.1016/j.rser.2016.11.083.

Ghosh, A., Parrish, K. & Chasey, A. D. 2015. Implementing a vertically integrated BIM curriculum in an undergraduate construction management program. *International Journal of Construction Education and Research*, 11, 121–139, https://doi.org/10.1080/15578771.2014.965396.

Glick, S. & Guggemos, A. April 2009. IPD and BIM: Benefits and opportunities for regulatory agencies. *In*: Proceedings of the 45th ASC National Conference, Gainesville, Florida.

Gubbi, J., Buyya, R., Marusic, S. & Palaniswami, M. 2013. Internet of Things (IoT): A vision, architectural elements, and future directions. *Future Generation Computer Systems*, 29, 1645–1660, https://doi.org/10.1016/j.future.2013.01.010.

HM Government 2012. *Industrial Strategy: Government and Industry in Partnership: Building Information Modelling*, London, HM Government.

HOCHTIEF-ViCon. 2020. *BIM Wiki* [Online]. Available: https://www.hochtief-vicon.com/vicon_en/BIM-Wiki/BIM-News-10.jhtml [Accessed 2020 14 April].

Holzer, D. 2016. *The BIM Manager's Handbook: Guidance for Professionals in Architecture, Engineering, and Construction*, Hoboken, NJ, Wiley.

Jiang, X. 2011. *Developments in Cost Estimating and Scheduling in BIM Technology*, Boston, MA, Northeastern University.

Khosrowshahi, F. & Arayici, Y. 2012. Roadmap for implementation of BIM in the UK construction industry. *Engineering, Construction and Architectural Management*, 19, 610–635, https://doi.org/10.1108/09699981211277531.

Ku, K. & Taiebat, M. 2011. BIM experiences and expectations: The constructors' perspective. *International Journal of Construction Education and Research*, 7, 175–197, https://doi.org/10.1080/15578771.2010.544155.

Lea, G., Ganah, A., Goulding, J. S. & Ainsworth, N. 2015. Identification and analysis of UK and US BIM standards to aid collaboration. *WIT Transactions on The Built Environment*, 149, 505–516, http://dx.doi.org/10.2495/BIM150411.

Lee, C. Y., Chong, H.-Y. & Wang, X. 2018. Enhancing BIM performance in EPC projects through integrative trust-based functional contracting model. *Journal of Construction Engineering and Management*, 144, 06018002, https://orcid.org/0000-0001-6222-2061.

NAFSA (National Association Of Foreign Student Advisers), COAA (Construction Owners Association Of America), APPA (American Public Power Association), AGC (American General Contractors) & AIA (American Institute of Architects) 2010. *Integrated Project Delivery for Public and Private Owners*. [Online]. Available: https://www.agc.org/sites/default/files/Files/Programs%20%26%20Industry%20Relations/IPD%20for%20Public%20and%20Private%20Owners_0.pdf/ [Accessed 13 April 2021].

National BIM Standard. 2020. *RE: About the National BIM Standard-United States® | National BIM Standard - United States*. [Online]. Available: https://www.nationalbimstandard.org/about/ [Accessed 13April 2021].

Preidel, C., Borrmann, A., Mattern, H., Markus König & Schapke, S.-E. 2015. Common Data Environment. *In*: Borrmann, A., Konig, M., Koch, C. & Beetz, J. (eds.) *Building Information Modeling Technology Foundations and Industry Practice*. Cham, Switzerland: Springer.

Sacks, R., Eastman, C., Lee, G. & Teicholz, P. 2018. *BIM Handbook: A Guide to Building Information Modeling for Owners, Designers, Engineers, Contractors, and Facility Managers*, Hoboken, NJ, Wiley.

Scanlin, J. 1998. The internet as an enabler of the Bell Atlantic project office. *Project Management Journal*, 29, 6–7, https://doi.org/10.1177/875697289802900202.

Schwabe, K., Dichtl, M., König, M. & Koch, C. 2015. COBie: A specification for the construction operations building information exchange. *Building Information Modeling Technology Foundations and Industry Practice*. Cham, Switzerland: Springer.

Scott, W. R. 1998. *Organizations: Rational, Natural, and Open Systems*, Englewood Cliffs, NJ, Prentice Hall.

Shih, R. 2020. *Parametric Modeling with Autodesk Inventor 2020*, Mission, TX, Sdc Publications.

Srivastava, L. 2006. Pervasive, ambient, ubiquitous: The magic of radio. *In*: European Commission Conference "From RFID to the Internet of Things", Brussels, Belgium.

Tang, S., Shelden, D. R., Eastman, C. M., Pishdad-Bozorgi, P. & Gao, X. 2019. A review of building information modeling (BIM) and the internet of things (IoT) devices integration: Present status and future trends. *Automation in Construction*, 101, 127–139, https://doi.org/10.1016/j.autcon.2019.01.020.

The American Institute of Architects. 2007. *Integrated Project Delivery: A Guide*, Sacramento, CA, The American Institute of Architects.

Uhm, M., Lee, G. & Jeon, B. 2017. An analysis of BIM jobs and competencies based on the use of terms in the industry. *Automation in Construction*, 81, 67–98, https://doi.org/10.1016/j.autcon.2017.06.002.

UK BIM Alliance. 2019. Guidance Part 1: Concepts. *Information Management According to BS EN ISO 19650*, London, UK BIM Alliance.

UK BIM Alliance. 2019. *Standards & Guidance* [Online]. Available: https://ukbimframework.org/standards-guidance/ [Accessed 14 April 2020].

Van Nederveen, S., Beheshti, R. & Gielingh, W. 2010. Modelling concepts for BIM. In: Underwood, J. & Isikdag, U. (eds.) *Handbook of Research on Building Information Modeling and Construction Informatics: Concepts and Technologies*, Hershey, IGI Global.

Walker, D. & Llyod-Walker, B. 2019. Characteristics of IPD: A collaboration framework overview. *In*: Walker, D. & Rowlinson, S. (eds.) *Routledge Handbook of Integrated Project Delivery*, New York, Routledge.

Wong, J., Wang, X., Li, H. & Chan, G. 2014. A review of cloud-based BIM technology in the construction sector. *Journal of Information Technology in Construction*, 19, 281–291.

4 Collaboration in BIM-based construction networks

Bimal Kumar and Benny Raphael

Introduction

BIM (building information modelling) is currently a huge topic of discussion in the construction industry in most parts of the world (Eastman et al., 2018). Several countries have mandated the use of BIM on all publicly procured projects, and others are considering this seriously. However, the nature of discourse around BIM seems to be quite confusing in many sections of the industry. This stems from a lack of understanding of the different aspects of BIM. That said, there is generally an awareness that BIM-driven project procurement and delivery is largely driven by promoting effective collaboration among all stakeholders in a typical construction project. As discussed in later sections, BIM is not just a technology (or tool), as perceived by many people, but a whole new way of procuring and delivering built assets driven by BIM processes, standards, and protocols. When one considers the BIM tools, processes, and standards as a unified package to implement effective collaboration through seamless information exchange, BIM is more akin to building information management than just building information modelling. This chapter examines the effectiveness of BIM tool(s) in implementing effective collaboration. Several key models of collaboration are first examined before being juxtaposed with BIM to assess how effective it is in relation to these collaboration models and the related technical requirements for implementing their key aspects.

Models of collaboration

Collaboration in teams – cognitive and psychological models

Any construction project involves synthesizing the work done by more than one party towards the production of a common deliverable. While in a construction project the final deliverable is a constructed artefact, at intermediate stages, deliverables are pieces of information in the form of drawings, specifications, reports, or structured information. Since practical applications of computers in construction projects are mostly limited to the production and processing of information, the scope of this paper is restricted to project tasks that aim to generate digital information. This includes tasks such as design, planning, management, and monitoring of the construction process.

The project teams are formed based on an understanding of the common top-level goals. These top-level goals are iteratively refined during the course of the project by the participants. During this process, they share a common interest in successful project completion because it ensures profit for all (Peña-Mora and Wang, 1998). This is the primary incentive for working together, and they cooperate so that the project will be completed on schedule and within the budget. At the same time, individual parties might be driven by self-interests such as maximizing individual profits, reducing the amount of work for oneself, or improving individual work at the expense of others so that they are able to obtain a better share of credit and fame from the project. It is these psychological factors that create competition among project members and make collaboration difficult.

Collaboration is inevitable in complex projects because a single party is unable to complete all the work on time or does not have the expertise to work on all aspects of the project. Ideally, the project should be broken down into tasks that can be executed in parallel and independently. That is, each party generates different pieces of information that are not interrelated. In this case, synthesizing the work done by all the parties becomes a trivial task. However, in practice, there is overlap in the information that is manipulated by multiple parties, and they cannot work in isolation. This is illustrated in Figure 4.1.

Each party in the project contributes bits of information which, when aggregated, become the project deliverable. As an analogy, if the project deliverable is viewed as a set of drawings, each piece of information contributed by a collaborator can be considered as one drawing. The drawings that are shown within the overlapping parts of the circles are developed by more than one party. Arrows linking the drawings are dependencies between information. In this analogy, if the content of one drawing is changed, related drawings have to be modified. The information generated by each party is a function of their private knowledge, assumptions, and interpretations and the resources available, as well as self-interests. The factors that influenced the decisions are not made public and cannot be easily inferred from the

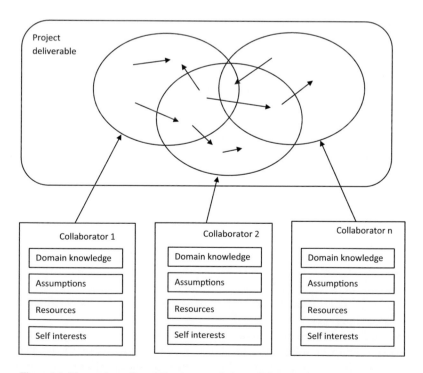

Figure 4.1 Illustration of contributions made by collaborators to the project deliverable.

information that is published in the project deliverable. Also, the dependencies are not explicitly stored and are known only to the parties that generated the information. The number of drawings and their structure are unknown at the start of the project. The project manager along with the team members have to ensure that the final set of drawings is logically consistent and meets the requirements of the client.

In the drawings analogy, the act of collaboration comes down to the following:

- exchanging relevant information (communicating drafts of relevant drawings);
- analyzing and interpreting new information (proofreading and reviewing);
- seeking explanations, critiquing, and proposing alternatives (discussions);
- synthesizing and incorporating new information into collaborator's task (redoing affected drawings).

All the above might require several cycles to reach successful completion. These activities might be carried out with or without computer support. For example, the proposal of solutions, analysis, and discussions are traditionally done in face-to-face meetings through verbal communications. Effective project managers make sure that the cycles are reduced and conclusions are reached quickly.

Due to more and more globalized projects in which collaborators are geographically separated, electronic communications media are becoming popular and are replacing face-to-face meetings. Teleconferencing, video conferencing, bulletin boards, and instant messaging are increasingly used. Electronic whiteboards and other online collaboration tools are currently available, but there is no evidence that these are widely used in construction projects. Among non-IT savvy team members, email and fax are still the preferred options when face-to-face meetings are not possible.

Whatever might be the mode of communication, the traceability of the information that is communicated is important. It might contain the reasoning that shaped the design, that is, the design rationale (Peña-Mora et al., 1995). This information is usually not available in the drawings, specifications, or CAD files which form part of the project deliverables. Such information is important to subsequent decision-making and, if used effectively, can create better optimal designs and prevent delays in the design and construction process. Traditionally, this information is captured in the minutes of project coordination meetings, which are circulated among all the project team members. When meetings are conducted over electronic media, new mechanisms have to be developed to capture such information.

The use of electronic communication media has exacerbated the difficulties with activities such as interpretation, critiquing, and synthesis of information. The lack of visual and nonverbal expressions to aid understanding is a handicap and could result in misunderstandings of intent, seriousness, and urgency.

Synthesizing new information and modifying solutions to adapt to it has always been a challenge even in traditional collaboration approaches. Lottaz et al. (2000) argue that generally, project partners do not welcome new ideas because it is a heavy burden to analyse their implications and determine inconsistencies. This is because very little knowledge is represented explicitly, and checking for inconsistencies is essentially a manual process.

Collaboration in teams – computational models

Work related to computational models for collaboration can be classified into groups depending on the issue that has been addressed. The following is a hierarchical decomposition of topics, issues, and solution approaches found in the literature on concurrent and collaborative engineering:

- Representation of information
    - What (Semantics): Product models, knowledge bases, design rationale
    - How (Syntax): Graphical databases, relational databases, conventional documents, CAD file formats, text-based structured representations such as XML
- Storage of information
    - Distribution: Centralized, decentralized
    - Access control, security
- Communication
    - Information loss, avoiding misunderstandings
- Use of information
    - Presentation of information
    - Interoperability
    - Search for information, querying
    - Maintenance
    Concurrency, version management, consistency management: Conflict identification, negotiation, conflict resolution, conflict mitigation
- Team interaction models
    - Shared workspaces
    - Recording communications
    - Design synthesis

A few of the above issues are discussed in the following sub-sections. Only critical issues that affect BIM-based collaboration are covered here.

Representation of information

Representation of information is of fundamental importance to all tasks in computer-aided engineering (Raphael and Smith, 2013). An exhaustive survey on this topic is outside the scope of this chapter. However, work specifically related to collaborative and concurrent engineering are briefly mentioned here.

The representation issue deals with *what* information needs to be represented and *how* this should be represented for effective collaboration. Since the construction industry is fragmented and involves many stakeholders, there has been no agreement on the answers to these questions for a long time. For example, architects and engineers could not agree on a common definition for everyday objects such as walls. Standard definitions and representations are essential for ensuring interoperability (See Chapter 5). The meaning of data, formally known as *semantics*, must be made explicit so that different systems can exchange information without ambiguity and all the stakeholders have a common understanding. The importance of standardization was recognized quite early on (Fischer and Froese, 1992; Björk and Penttilä, 1978, Björk 1989). Considerable effort was put into the development of standard representations known as product models. Industry Foundation Classes (IFC) has emerged as a standard that has achieved universal recognition.

The issue of how to represent information is equally important. From the early days of computing, drawings were used to represent engineering information. CAD file formats such as DXF were developed for representing drawings. However, computers could not extract useful information from drawings because the knowledge to interpret the geometric information was not available. For example, a set of lines might represent a beam, but the knowledge to extract the properties of the beam is not stored in the drawing. In a way, this limitation of conventional drawings gave rise to the developments in BIM. Structured representations are needed for automating information processing by computers. Relational databases are a popular method for structuring data. In a relational database, all the data is decomposed into tables with a well-defined structure. The idea of decomposing data into smaller parts is crucial from the point of view of collaboration, because it enables multiple parties to work independently on the different parts. However, relational databases are not ideal for many applications. For example, data within an image may not be easily converted into tables such that it can be efficiently processed. Semi-structured representations involving text, images, and structured meta-data are becoming increasingly popular. Another requirement that has come about with the adoption of standards such as IFC is a standard format to represent the decomposition of objects into parts and sub-parts. Text-based formats that are capable of capturing this information and efficiently processing it have been developed. Examples include STEP (STEP Tools 2007) and XML (extensible markup language).

While product models focus on standard representations of commonly used engineering objects, collaboration involves sharing more than product information. Key components that are necessary to facilitate a comprehensive level of collaboration have been identified by Scherer and Katranuschkov (2006). These include a standard global schema such as IFC, ontologies providing commonly agreed interpretations for concepts that exist within and beyond the global schema, standardized constructs extending the global schema to support domain- and discipline-specific needs, model views defining reusable units of information from the global schema or its domain extensions, and rules and methods to ensure consistency of the data throughout the life cycle of the building. It is important to note that a standard data model such as IFC alone does not enable effective collaboration. Several types of knowledge and tools are required. Many researchers have articulated the need to shift emphasis from data to knowledge. Wong and Sriram (1993) argue that the representation of product information which supports communication and coordination should include not only the geometric and physical properties of the product and its parts, but also information about functions, constraints, and design rationale. Work related to the use of design rationale in collaboration include Peña-Mora et al. (1995). Another knowledge item that is essential for consistency checking is the relationships between design objects. Rosenman et al. (2007) categorize relationships as intra-disciplinary and inter-disciplinary. They argue that inter-discipline relationships are an essential part of multi-disciplinary collaboration, and such relationships need to be explored and included in IFC standards. Even though modelling relationships has not been given sufficient attention by the product modelling community, this is a widely researched topic in concurrent engineering. Representing relationships as constraints and the use of design structure matrices are described in later sections.

Storage of information and access control

Early works on collaboration were based on the concept of a central database hosting all the information. Examples of such systems can be found in Fenves et al. (1994), Faraj et al. (2000), Lee et al. (2003), etc. This model facilitates simultaneous modification of the model by collaborators in a project, and each party is able to access the latest data at any time. Distributed storage of project data (Scherer and Schapke, 2011, etc.) and the use of cloud services are a more recent development. This improves the reliability of data access and helps to distribute the computational load among several machines.

Lots of work on computer-assisted collaboration has been centred on document management. Electronic document management systems (EDMSs) provide shared access to documents among team members. The validation and distribution of documents are based on rules set up on the server. Apart from controlling access to documents, EDMSs also provide different ways of communication, including instant messages, emails, and video conferencing. Project extranets (Wilkinson, 2005) were developed as collaboration-driven EDMSs and have been quite successfully adopted, particularly in larger projects, since about 2000.

When data is stored on a remote server which can be potentially accessed by anybody across the globe, security is a major concern. Even though security systems are getting stronger with advanced encryption and authentication methods, enterprises are hesitant to host their data outside their intranet because of the many reports of security breaches that are emerging almost daily. Collaboration tools will never be widely adopted for professional work until these security concerns are fully addressed.

Presentation of information

Some collaboration models are based on the idea that only the data is stored in a central location, and the presentation of information is done locally by downloading the required information. Visualization software provides 3D renderings and animations of the design individually for each project partner, without any interaction with other collaborators. A recent development is the use of mobile devices for visualization, which has the potential to extend the use of BIM to the construction stage for applications such as progress monitoring.

Rosenman et al. (2007) argue for a collaborative virtual environment that provides 3D visualization, walkthroughs, and rendering to allow for communication of the various views of the design as modelled by the different disciplines. According to them, the main advantage of a virtual world environment is that it allows users to be immersed in the environment, allowing for real-time walkthroughs and collaboration. The idea of a shared workspace is also presented by other authors (Maher and Rutherford, 1997). The shared workspace is a means of sharing all or part of a desktop with other users and facilitates concurrent or synchronous problem solving.

Interoperability

Scherer and Schapke (2011) identify three types of interoperability problems that are prevalent in construction projects. Firstly, *horizontal* interoperability problems exist between management and various engineering teams involved in a project. This is due to the different ways in which multiple disciplines model the same information. Different project partners will continue to use distinct engineering and management models optimized for their specific application domain. Secondly, *vertical* interoperability problems result during adding or aggregating details for supporting decision-making at different management levels. That is, different amounts of detail are needed for different activities, from conceptual design to construction implementation. Thirdly, *longitudinal* interoperability problems might exist among model versions generated at different points in time.

Current efforts in BIM development seem to focus mainly on the horizontal interoperability problem. The emphasis is on the exchange of model and other information between software applications used by architects and engineers for visualization and simulations.

As the project progresses, several versions of the model might be developed. To further complicate matters, each part of the model might have multiple versions as a separate team might be working on it. Keeping track of the versions of parts and sub-parts is not easy when the designs are changing dynamically with input from many people simultaneously. Construction projects are normally executed concurrently, meaning that tasks are executed in parallel. In this situation, computational support is important for ensuring that the correct versions are used for all the parts.

What makes collaborative work challenging is the interdependencies between solution parts which are developed by separate teams, as shown in Figure 4.1. Since each construction project is unique, the dependencies cannot be identified a priori and enumerated exhaustively. Many dependencies are "discovered" during the course of project execution through communication between collaborators. The hidden dependencies result in inconsistencies in design during changes. The topic of methods to ensure data consistency does not seem to have generated sufficient attention from BIM researchers and practitioners. However, a huge body of literature exists on this topic within the context of concurrent engineering.

In this discussion, a distinction is made between checking for inconsistencies and resolving inconsistencies, since the former is relatively easy compared to the latter. Consistency checks can be performed using many approaches, including rule-based systems (Raphael and Smith, 2013, Chapter 9), models based on physical principles (Raphael and Smith, 2013, Chapter 9), and case-based reasoning systems (Kumar and Raphael, 2001). In this respect, work related to the representation of codes of practice and standards (McGibney and Kumar, 2013) are important, since they are knowledge sources for consistency checks.

The type of inconsistencies might be structural or semantic. Structural inconsistencies are easier to detect as rules governing these are usually well formalized in product data models. For example, if a beam and a part of ductwork overlap in a model, it can be easily detected by checking their geometry. Semantic conflicts are more difficult to locate, and this requires additional knowledge. Automatic consistency checking requires explicit representation of this knowledge. In an open world scenario, where it is impossible to encode all the knowledge involved, human input is necessary to perform consistency checks. Decision support systems that assist human experts in this task are relevant.

Factors that cause data to be in an inconsistent state are called *conflicts*. Constraint satisfaction methods have been proposed for managing conflicts (Raphael and Smith, 2013, Chapter 7; Panchal et al., 2007). Constraints capture the relationship between variables such that the solution is valid and satisfies all the requirements. There have been other approaches to capturing dependencies between project activities. Hegazy et al. (2001) represent dependencies between building components within an information flow model. Peña-Mora et al. (1995) present an approach to conflict mitigation by representing design rationale. This research is based on the view that the designers' perspectives are expressed in their design rationale and a system for capturing the design rationale needs to represent and manage design intent evolution, artefact evolution, and relationships between intents and between intent and artefact. They propose an ontology for design rationale and a system for conflict mitigation based on this ontology. In a similar work, Peña-Mora and Wang (1998) present a methodology for facilitating the negotiation of conflicts during the development of large-scale civil engineering projects based on game theory and negotiation theory.

An approach that has been shown to have good potential in conflict management is the design structure matrix (DSM) (Steward, 1981; Zhao et al, 2010). DSM presents dependency relationships between activities by means of a matrix. This matrix captures the input and output information of all activities, that is, what information is needed to start a particular activity and which other activities utilize its output. The advantage of DSM is the ability to provide a systematic mapping among elements compactly (Maheswari et al., 2006). Senthilkumar et al. (2010) present an approach to using DSM for design information modelling of construction projects. The approach is based on conventional design drawings for exchanging information. Drawings are considered as the input and output of design activities in order to identify interface issues and relationships.

Collaborative design synthesis

There are two strategies that are commonly employed for effectively synthesizing contributions of interacting designers. These are based on centralized and decentralized decision-making. Centralized decision-making requires the transfer of knowledge from various domain experts to a single decision-maker. The central decision-maker identifies good solutions through simultaneous consideration of system-level tradeoffs. Decentralized decision-making requires a two-way flow of information between decision-makers as well as the active involvement of domain experts throughout the decision-making process. In multi-functional design, different designers and domain experts control a common set of design variables and share responsibility for achieving different objectives.

Point-based methods are conventionally adopted in decentralized decision-making. A solution point is initially proposed, which is evaluated and critiqued by domain experts. The point is iteratively modified to account for one defect at a time until an acceptable solution is found. A critical problem with point-based methods is that the process may never converge and might result in sub-optimal solutions. To avoid the problem of divergence, set-based approaches have been proposed. Instead of working with a single point at a time, feasible ranges of values of design variables are evaluated by domain experts (Toche et al., 2020).

The table below shows the relationship between the different aspects of collaboration with other key dimensions for their implementation. Implementation details are discussed in the "Approaches and implementations" section. Acronyms used in the table: PE – project extranet, CDE – common data environment.

	Hardware	Software	Standards	PE	CDE	BIM Server
Interoperability			✓		✓	✓
Communication	✓	✓	✓	✓	✓	✓
Storage						✓
Access management		✓		✓	✓	✓
Version management		✓		✓	✓	✓
Shared workspaces		✓			✓	✓
Collaborative design synthesis	✓	✓	✓			
Consistency management		✓	✓		✓	✓
Conflict management		✓	✓		✓	

Collaboration in construction

Due to the complexities and the large volume of work in construction projects, separate engineering and architecture teams have to work in parallel to produce parts of the design that have to be integrated into a consistent product. The parts of the design solution are shared with other team members only at critical stages. This process involves long periods of working in isolation while searching for an acceptable solution. The initial conceptual design is developed by the architect and is shared by the rest of the team. For a commercial building, this usually consists of information such as the orientation of the building, number of building blocks, number of storeys, height of each storey, spacing of structural grid, etc. Details are added to the conceptual design by various parties in order to develop the final design drawings. Architectural design teams prepare detailed drawings of room layouts with other architectural elements such as façade, shading, doors, windows, etc. Structural engineers perform the design of the structural system and the sizing of structural members. Mechanical and electrical (M&E) engineers prepare drawings showing the location of plants, ducts, electrical wiring, and equipment. During this process, a number of specialist software tools are used by each party. Structural engineers might use a finite element analysis package, while M&E engineers might perform energy simulations and HVAC plant simulations. The type of data and the level of detail required by these software tools are different, and frequently assumptions have to be made. For example, HVAC engineers might have to assume the slab thickness to estimate the available space above the false ceiling since the structural design has not been completed. Inconsistencies in assumptions and conflicts in the design are detected only when the completed designs from all the parties are compared. This results in iterative loops in the design process. The completed design drawings are given to the construction contractors, and it is not uncommon to make design changes after the construction has started.

Early work on computer-based collaboration focused mainly on the exchange of information between software applications, that is, they primarily addressed the interoperability problem. The integration of software for different activities in the construction project was the main concern. The use of shared product models was identified to be a key element in integration. Several publications that use this approach can be found in the last two decades. Lee et al. (2003) present an integrated information system called a design information management system (DIMS), which uses a representation of design information based on IFC along with schedule, estimation, and structural engineering information. Roddis (1998) describes an integrated CADD environment called SteelTeam (Pasley et al., 1996; Pasley and Roddis, 1994), which uses an object-based building data model for downstream communication from designers to the parties involved with detailing, fabrication, and construction.

In spite of great efforts towards improving the interoperability of software applications used in the construction industry, seamless integration is far from being achievable. Since the number of specialist domains and applications are increasing, it seems that complete integration can never be achieved. In order to avoid the complexity of achieving true interoperability, process models have been developed for integrating selected applications used by designers, consultants, and contractors.

Multi-models have been proposed for integrating heterogeneous application data (Scherer and Schapke, 2011). This approach is based on the assumption that project collaboration in the construction phases cannot be enabled through one central project model or database. Different project partners will continue to use distinct engineering and management models optimized for their specific application domain. To allow for the modification, re-use, and synchronization of these models, they propose a distributed process-centric approach based on multi-models. The main idea of a multi-model is to combine distributed application models, or selected views of them, in a single exchangeable information resource. Within the resource, the application models are bound together by link models. The link models explicitly specify the interdependencies among the application models referencing the respective model elements by their identifiers (ID). The resulting compound model is stored in a persistent data store or multi-model container.

Lottaz et al. (1999) propose the use of solution spaces instead of single solutions to support collaborative tasks during design and construction. In the web-based platform they have developed, collaborating partners specify constraints instead of single solutions to their subtasks. Solutions to the complete project task are obtained through solving the constraint satisfaction problem formed by all the constraints. The application of this approach is illustrated using a real construction case involving a steel-framed building in Lottaz et al. (2000). The concept of using solution spaces for collaborative design can be found in more recent publications, for example, Rempling et al. (2019) and Raphael (2011).

A number of researchers have proposed the use of knowledge-based systems (KBSs) containing heuristics for improving collaboration and communication. SteelTeam (Roddis, 1998) contains a KBS used by the designer as a decision support tool representing upstream communication of the concerns of the detailer, fabricator, and constructor.

Defining BIM from the viewpoint of collaboration

Other chapters in this book discuss the meaning of BIM at great length. Here some concepts are re-visited from the viewpoint of collaboration. According to accepted definitions, BIM stands for *building information modelling*, but although when the *M* in BIM stands for *modelling*, we are essentially referring to the BIM technology, in the context of collaboration, it is more appropriate for *M* to stand for *management*.

It should be mentioned that one of the problems with the whole BIM "thing" is that BIM technology hit the scene way before BIM processes, protocols, and standards. The upshot of this is that someone familiar with BIM technologies takes it as such and no more. In fact, it would arguably be more appropriate if the *B* in BIM should stand for *building* as a verb rather than a noun, implying its scope to be vertical (building) structures as well as horizontal (civil infrastructure) structures like bridges, tunnels, etc. Similarly, the *IM* in BIM should more appropriately stand for *information management* rather than *information modelling*. The wider connotation of BIM is much bigger than just modelling and has huge implications for overall information management for the entire asset lifecycle, from inception to demolition. This includes modelling but is not restricted to it. In fact, a lot of the misconceptions about BIM arise because of this historical fact more than anything else. Therefore, BIM normally means not just the technology but also the associated set of processes, standards, and protocols for information exchange. All these aspects of BIM have to work together as illustrated by the diagram in Figure 4.2. However, as the diagram also suggests, despite tools (technologies), standards, and processes working perfectly, the biggest influence as always is that of people, economics, and political issues, which may make or break a perfectly good and effective way of procuring projects using BIM.

It should be pointed out that there are several examples of BIM uses that relate to the use of technology almost in complete isolation from the processes and standards and protocols for information exchange that are so vitally important for getting the real benefits of BIM in a project. These uses are sometimes referred to as *lonely BIM* uses. It is important to appreciate that although an organization can gain significant benefits internally through the use of BIM technologies, unless associated processes, standards, and protocols are in place for the whole project team to sign up to, the real benefits to the project, and hence the client, are not going to accrue. This point cannot be over-emphasized, it is so important. In order for the client

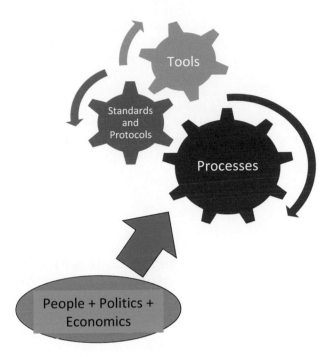

Figure 4.2 Key elements of BIM implementation in AEC projects.

organizations to get real benefits from BIM, *collaborative BIM*, which is more akin to building information management, needs to be implemented.

In summary, in the context of this chapter, perhaps a more apt definition of BIM could be the following:

> *BIM provides a framework for facilitating collaboration between all stakeholders in a construction project. This platform is a collection of processes, standards, and contractual protocols underpinned by hardware and software technologies as enablers of collaboration.*

BIM-based collaboration

In order for a BIM-based collaboration to be implemented, there are a number of pre-requisites that need to be in place. There are now a number of guidelines available for this, most notably the ones published by the UK BIM Task Group (2013).

A high-level workflow for effective BIM-based collaboration should consist of

- Processes for capturing information requirements along with format and schema specification by the owner's team.
- Processes for capturing and evaluating information generated at each stage of the asset design and construction lifecycle. These are typically called data drop-points in some countries. These data drops have to be evaluated against the initial information requirement specification mentioned above to ensure that the client is getting what they asked for.

- BIM project execution plans (Wilkinson, 2005) for information delivery along the whole asset life cycle.
- Protocols and contracts to bind all stakeholders to the required information deliverables in the required formats at the different data points. These should require the appointment of an *information manager* for the project, which is typically not the case in this industry traditionally.

If these pre-requisites are not in place, any use of BIM technologies for modelling may well deliver localized benefits to the stakeholders using the technology, but any project-level and asset life cycle benefits will not be achieved. More recently, these processes are being adapted for international application through the publication of the ISO19650 series of guidance documents (UK BIM Alliance, 2020).

Approaches and implementations

BIM servers

Exchange of information among applications in the project life cycle can be facilitated by centralized servers hosting BIM data, called *BIM servers*. BIM servers provide several services such as supplying relevant partial models, merging and integrating changed data into the central model, searching and querying, access control and security, version control, transaction processing, etc.

Since integrated BIM data might be voluminous, it is not advisable to transfer the complete model each time a project team member needs to work on his solution. Therefore, the BIM server should be able to supply a partial model instead of a complete IFC file. In the commonly adopted terminology, the team member "checks out" a partial model. After the team member has finished working on his design, he "checks in" the modified partial model. The modified model is verified by the server for consistency; the changes are identified and finally merged into the central model. Version control and transaction processing capabilities are important for ensuring data integrity and consistency.

The BIM server should also allow easy searches for information, and query languages should be provided for locating information precisely, similar to retrieving data from traditional relational databases. Since navigating through BIM data is most naturally done visually, elementary visualization capabilities should be built into BIM servers.

Rights and permissions are granted to users based on their roles in the project, and these are assigned by a system administrator representing the project manager. Since roles and responsibilities change during the course of the project, it is advisable to have a flexible system for granting access permissions so that individuals are able to easily share information with team members.

Workflow

Several workflow configurations and scenarios are possible with a central BIM server. For example, the initial architectural design is used to develop the first model which is uploaded into the server. Electrical and mechanical design teams download this model as an IFC file and create their designs using their specialized software. The newly generated BIM elements are then merged into the integrated model on the server. Documents for design reviews are generated from the integrated model, which is studied by all the team members and problems are reported. The process continues until all the issues are resolved and detailed design drawings are generated. Several other workflow scenarios can be considered. It is the responsibility of the project management team to develop an efficient BIM-based workflow that is adapted to the particular characteristics of each project.

Version management

Versioning is one of the techniques adopted for ensuring the consistency of models in a distributed environment. Koch and Firmenich (2011) make a distinction between state-oriented and change-oriented versioning approaches. In state-oriented versioning, snapshots of different versions of the evolving design solution are stored. On the other hand, in change-oriented versioning, actions that produce the changes to the solutions are stored. The authors argue that by storing the design operations, design intents can be represented, and there is better capability to ensure the consistency of the model by easily identifying changes. It should be remarked that the change-oriented versioning resembles a derivational analogy approach (Kumar and Raphael, 2001) to case-based design.

Project extranets

As mentioned earlier, project extranets (PE) were devised around the turn of the century to facilitate communication between project stakeholders through a central project repository. A lot of the issues mentioned earlier like version management, file sharing etc. were implemented in project extranets. The hardware and other IT infrastructure required to implement these repositories were just becoming readily available at the time, which facilitated their implementations. However, a number of key components for the seamless information exchange that is critically required for true collaboration were still not in place. Therefore, although in one sense PEs served a great purpose and can be regarded as an important milestone towards facilitating collaboration, they still left a lot of unanswered questions.

Figure 4.3 above highlights the challenges of communication without having a common language. It is quite common in large projects in the architecture/engineering/construction (AEC) sector to have key stakeholders from different parts of the world often speaking a different *language*. It is easy to see that if the uploaded project documents in the central repository are in different languages; the challenges to effective communication are obvious. Extended to real project scenarios, this analogy means that the *language* of communication relating to different elements of an asset should be common to all the stakeholders, i.e., the parameters used to convey information about any object should be the same for all stakeholders. It is well known that the genesis behind industry-wide common data models like IFC (Industry Foundation Classes) is exactly this. IFC has contributed greatly to streamlining the exchange of information with the AEC sector and is a key factor in the emergence of BIM in the industry. However, IFC also suffers from several limitations. Data loss can happen both in importing from and exporting to the IFC format. In addition, for the IFC model to facilitate full interoperability between applications, it would have to be a superset of all their data models, which would be a near-impossible task. It is important to keep this in mind so that expectations from the IFC do not exceed what is realistically possible. Due to problems with this relatively uncontrolled (apart from access controls) approaches like project extranets, a more advanced project repository system was proposed in BS1192:2007 called the *common data environment* (CDE). ISO19650 Part 1 and 2, which are largely based on BS1192:2007 and PAS1192:Part 2, have now superseded the UK standards, and these

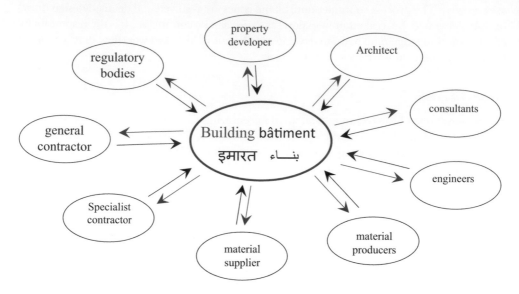

Figure 4.3 A federated project communication model without a *common language*.

international standards are to be adopted worldwide with national annexes added to comply with any local national and local issues. But the concept of CDE has been largely left intact, as proposed in BS1192:2007 (Figure 4.4). CDE has a number of internal checks and balances in place within the repository, as shown above, implemented as gateways to ensure that the information being shared satisfies certain constraints before being shared with other project stakeholders. CDE also implements a life cycle for the shared information by categorizing them as *WIP*, *Shared*, *Published*, and *Archived*. These enhancements to the traditional project extranet have had a profound effect on streamlining information exchange in the construction industry.

CDE has been adopted as a key component in ISO19650 series of documents and forms one of the main mandatory elements in its implementation in projects. There are several examples of the use of CDE (Pathfinder report, 2018) in real projects in the United Kingdom and elsewhere, and it would appear that project extranets have largely been replaced by CDEs in large projects.

Discussion, summary, and conclusions

It is easy to consider the *sharing* of files as *collaboration*. In one sense, it is true that sharing information like one might do through several file sharing technologies (like MS Sharepoint) may be considered to be a necessary technical requirement for collaboration, but as mentioned earlier, a true collaboration between stakeholders is a lot more than simply file sharing. As Figure 4.3 indicates, unless the *sharing* has a *common language* underpinning the collaborative activities, it may not be very effective. BIM-based collaboration has its main objective that *all stakeholders in any project share and exchange information in a pre-agreed standard format which is enabled by appropriate technologies, processes, and standard protocols*. BIM attempts to achieve this by having all these ingredients of effective collaboration in place as indicated in Figure 4.2. Therefore, it can be concluded that BIM-based collaboration facilitates much more than just file sharing

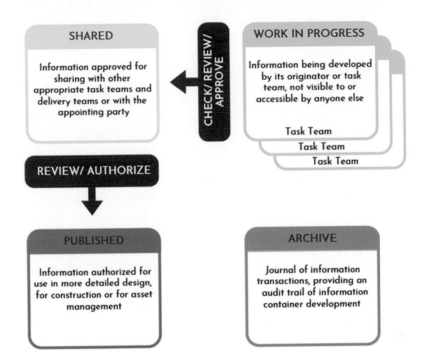

Figure 4.4 Common data environment (CDE) as envisioned in ISO19650 Part 1 Figure 10.

as project extranets (for example) and goes a long way towards true collaboration. That said, it does have some limitations, and these could be categorized as technological and legal as well as professional.

On the *professional* level, the limitations relate to the challenges around a truly integrated model development, which needs greater collaboration and communication across disciplines (Singh et al., 2011). New communication modes and tools are needed for collaborative model development where multiple parties contribute to a centralized model. This will inevitably lead to new roles and relationships within project teams. In general, dedicated roles, such as BIM model manager and BIM server manager, will be inevitable for complex projects (Singh et al., 2011).

A survey conducted by Howard and Björk (2008) among experts in the construction industry indicated that BIM solutions appear too complex for many and may need to be applied in limited areas initially. There is general agreement that knowledge about building requirements should be modelled in a way that integrates as much of the data as possible. However, current solutions which are complex fall short of the ideals that have been sought since the 1970s, and special information-manager roles are needed in project teams to coordinate the use of models throughout the project. This ideal, at least to an extent, did become a reality by the introduction of the role of *information manager* as proposed by the UK BIM Task Group (2013) for all construction projects from 2016. However, this is not the case elsewhere in the world. Finally, it should be noted that the future of BIM-based collaboration will most likely be based on Industry 4.0 ideas by linking it to IoT and Big Data, working as a network of systems feeding data to each other to enable critical decision-making in real time right across the spectrum of design, construction, and facilities management.

Summary

For effective collaboration using BIM, several issues need to be addressed. These are related to the representation, storage, use, and communication of the information that is needed by the different stakeholders in a project. While several solutions have been proposed for addressing these issues, maintaining consistency in data remains a challenge because the rationale for choosing particular options is not explicitly stored in the project database, and it is not possible to automatically check whether changes are consistent with the assumptions made by the consultants. Since human involvement is inevitable for checking logical consistency, the software architecture and workflow should be carefully designed to facilitate this.

Questions

1. What are the options available for capturing dependencies between variables in a project database?
2. What additional information is needed for effective collaboration, which is currently not represented in standards such as IFC?
3. Discuss the difference between structural and semantic inconsistencies.
4. What are the types of consistency checks that a BIM server might be able to execute?
5. Give examples of structured, unstructured, and semi-structured representations of project information.

References

Björk B-.C. (1989), Basic structure of a building product model. *Computer-Aided Design*, 21 (2), 71–78.

Björk, B.-C., and Penttilä, H. (1978), A scenario for the development and implementation of a building product model standard (1989). *Advances in Engineering Software*, 11 (4), 176–187.

British Standard Institution. (2007), BS1192: 2007, *Collaborative Production of Architectural, Engineering and Construction Information: Code of Practice*, BSI.

Eastman, C., Teicholz, Paul M., Sacks, R., and Lee, G. (2018), *BIM Handbook: A Guide to Building Information Modelling for Owners, Managers, Designers, Engineers and Contractors*, Third edition, Wiley.

Faraj, I., Alshawi, M., Aouad, G., Child, T., and Underwood, J. (2000), Industry foundation classes Web-based collaborative construction computer environment: WISPER, *Automation in Construction*, 10 (1), 79–99.

Fenves, S., Flemming, U., Hendrickson, C., Maher, M.L., Quadrel, R., Terk, M., and Woodbury, R. (1994) *Concurrent Computer-Integrated Building Design*, Prentice-Hall, Englewood Cliffs, N.J.

Fischer, M., and Froese, T. (1992), Integration through standard project models. In Proceedings CIB W78 Workshop on Computer-Integrated Construction, pp. 189–205. Montreal, 12–14 May.

Hegazy, T., Zaneldin, E., and Grierson, D. (2001), Improving design coordination for building projects. I: Information model. *Journal of Construction Engineering and Management*, 127 (4), 322–329.

Howard, R. and Bjork B.-C. (2008), Building information modeling: Experts' views on standardisation and industry development. *Advanced Engineering Informatics*, 22 (2), pp. 271–280.

Koch, C., and Firmenich, B. (2011). An approach to distributed building modeling on the basis of versions and changes. *Advanced Engineering Informatics*, 25 (2), 297–310.

Kumar, B., and Raphael, B. (2001), *Reconstructive Memory in Design Problem Solving*, Saxe-Coburg Publications.

Lee, K., Chin, S., and Kim, J. (2003), A core system for design information management using industry foundation classes. *Computer-Aided Civil and Infrastructure Engineering*, 18, 286–298.

Lottaz, C., Clément, D.E., Faltings, B.V., and Smith, I.F.C. (1999), Constraint-based support for collaboration in design and construction. *Journal of Computing in Civil Engineering*, 13 (1), 23–35.

Lottaz, C., Smith I.F.C., Robert-Nicoud Y., and Faltings B.V. (2000), Constraint-based support for negotiation in collaborative design. *Artificial Intelligence in Engineering*, 14, 261–280.

Maher, M.L., and Rutherford J. (1997), A model for collaborative design using CAD and database management. *Research in Engineering Design*, 9 (2), pp. 85–98.

McGibbney Lewis J, and Kumar, B. (2013), A comparative study to determine a suitable representational data model for UK building regulations. *Electronic Journal of IT in Construction*, 18, 20–39.

Ouertani M.Z. (2008), Supporting conflict management in collaborative design: An approach to assess engineering change impacts, *Computers in Industry*, 59, 882–893.

Panchal Jitesh H., Fernandez Marco Gero, Paredis Christiaan J. J., Allen Janet K., and Mistree Farrokh (2007), An interval-based constraint satisfaction (IBCS) method for decentralized, collaborative multifunctional design, *Concurrent Engineering Research and Applications* 15 (3), 309–323.

Pasley, G. P., and Kim Roddis, W. M. (1994). Using artificial intelligence for concurrent design in the steel building industry. *Concurrent Engineering* 2 (4), 303–310.

Pasley, G. P., and Roddis, W. M. K. (1996). Decision support tool for the steel building industry. In *Computing in Civil Engineering*, New York, pp. 725–731.

Pathfinder Report (2018). https://www.scottishfuturestrust.org.uk/storage/uploads/pathfindersrep ortfinaljuly2017.pdf, Scottish Futures Trust, pages 85.

Peña-Mora, F., and Wang Chun-Yi (1998), Computer-supported collaborative negotiation methodology, *Journal of Computing in Civil Engineering*, 12 (2), 64–81.

Peña-Mora, Feniosky, Sriram, Duvvuru, and Logcher, Robert (1995), Design rationale for computer-supported conflict mitigation. *Journal of Computing in Civil Engineering*, 9 (1), pp. 57–72.

Raphael B. (November, 2011), Multi-criteria decision making for collaborative design optimization of buildings. *Built Environment Project and Asset Management*, 1 (2), 122–136.

Raphael, B. and I.F.C. Smith (2013), *Engineering Informatics: Fundamentals of Computer Aided Engineering*, Second edition, John Wiley.

Rempling, R., Mathern, A., Tarazona Ramos, D., and Luis Fernández, S. (2019), Automatic structural design by a set-based parametric design method. *Automation in Construction*, 108, 102936.

Roddis W.M. Kim (1998), Knowledge-based assistants in collaborative engineering lecture notes in computer science. *Artificial Intelligence in Structural Engineering*, 1454, 320–334.

Rosenman, M.A., Smith, G., Maher, M.L., Ding, L., and Marchant, D. (2007), Multidisciplinary collaborative design in virtual environments. *Automation in Construction*, 16 (1), 37–44.

Scherer, R.J., and Katranuschkov, P. (2006). From data to model consistency in shared engineering environments, Lecture Notes in Computer Science (including subseries Lecture Notes in Artificial Intelligence and Lecture Notes in Bioinformatics), 4200 LNAI, pp. 615–626.

Scherer R.J., and Schapke S.-E. (2011), A distributed multi-model-based management information system for simulation and decision-making on construction projects, *Advanced Engineering Informatics*, 25, 582–599.

Senthilkumar V., Varghese K., and Chandran A. (2010), A web-based system for design interface management of construction projects. *Automation in Construction*, 19, 197–212.

Singh V., Gu N., and Wang X. (2011), A theoretical framework of a BIM-based multi-disciplinary collaboration platform. *Automation in Construction*, 20, 134–144.

STEP Tools (2007), *STEP ISO10303*, STEP Tools Inc., http://www.steptools.com/library/standard /2007.

Steward, D.V. (1981). *Analysis and Management: Structure, Strategy and Design*, Petrocelli Books.

Toche, B., Pellerin, R., and Fortin, C. (2020), Set-based design: A review and new directions. *Design Science*, e18.

UK BIM Alliance (2020), https://www.ukbimframework.org/resources/.

UK BIM Task Group (2013), *PAS1192:Part 2-2013*, British Standards Institution.

Uma Maheswari, J., Varghese, K., and Sridharan, T. (2006). Application of dependency structure matrix for activity sequencing in concurrent engineering projects. *Journal of Construction Engineering and Management*, 132 (5), 482–490.

Wilkinson, Paul (2005), *Construction Collaboration Technologies: The Extranet Evolution*, Taylor & Francis.

Wong, A., and Sriram, D. (1993). SHARED: An information model for cooperative product development. *Research in Engineering Design*, 5 (1), 21–39.

Zhao Zhen Yu, Lv Qian Lei, Zuo Jian, and Zillante George (2010), Prediction system for change management in construction project, *Journal of Construction Engineering and Management*, 136(6), 659–669.

Section 1–2

BIM applications

5 Towards adopting 4D BIM in construction management curriculums

A teaching map

*Faris Elghaish, Sepehr Abrishami, Salam Al-Bizri,
Saeed Talebi, Sandra Matarneh, and Song Wu*

Project planning and scheduling

Construction planning and scheduling are the main processes of construction management which have been developed through the last few decades (Gould & Joyce, 2003). Construction planning includes defining the project activities, estimating the resources, and determining the required durations to perform the required activities and to define the interrelationships among project tasks (Ritz, 1994). On the other hand, project scheduling is about determining the sequence of activities and defining resources needed to execute each activity by identifying the critical/non-critical paths (Illingworth, 2017). With the growing use of computers in computational processes, the construction planning and scheduling process has been enhanced by reducing the required time, minimizing errors, and improving the visualization of presented data through the use of such programmes as Autodesk Microsoft Project (Baldwin & Bordoli, 2014). Similarly, building information modelling (BIM) is expected to significantly improve the lifecycle of construction projects (Abrishami et al., 2014) including construction planning and scheduling, which are known as 4D BIM (Han & Golparvar-Fard, 2015).

The optimization of the project schedule is vital to exploit all resources and deliver the project with a minimum cost and according to the agreed time. Therefore, optimization tools could be used to optimize the project duration and levelling the resources.

Research on evolutionary schedule optimization

Hegazy (1999) developed a model to reach the optimal schedule path based on resource levelling and allocation simultaneously by embedding a genetic algorithm; however, this method did not consider the cost factor. Furthermore, Hegazy and Ersahin (2001) created spreadsheets to optimize the project schedule based on resource levelling, cost, and time. Another model has been developed by Senouci and Eldin (2004) to articulate a model which considers the time–cost trade-off to minimize project cost. However, this model did not consider various possible methods that could be used to perform the activity. Regarding multi-objective optimization by genetic algorithm, Leu and Yang (1999) developed a model that considers the time–cost trade-off, resource-constrained allocation, and unlimited resource-levelling models. This model was implemented in two stages: the first stage is to reach the optimal cost regardless of the resource-levelling constraints, and the second stage focuses on levelling the resources. The shortcoming of this model is that the final results can be adversely affected because the second stage can increase the selected cost.

Ghoddousi et al. (2013) developed a model which considers multi-objectives, namely, the multi-mode resource-constrained project scheduling problem (MRCPSP), the discrete time–cost trade-off problem (DTCTP), and the resource-allocation and resource-levelling problem (RLP). Despite the fact that the results of this research presented a high level of multi-objective optimization, the model did not appear to be applicable in a complex project which usually includes many activities with unlimited possible solutions for performing each activity. Furthermore, Elbeltagi et al. (2016) developed a model to optimize the schedule based on multi-criteria, which are the cost, resources, and cash flow. A particle swarm optimization method was used to determine the optimal path of activities based on several possible solutions to execute each activity (Elbeltagi et al., 2016). However, the proposed model has the following shortcomings: (1) it relies on collecting data manually to enable implementing the optimization model, and (2) it cannot be applied in a complex project in which activities are linked manually in the presented model.

After presenting the different attempts throughout the last few decades to use the genetic algorithm and other algorithm to optimize the project duration and resources, it is obvious that all proposed models can be effective for small projects and optimization models cannot be applied to large-scale projects. Therefore, BIM, as we will see, can help to optimize project duration and resources by enabling you to choose the proper construction method.

What is BIM?

In this section, an introduction to the BIM process will be presented to enable you to move to 4D BIM with a good understanding of the BIM process.

Due to the increase in the complexity of construction projects, it has become more complicated to manage them because of the reciprocal interdependencies between different stakeholders (Alshawi & Ingirige, 2003; Qureshi & Kang, 2015). Therefore, utilizing technologies to enhance management processes has been considered in the management of project communication and information data (Taxén & Lilliesköld, 2008). During the last few years, enhancements in managing project information have been developed, and building information modelling (BIM) is one of these developments; BIM can be identified as "a set of interacting policies, processes and technologies generating a methodology to manage the essential building design and project data in digital format throughout the building's lifecycle" (Succar, 2009). Consequently, beginning in 2014, the United Kingdom began to recommend BIM as a way to approach projects to reduce transaction costs, as well as minimize design errors. The government required the adoption of BIM as a condition for many projects, which were to deliver project documentation as fully collaborative BIM (Smith, 2014). BIM is not only a geometric model which includes all design elements, but also a holistic process that contains several tasks such as project management tools and techniques, 3D design, contractual issues, and facility management. It helps to streamline construction activities and maximize the value of a constructed asset (Bryde et al., 2013).

After explaining the BIM process, the next section presents the origin, definition, and characteristics of 4D BIM.

4D BIM

The origin of the 4D BIM process goes back to the 1980s, when Bechtel and Hitachi Ltd collaborated to generate a 4D visual model (Rischmoller & Alarcón, 2002). However, the core of 4D techniques has been developed by Fischer and Associates from Stanford University to create visually supported planning and scheduling (Dawood & Mallasi, 2006). Currently, 4D BIM can integrate several models within the project schedule while loading multiple resources as well as creating logical relationships between project activities (Gledson & Greenwood, 2016). The main function of 4D BIM is to link the 3D BIM model with the project schedule (Gledson & Greenwood, 2016). This function includes several capabilities, such as visualizing model spaces and the time required to create the design elements (Büchmann-Slorup & Andersson, 2010; Heesom & Mahdjoubi, 2004; Liston et al., 2003), considering the constructability methods of performing each activity (Koo & Fischer, 2000), and supporting communication between all stakeholders, which minimizes errors (Dawood, 2010).

In recent years, the application of 4D BIM in different construction contexts has been explored. For example, Gledson and Greenwood (2016) conducted a survey within UK industry firms to measure the applicability of the 4D BIM in the UK. This study shows positive results regarding the level of awareness and experience. On the other hand, most of the research studies that explored the application of 4D BIM and the improvement of its features and processes were ignored. Thereby, there is a need to further explore how to improve the functionality of 4D BIM process regarding (1) proposing a new method to create the list of activities (processes), (2) proposing an integration way to incorporate genetic algorithms into BIM platforms using API, and (3) articulating a new approach to understanding/utilizing the output of 4D BIM.

4D BIM is defined as a way to improve the functions of the planning process (Sloot et al., 2019). These functions can be defined as follows: (1) the function of extracting the needed planning information from BIM 3D design model (Turkan et al., 2012) ; (2) the function of identifying the activities by analyzing the extracted design elements via specific constructability methods (Hartmann et al., 2008); (3) estimation and process interdependency functions (Heesom & Mahdjoubi, 2004); (4) and planning project resources and site logistic data functions (Gledson & Greenwood, 2016). 4D planning is mainly related to linking the project schedule to BIM 3D design elements to improve the buildability of construction activities. Moreover, it has other capabilities, such as visualizing the time and construction process (Büchmann-Slorup & Andersson, 2010), analysing the project schedule to determine the suitable buildability method (Koo & Fischer, 2000), minimizing construction errors by exploiting virtual simulation before entering the construction phase, and improving collaboration and communication between project parties (Dawood, 2010).

4D BIM is characterized by the following factors:

- It has a visualization feature that can help the non-specialized employer to integrate and involve in the construction process within different stages (Heesom & Mahdjoubi, 2004). This is particularly important because decision-making needs visualization to clarify the information required to build an effective argument to get an optimum decision (Dawood, 2010).
- It enables efficient communication by building an information channel, which facilitates integrating and combining all project stakeholders in a dynamic panel (Hartmann et al., 2008). The dynamic panel begins to be developed from the conceptualization stage by integrating the owner with the architect to set the project outlines. This process requires information from trade contractors and other specialists (Elghaish et al., 2020a).
- It allows collaborative planning and scheduling (Gledson & Greenwood, 2016).
- It facilitates claims and dispute resolution by utilizing the clash detection feature in the 4D BIM (Sloot et al., 2019).

After this discussion of the origin of 4D BIM and its characteristics, the next section examines the development of 4D BIM for construction projects. Given we are in the era of automation, the next section will show how 4D BIM can be automated regarding the development of the model as well as the collection of data from construction sites.

4D BIM automation process

For 4D BIM automation, Montaser & Moselhi (2015) developed a model which allows users to import data from MS Project to the developed BIM model using the Revit application programming interface (API), coded by C#.NET. The main feature of this model was its ability to correlate between the design elements' implementation and the activity start and end dates. Furthermore, the study designed a project progress control methodology through process-based colour coding. For instance, the completed activities are highlighted in green, and the ones under construction are highlighted in another colour. Moreover, the implemented activities for specific construction operations will be hidden once finished to allow the planner to easily follow the progression of the project (Ali Montaser, 2013).

Omar and Dulaimi (2015) reported that embedding BIM in daily construction activities will help overcome all persistent problems. For example, updating all site-related information in BIM will enhance productivity and strengthen the relationship between all stakeholders. As such, El-Omari and Moselhi (2011) asserted that using unsystematic procedures to collect site data leads to a huge loss of information and will lead to unreliable results. Thus, 4D BIM automation will enhance the quality of the collected data and reduce human interference in the data collection process (Boton et al., 2015; Hakkarainen et al., 2009).

The progress of the collection of construction data has been intensively improved through various technologies such as barcoding, radio-frequency identification, 3D laser scanning, photogrammetry, multimedia, and pen-based computers (El-Omari & Moselhi, 2011; Kim et al., 2013; Talebi et al., 2018; Turkan et al., 2012, 2013). At the moment, however, the collected data cannot be fully exploited to update cost information. Consequently, Hamledari et al. (2017) advised that progress data must be automatically analyzed through advanced information technology. Furthermore, Wang et al. (2016) developed a model that utilizes BIM to create a project budgeting curve, namely an S curve. This model generates an optimized cost–budget curve based on multiple criteria, making it more reliable in implementation and giving a realistic indication concerning cost/schedule cases.

Elghaish and Abrishami (2020) developed a new philosophy to develop planning and scheduling in the 4D BIM model. A BIM library of the project activities is developed to enable the automation of the creation of the project schedule with respect to the 3D BIM design sequence. The optimization of the project duration is considered to be automated within the creation process using the proposed genetic algorithm model.

Steps for implementing the proposed constructability optimization method according to (Elghaish & Abrishami, 2020):

1. The user defines activities that require specific constructability methods, in other words, the activities that can be executed using different tools.
2. The appropriate resources (i.e. different types of equipment) are assigned to the selected activities.
3. The optimization criteria are selected from the designed panel (i.e. complexity, degree of uncertainty, etc.)
4. The optimization process is run, and the proper list of activities corresponding to the optimized cost is produced.

Figure 5.1 shows the differences between the traditional path of formulating a project schedule as well as the proposed path.

According to (Elghaish & Abrishami, 2020), Figure 5.2 includes all tasks (1, 2, 3 and 4) to show how the proposed framework can be implemented. The process begins by importing the 3D BIM model to Navisworks and thereafter creating the list of activities and conducting the optimization process for different construction methods. All the mentioned tasks will be implemented in a single platform (Navisworks). Figure (5.2) shows the configuration of the Navisworks hierarchy level, which helps the 4D planner to track all consumed resources in the project as well as to assign the right responsibilities to all project stakeholders. On the other hand, when the animation option works, each type of activity has two distinct colours: the appearance colour during the execution, and another colour when the activity has been accomplished. Therefore this configuration could help to check the performance of each resource in the project by measuring the duration of its appearance in the animation video. The price of materials, equipment, and labour are updated to the library.

Figure 5.2 shows the proposed library that was embedded into Navisworks by using the application programming interface (API) which has been coded by C# .NET. This can support the dynamic/single automation process by using a single platform, rather than exporting the data to several platforms in order to perform each task, such as importing the list of activities from Microsoft Project to generate a 4D model as well as exporting the 4D BIM model back to Microsoft Project in order to extract the budgeted cost of work schedule (BCWS), which represents the project budget. Currently, by adopting the proposed model, the planner will be able to finish all planning and scheduling tasks on the same platform. On the other

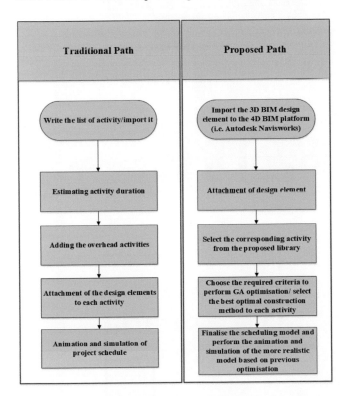

Figure 5.1 Comparison of traditional/proposed scheduling paths (Elghaish & Abrishami, 2020).

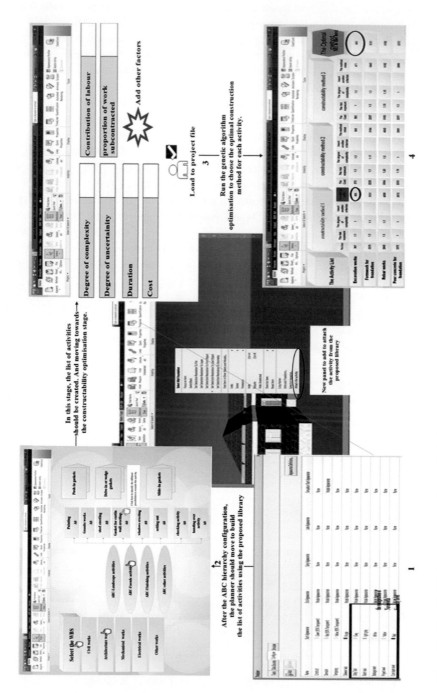

Figure 5.2 The 4D BIM process according to Elghaish & Abrishami, 2020.

hand, when the construction process starts, the 4D/5D BIM manager will be able to track the project by using the same platform as well. The criteria to enable the genetic algorithm (GA) to work can be selected from the proposed browser as can be seen in Figure 5.2, section 3.

Figure 5.2, section 4 shows the output of the genetic algorithm optimization process so that each activity has three construction methods; these methods can survive several iterations, and the successful method will achieve the minimum required value to perform the construction process as seen in Figure (5.2).

Given that 4D BIM can be used in integration with other technologies such as unmanned aerial vehicle technologies and immersive technologies, the next sections therefore present the application of these technologies with 4D BIM.

Unmanned aerial vehicle technologies and 4D BIM

Unmanned aerial vehicles (UAVs), through their advanced data collection capabilities, are revolutionizing a wide range of construction-related activities (Asadi et al., 2020). In terms of accuracy, efficiency, and cost-effectiveness, UAV-based applications surpass conventional methods on construction sites (Greenwood et al., 2019).

UAVs are becoming an essential component of virtual design and construction (VDC), giving architects and engineers new and efficient ways to visualize and analyze structural requirements from the ground upwards. During the past two years, integrating UAV technology to enhance information management and visualization has drawn much attention. Lu and Davis (2018) proposed a framework to integrate unordered images, geometric models, and the surrounding environment on Google Earth using three major components: UAV-centric image alignment and processing, keyhole markup language-based (KML) imagery, and a 3D model-management system. The proposed system is aimed at providing construction engineers with a low-cost and low technology-barrier solution to represent a dynamic construction site through information management, integration, and visualization. Puppala et al. (2018) developed the 3D models using UAV-based photogrammetry studies to provide the health condition of the structure, with a focus on material performance. Different image datasets about the conditions of various infrastructure assets were collected using a visual range camera mounted on a UAV. The 3D models were then developed to visualize the collected data and analyze the health condition of the structure. Ajibola et al. (2019) developed a model that integrates a weighted averaging and additive median filtering algorithms to improve the accuracy and quality of the digital elevation model (DEM) produced by UAV. Analysis of the result shows a remarkable increase of 88% in the accuracy of the fused DEM. Li and Liu (2019) presented improved neural networks to extract road information from remote-sensing images using a camera sensor equipped with a UAV. Kim et al. (2019) proposed a UAV-assisted robotic approach that can significantly reduce human intervention in, as well as the time required for, data collection and processing. This approach is to enable cluttered environments to be frequently monitored, updated, and analyzed to support timely decision-making. Ham and Kamari (2019) proposed a new method of automatically retrieving photo-worthy frames containing construction-related content that are scattered in collected video footages or consecutive images. The proposed automated method enables practitioners to assess the as-is status of construction sites efficiently through selective visual data, thereby facilitating data-driven decision-making at the right time.

Liu et al. (2019) proposed a safety-inspection method that integrates UAV and dynamic BIM. A dynamic BIM model is created by aggregating timely updated safety information with a BIM model in the web environment. The synchronous navigation of UAV video and dynamic BIM is realized by matching the virtual camera parameters with the real ones. The proposed method enables off-site managers to view the inspection video and make timely and comprehensive safety evaluations with the support of dynamic BIM. De Melo & Costa (2019) developed a conceptual framework for integrating resilience engineering (RS) and UAS technology into construction projects to support the safety planning and control (SPC) process. This framework highlights the fact that UAVs can be used to perform regular safety inspections. Such inspections provide information to help managers' decision-making, especially in tasks which involve a high risk of accidents. The visual assets collected with UAVs can also be used for feedback about the SPC and to increase workers' awareness through safety training. Finally, Gheisari and Esmaeili (2019) conducted a survey study to determine the effectiveness and frequency of using UAVs in improving safety operations in hazardous situations. The results indicated that the most important safety activities that could be improved using UAVs were the monitoring of boom vehicles or cranes in the proximity of overhead power lines, the monitoring of activities in the proximity of boom vehicles or cranes, and the monitoring of unprotected edges or openings. In terms of the UAV technical features required for safety-inspection applications, the most important features were camera movability, sense-and-avoid capability, and a real-time video communications feed.

After this discussion on how UAVs can be used to collect data from the site and compare the collected data with the 4D BIM, immersive technologies will be presented in the next section to illustrate how BIM can be benefited from mixed reality (MR) and relevant technologies to enhance the understanding of construction processes for all project parties.

Immersive technologies and 4D BIM

Meža et al. (2014) highlighted that augmented reality (AR) on a tablet, personal computer, or mobile is the best option for monitoring and tracking a construction project. They also clarified how AR technology can facilitate the visualization and estimation of the work performed on-site and compare it with the proposed schedule of construction projects. Park and Kim (2013) added another application to schedule monitoring by connecting AR material tracking to ensure that the necessary materials are located on the project site.

Combined immersive technologies and 4D BIM can be also used for safety monitoring by improving the situational awareness of construction workers (Cheng & Teizer, 2013; Kim et al., 2017). Few studies were found that focused on project scheduling. One was that by Kim et al. (2018), who developed an AR-based 4D CAD system which connects 4D and 5D (cost-management dimension) objects with a real field image and an AR object to implement several types of schedule information, and enables the use of constantly changing schedule information through AR objects. Another study was by Ratajczak et al. (2019), who developed a unique field application that integrates a location-based management system (LBMS) into BIM and an AR platform. This was to (1) detect scheduling deviations easily by visualizing construction progress in AR, (2) provide daily progress information, (3) provide performance data regarding construction activities, and (4) provide context-specific information/documents on scheduled tasks.

Construction projects involve collaboration between several project disciplines, including contractors, designers, managers, and more. A successful partnership confirms that a project will be completed on time, as per the proposed budget. However, not all project teams involved in a project are always present on a job site. If any error occurs that requires immediate action to be agreed upon by all parties involved, immersive technologies allow users to take notes and share views of an error and to send information to remote teams in real-time (Elghaish et al., 2020b). Pejoska et al. (2016) realized that in comparison with more traditional information sources, the accessibility of on-site project information and effective communication are significantly improving with the utilization of immersive technologies. However, some studies appeared to focus on collaboration and communication in construction projects. For example, some studies focused on using immersive technologies to facilitate collaboration and communication between the design team. A study by Goulding, Nadim, Petridis, and Alshawi (2012) demonstrated the need for integrating collaborative design teams to facilitate project integration and interchange by applying a game environment supported by a web-based VR cloud platform to facilitate collaboration and decision-making during the design process. Another is by Chalhoub and Ayer (2018), who examined the application of MR technologies in communicating electrical designs by comparing the performance of 18 electrical construction personnel who were tasked with building similar conduit assemblies using traditional paper. Du et al. (2018), meanwhile, developed a real-time synchronization system of BIM data in virtual reality (VR) for collaborative decision-making. The system is based on an innovative cloud-based BIM metadata interpretation and communication method to allow users to update BIM model changes in VR headsets automatically and simultaneously.

Other studies focused more on facilitating communication between project parties. For instance, Lin et al. (2015) proposed a visualized environment to facilitate the discussion among parties by using a stationary display called BIM Table, which displays public information. The proposed visualized environment uses AR technologies to connect the BIM Table and the mobile devices. Zaker and Coloma (2018) investigated the application of a VR-based workflow in a real project. A case study of VR integrated collaboration workflow was used to serve as an example of how AEC firms could overcome the challenge of collaboration between a project's teams. Du et al. (2018) also developed a cloud-based multi-user VR headset system called collaborative virtual reality (CoVR), which facilitates interpersonal project communication in an interactive VR environment. Another study, conducted by Boton (2018), proposed an immersive VR-based collaborative 4D BIM simulation to provide a supportive environment for conducting constructability analysis meetings.

The capabilities of immersive technologies to pool digital data and documentation with the physical view are a game-changer. Examples include Yeh et al. (2012), who presented a wearable device that could project the construction drawings and related information to help engineers to avoid carrying bulky construction drawings to the site and to reduce the effort required in looking for the correct drawings to obtain the information needed. Zhang et al. (2009) developed a system to facilitate the accurate exchange of project information among field personnel, using existing and already available camera-equipped mobile devices. Kim et al. (2013) developed a comprehensive system using mobile computing technology to provide construction stakeholders with a sufficient level of project information required for task management, including the visualization of task location in an AR environment. Chu et al. (2018) evaluated the effectiveness of BIM and AR system integration to enhance task efficiency through the improvement of the information retrieval process during construction by developing a mobile BIM AR system with cloud-based storage capabilities.

Internet of Things (IoT) and 4D BIM

We are currently in the era of industry 4.0, associated with utilizing IoT to automate all processes and reduce human interference to complete tasks. As such, in this section, the utilization of IoT with 4D BIM will be highlighted.

The IoT concept was earlier utilized in measuring project progress by employing radio-frequency identification (RFID) technologies (readers and sensors), particularly for health and safety and for evacuation planning during the construction stage (Kiani et al., 2014). This research proposed valuable extensions for the developed system to enhance and support the visualization and reliable data acquiring for construction health and safety management tasks. Subsequently, significant research has been conducted to cover a wide range of IoT applications for measuring construction project progress. Zhou and Ding (2017) utilized the IoT to provide automated warning systems as well as safety-barrier strategies for underground sites to avoid accidents. However, even though the system was tested using a case study – the Yangtze River-Crossing Metro Tunnel – health and safety regulations are different from one country to another. Therefore, more applications and extensions to this system are still needed to maximize the benefits.

Kochovski and Stankovski (2018) explored how edge computing applications, such as video communications and construction process documentation, can support the movement to smart construction with high quality of service (QoS). However, the security of the data was an issue, and this is why the researchers recommended the integration of the presented applications and blockchain technology. Further case studies have been conducted to measure the significance of the IoT in managing smart buildings. Zhou et al. (2019) developed a cyber-physical-system-based safety monitoring system for metro and underground construction, particularly for blind hosting. A case study was conducted to measure the validity of the system in a complex site environment. The findings show that the integration of BIM models and physical activities can provide real-time feedback information for all movements of equipment on-site, enabling risks to be identified automatically. However, the authors recommended that studying and optimizing the relationships between safety issues and construction conditions could enable the development of simulations in future to predict similar issues. More utilizations of the IoT in health and safety have been presented, such as the use of IoT-based architecture to automate non-hard-hat use (NHU) testing (Zhang et al., 2019). The researchers proposed a system that relied on an infrared beam detector and a thermal infrared sensor for non-intrusive NHU detection to deal with the problems of employing traditional sensors, which were not efficient enough to detect human movements.

Construction mobility for industry 4.0 requires an ecosystem to utilize the IoT in the entire construction operation rather than utilizing it in a single operation (Woodhead et al., 2018). This research revealed that there was a contradiction between acquiring a highly secure IoT environment and sharing data and a set of new processes and systems should be developed to enable the utilization of IoT in the construction industry, such as new information workflow and new business models.

Teaching map of 4D BIM

To teach 4D BIM to undergraduate or postgraduate students, the teaching process should be systematic to enable students to absorb the required theoretical knowledge before moving on to the practical and sophisticated applications. The following steps should be followed to prepare/adopt 4D BIM in any educational institute:

- *Teaching planning and scheduling process*: It is suggested that the project planning and scheduling tools and techniques (e.g. the critical path method) should be taught before moving on to 4D BIM implementation procedures. This is particularly important because 4D BIM is based on project planning and scheduling concepts.

- *Teaching the clash detection process using traditional approaches*: 4D BIM includes clash detection to investigate coordination problems between different disciplines. It is expected that students will be aware of techniques that were used traditionally for such a purpose so students will better understand the value of 4D BIM.

- *Teaching 4D BIM implementation process*: After students understand all essential theoretical knowledge, 4D BIM then can be introduced systematically through (1) teaching the process of developing the list of activities, (2) teaching how the durations of activities can be estimated in a different method (i.e. parametric estimation), (3) teaching how activities can be linked with corresponding design elements to create a simulation of project works, (4) teaching students how to create scripts and viewpoints to enhance the animation of the project works in the virtual environment, and (5) teaching how to create a clash detection report as well as the mechanism of writing a report and solving the discovered clashes in the 3D BIM model.

- *Teaching advanced 4D BIM integrations with other technologies*: Given that BIM can be integrated with several technologies such as immersive technologies and blockchain, it will be useful to show how 4D BIM can be enhanced by using these technologies; for example, mixed reality can be used to help project parties to compare performed works and designed works, using drones to collect data from the site and subsequently comparing this data with the 4D BIM model to evaluate the progress of the project.

Figure 5.3 shows the process of teaching 4D BIM to students, and all teaching procedures, tools, and techniques are also highlighted. The process of teaching was divided into three main stages as seen in Figure 5.3.

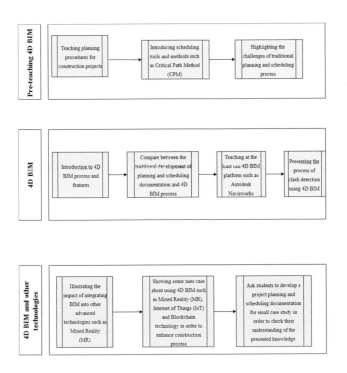

Figure 5.3 4D BIM teaching map.

Summary

The AEC industry is currently moving towards relying heavily on various BIM services. BIM services like clash detection and coordination services are in high demand, and they can be utilized throughout the project life cycle. This chapter presented a brief overview of 4D BIM uses in the various phases of construction and shed some light on its benefits and challenges. During the pre-design phase, 4D BIM supports strategic planning. When all the required information is available, the best strategies can be successfully shaped. 4D BIM can also be implemented to determine sequences and to make the best use of construction schedules. During the design development phase, with the support of 4D BIM, A/E can enhance the design constructability. 4D BIM can also optimize construction schedules with the help of the model, which can also benefit the contractors by showing the phasing plans to the owners.

4D BIM can also be integrated with other technologies such as immersive technologies (e.g. virtual reality) to help project parties visually observe differences between performed works and designed works. Also, 4D BIM can be integrated with data collected from drones to monitor the progress of a project by comparing data collected from the site with BIM.

Finally, this chapter proposed a map that includes steps to facilitate 4D BIM teaching and to enable students to absorb the required theoretical knowledge before moving on to practical and sophisticated BIM applications.

Questions

1. What are the main features of 4D BIM?
2. What are the differences between the traditional planning and scheduling process and 4D BIM?
3. How can virtual reality (VR) and mixed reality (MR) be used with 4D BIM?
4. How can UAV be used to collect data from construction sites?
5. Draw a flowchart to describe the features of 4D BIM and how it can be implemented in a construction project.
6. How can IoT be integrated into 4D BIM?
7. List various technologies that can be integrated with 4D BIM.

References

Abrishami, S., Goulding, J., Rahimian, F. P., & Ganah, A. (2014). Integration of BIM and generative design to exploit AEC conceptual design innovation. *Information Technology in Construction, 19*, 350–359. Retrieved from http://www.itcon.org/2014/21.

Ajibola, I. I., Mansor, S., Pradhan, B., & Shafri, H. Z. M. (2019). Fusion of UAV-based DEMs for vertical component accuracy improvement. *Measurement, 147*, 106795. doi:10.1016/j.measurement.2019.07.023

Alshawi, M., & Ingirige, B. (2003). Web-enabled project management: An emerging paradigm in construction. *Automation in Construction, 12*(4), 349–364.

Asadi, K., Suresh, A. K., Ender, A., Gotad, S., Maniyar, S., Anand, S., . . . Wu, T. (2020). An integrated UGV-UAV system for construction site data collection. *Automation in Construction, 112*, 103068. doi:10.1016/j.autcon.2019.103068

Baldwin, A., & Bordoli, D. (2014). *Handbook for Construction Planning and Scheduling*: Wiley.

Boton, C. (2018). Supporting constructability analysis meetings with immersive virtual reality-based collaborative BIM 4D simulation. *Automation in Construction, 96*, 1–15. doi:10.1016/j.autcon.2018.08.020

Boton, C., Kubicki, S., & Halin, G. (2015). The challenge of level of development in 4D/BIM simulation across AEC project lifecycle. A case study. *Procedia Engineering, 123*, 59–67.

Bryde, D., Broquetas, M., & Volm, J. M. (2013). The project benefits of building information modelling (BIM). *International Journal of Project Management, 31*(7), 971–980.

Büchmann-Slorup, R., & Andersson, N. (2010). BIM-based scheduling of Construction – A comparative analysis of prevailing and BIM-based scheduling processes. In Paper presented at the Proceedings of the 27th International Conference of the CIB W78, Cairo, Egypt, 16–18 November.

Chalhoub, J., & Ayer, S. K. (2018). Using mixed reality for electrical construction design communication. *Automation in Construction, 86*, 1–10. doi:10.1016/j.autcon.2017.10.028

Cheng, T., & Teizer, J. (2013). Real-time resource location data collection and visualization technology for construction safety and activity monitoring applications. *Automation in Construction, 34*, 3–15. doi:10.1016/j.autcon.2012.10.017

Chu, M., Matthews, J., & Love, P. E. (2018). Integrating mobile building information modelling and augmented reality systems: An experimental study. *Automation in Construction, 85*, 305–316. doi:10.1016/j.autcon.2017.10.032

Dawood, N. (2010). Development of 4D-based performance indicators in construction industry. *Engineering, Construction and Architectural Management, 17*(2), 210–230.

Dawood, N., & Mallasi, Z. (2006). Construction workspace planning: Assignment and analysis utilizing 4D visualization technologies. *Computer-Aided Civil and Infrastructure Engineering, 21*(7), 498–513.

de Melo, R. R. S., & Costa, D. B. (2019). Integrating resilience engineering and UAS technology into construction safety planning and control. *Engineering, Construction and Architectural Management, 26*(11), 0969–9988. doi:10.1108/ECAM-12-2018-0541

Du, J., Shi, Y., Zou, Z., & Zhao, D. (2018). CoVR: Cloud-based multiuser virtual reality headset system for project communication of remote users. *Journal of Construction Engineering and Management, 144*(2), 04017109. doi:10.1061/(ASCE)CO.1943-7862.0001426

El-Omari, S., & Moselhi, O. (2011). Integrating automated data acquisition technologies for progress reporting of construction projects. *Automation in Construction, 20*(6), 699–705. doi:https://doi.org/10.1016/j.autcon.2010.12.001

Elbeltagi, E., Ammar, M., Sanad, H., & Kassab, M. (2016). Overall multiobjective optimization of construction projects scheduling using particle swarm. *Engineering, Construction and Architectural Management, 23*(3), 265–282.

Elghaish, F., & Abrishami, S. (2020). Developing a framework to revolutionize the 4D BIM process: IPD-based solution. *Construction Innovation, 20*(3), 1471–4175.

Elghaish, F., Abrishami, S., Hosseini, M. R., & Abu-Samra, S. (2020a). Revolutionizing cost structure for integrated project delivery: A BIM-based solution. *Engineering, Construction and Architectural Management*, 0969–9988.

Elghaish, F., Matarneh, S., Talebi, S., Kagioglou, M., Hosseini, M. R., & Abrishami, S. (2020b). Toward digitalization in the construction industry with immersive and drones technologies: a critical literature review. *Smart and Sustainable Built Environment.*

Gheisari, M., & Esmaeili, B. (2019). Applications and requirements of unmanned aerial systems (UASs) for construction safety. *Safety Science, 118*, 230–240. doi:10.1016/j.ssci.2019.05.015

Ghoddousi, P., Eshtehardian, E., Jooybanpour, S., & Javanmardi, A. (2013). Multi-mode resource-constrained discrete time–cost–resource optimization in project scheduling using non-dominated sorting genetic algorithm. *Automation in Construction, 30*, 216–227.

Gledson, B., & Greenwood, D. (2016). Surveying the extent and use of 4D BIM in the UK. *Journal of Information Technology in Construction, 21*, 57–71.

Gould, F. E., & Joyce, N. E. (2003). *Construction Project Management*: Prentice Hall.

Goulding, J., Nadim, W., Petridis, P., & Alshawi, M. (2012). Construction industry offsite production: A virtual reality interactive training environment prototype. *Advanced Engineering Informatics, 26*(1), 103–116. doi:10.1016/j.aei.2011.09.004

Greenwood, W. W., Lynch, J. P., & Zekkos, D. (2019). Applications of UAVs in civil infrastructure. *Journal of Infrastructure Systems, 25*(2), 04019002. doi:10.1061/(ASCE)IS.1943-555X.0000464

Hakkarainen, M., Woodward, C., & Rainio, K. (2009). Software architecture for mobile mixed reality and 4D BIM interaction. Paper presented at the Proceedings of the 25th CIB W78 Conference, Santiago, Chile, July 15–17 2008.

Ham, Y., & Kamari, M. (2019). Automated content-based filtering for enhanced vision-based documentation in construction toward exploiting big visual data from drones. *Automation in Construction, 105*, 102831. doi:10.1016/j.autcon.2019.102831

Hamledari, H., McCabe, B., Davari, S., & Shahi, A. (2017). Automated schedule and progress updating of IFC-based 4D BIMs. *Journal of Computing in Civil Engineering, 31*(4), 04017012. doi:https://doi.org /10.1061/(ASCE)CP.1943-5487.0000660

Han, K. K., & Golparvar-Fard, M. (2015). Appearance-based material classification for monitoring of operation-level construction progress using 4D BIM and site photologs. *Automation in Construction, 53*, 44–57.

Hartmann, T., Gao, J., & Fischer, M. (2008). Areas of Application for 3D and 4D Models on Construction Projects. *Journal of Construction Engineering and Management, 134*(10), 776–785. doi:https ://doi.org/10.1061/(ASCE)0733-9364(2008)134:10(776)

Heesom, D., & Mahdjoubi, L. (2004). Trends of 4D CAD applications for construction planning. *Construction Management and Economics, 22*(2), 171–182.

Hegazy, T. (1999). Optimization of resource allocation and leveling using genetic algorithms. *Journal of Construction Engineering and Management, 125*(3), 167–175.

Hegazy, T., & Ersahin, T. (2001). Simplified spreadsheet solutions. II: Overall schedule optimization. *Journal of Construction Engineering and Management, 127*(6), 469–475.

Illingworth, J. R. (2017). *Construction Methods and Planning*: CRC Press.

Kiani, A., Salman, A., & Riaz, Z. (2014). Real-time environmental monitoring, visualization, and notification system for construction H&S management. *Journal of Information Technology in Construction, 19*, 72–91.

Kim, C., Park, T., Lim, H., & Kim, H. (2013). On-site construction management using mobile computing technology. *Automation in Construction, 35*, 415–423. doi:10.1016/j.autcon.2013.05.027

Kim, C., Son, H., & Kim, C. (2013). Automated construction progress measurement using a 4D building information model and 3D data. *Automation in construction, 31*, 75–82.

Kim, H., Lee, J., Ahn, E., Cho, S., Shin, M., & Sim, S.-H. (2017). Concrete crack identification using a UAV incorporating hybrid image processing. *Sensors, 17*(9), 2052. doi:10.3390/s17092052

Kim, H. S., Kim, S.-K., Borrmann, A., & Kang, L. S. (2018). Improvement of realism of 4D objects using augmented reality objects and actual images of a construction site. *KSCE Journal of Civil Engineering, 22*(8), 2735–2746. doi:10.1007/s12205-017-0734-3

Kim, P., Park, J., Cho, Y. K., & Kang, J. (2019). UAV-assisted autonomous mobile robot navigation for as-is 3D data collection and registration in cluttered environments. *Automation in Construction, 106*, 102918. doi:10.1016/j.autcon.2019.102918

Kochovski, P., & Stankovski, V. (2018). Supporting smart construction with dependable edge computing infrastructures and applications. *Automation in Construction, 85*, 182–192.

Koo, B., & Fischer, M. (2000). Feasibility study of 4D CAD in commercial construction. *Journal of Construction Engineering and Management, 126*(4), 251–260.

Leu, S.-S., & Yang, C.-H. (1999). GA-based multicriteria optimal model for construction scheduling. *Journal of Construction Engineering and Management, 125*(6), 420–427.

Li, Y., & Liu, C. (2019). Applications of multirotor drone technologies in construction management. *International Journal of Construction Management, 19*(5), 401–412. doi:10.1080/15623599.2018.1452101

Lin, T.-H., Liu, C.-H., Tsai, M.-H., & Kang, S.-C. (2015). Using augmented reality in a multiscreen environment for construction discussion. *Journal of Computing in Civil Engineering, 29*(6), 04014088. doi:10.1061/(ASCE)CP.1943-5487.0000420

Liston, K., Fischer, M., & Winograd, T. (2003). Focused sharing of information for multidisciplinary decision making by project teams. *Journal of Information Technology in Construction, 6*(6), 69–82.

Liu, D., Chen, J., Hu, D., & Zhang, Z. (2019). Dynamic BIM-augmented UAV safety inspection for water diversion project. *Computers in Industry, 108*, 163–177. doi:10.1016/j.compind.2019.03.004

Lu, X., & Davis, S. (2018). Priming effects on safety decisions in a virtual construction simulator. *Engineering, Construction and Architectural Management, 25*(2). doi:10.1108/ECAM-05-2016-0114

Meža, S., Turk, Ž., & Dolenc, M. (2014). Component based engineering of a mobile BIM-based augmented reality system. *Automation in Construction, 42*, 1–12. doi:10.1016/j.autcon.2014.02.011

Montaser, A. (2013). *Automated Site Data Acquisition for Effective Project Control*: Concordia University,

Montaser, A., & Moselhi, O. (2015). Methodology for Automated Generation of 4d BIM. In Paper presented at the Proceedings of the 11th Construction Specialty Conference of the Canadian Society of Civil Engineering, Vancouver, British Columbia, June 8–10 2015.

Omar, H., & Dulaimi, M. (2015). Using BIM to automate construction site activities. *Building Information Modelling (BIM) in Design, Construction and Operations, 149*, 45. doi:http://doi.org/10.2495/BIM150051

Park, C.-S., & Kim, H.-J. (2013). A framework for construction safety management and visualization system. *Automation in Construction, 33*, 95–103. doi:10.1016/j.autcon.2012.09.012

Pejoska, J., Bauters, M., Purma, J., & Leinonen, T. (2016). Social augmented reality: Enhancing context-dependent communication and informal learning at work. *British Journal of Educational Technology, 47*(3), 474–483. doi:10.1111/bjet.12442

Puppala, A. J., Congress, S. S., Bheemasetti, T. V., & Caballero, S. R. (2018). Visualization of civil infrastructure emphasizing geomaterial characterization and performance. *Journal of Materials in Civil Engineering, 30*(10), 04018236. doi:10.1061/(ASCE)MT.1943-5533.0002434

Qureshi, S. M., & Kang, C. (2015). Analysing the organizational factors of project complexity using structural equation modelling. *International Journal of Project Management, 33*(1), 165–176.

Ratajczak, J., Riedl, M., & Matt, D. T. (2019). BIM-based and AR application combined with location-based management system for the improvement of the construction performance. *Buildings, 9*(5), 118. doi:10.3390/buildings9050118

Rischmoller, L., & Alarcón, L. F. (2002). 4D-PS: Putting an IT new work process into effect. In Paper presented at the Proceedings of CIB w78 2002 Conference on Construction Information Technology, Aarhus, Denmark.

Ritz, G. J. (1994). *Total Construction Project Management*: New York, NY, United States: McGraw-Hill, Incorporated.

Senouci, A. B., & Eldin, N. N. (2004). Use of genetic algorithms in resource scheduling of construction projects. *Journal of Construction Engineering and Management, 130*(6), 869–877.

Sloot, R., Heutink, A., & Voordijk, J. (2019). Assessing usefulness of 4D BIM tools in risk mitigation strategies. *Automation in Construction, 106*, 102881.

Smith, P. (2014). BIM implementation: Global strategies. *Procedia Engineering, 85*, 482–492.

Succar, B. (2009). Building information modelling framework: A research and delivery foundation for industry stakeholders. *Automation in Construction, 18*(3), 357–375.

Talebi, S., Koskela, L., & Tzortzopoulos, P. (2018). Tolerance compliance measurement using terrestrial laser scanner. Paper presented at the 26th Annual Conference of the International Group for Lean Construction: Evolving Lean Construction – Towards Mature Production Across Cultures and Frontiers, Chennai, India.

Taxén, L., & Lilliesköld, J. (2008). Images as action instruments in complex projects. *International Journal of Project Management, 26*(5), 527–536.

Turkan, Y., Bosche, F., Haas, C. T., & Haas, R. (2012). Automated progress tracking using 4D schedule and 3D sensing technologies. *Automation in Construction, 22*, 414–421. doi:https://doi.org/10.1016/j.autcon.2011.10.003

Turkan, Y., Bosché, F., Haas, C. T., & Haas, R. (2013). Toward automated earned value tracking using 3D imaging tools. *Journal of Construction Engineering and Management, 139*(4), 423–433. doi:https://doi.org/10.1061/(ASCE)CO.1943-7862.0000629

Wang, K.-C., Wang, W.-C., Wang, H.-H., Hsu, P.-Y., Wu, W.-H., & Kung, C.-J. (2016). Applying building information modeling to integrate schedule and cost for establishing construction progress curves. *Automation in Construction, 72*, 397–410. doi:https://doi.org/10.1016/j.autcon.2016.10.005

Woodhead, R., Stephenson, P., & Morrey, D. (2018). Digital construction: From point solutions to IoT ecosystem. *Automation in Construction, 93*, 35–46.

Yeh, K.-C., Tsai, M.-H., & Kang, S.-C. (2012). On-site building information retrieval by using projection-based augmented reality. *Journal of Computing in Civil Engineering, 26*(3), 342–355. doi:10.1061/(ASCE)CP.1943-5487.0000156

Zaker, R., & Coloma, E. (2018). Virtual reality-integrated workflow in BIM-enabled projects collaboration and design review: A case study. *Visualization in Engineering, 6*(1), 4. doi:10.1186/s40327-018-0065-6

Zhang, H., Yan, X., Li, H., Jin, R., & Fu, H. (2019). Real-time alarming, monitoring, and locating for non-hard-hat use in construction. *Journal of Construction Engineering and Management, 145*(3), 04019006.

Zhang, X., Arayici, Y., Wu, S., Abbott, C., & Aouad, G. (2009). Integrating BIM and GIS for large-scale facilities asset management: A critical review. Paper presented at the 12th International Conference on Civil, Structural and Environmental Engineering Computing, Funchal, Portugal.

Zhou, C., & Ding, L. (2017). Safety barrier warning system for underground construction sites using Internet-of-Things technologies. *Automation in Construction, 83*, 372–389.

Zhou, C., Luo, H., Fang, W., Wei, R., & Ding, L. (2019). Cyber-physical-system-based safety monitoring for blind hoisting with the Internet of Things: A case study. *Automation in Construction, 97*, 138–150.

6 Cost management–based BIM

Skills, implementation, and teaching map

Faris Elghaish, Saeed Talebi, and Song Wu

Background of the cost management process

The cost management process is a system of managing all cost tasks within the different stages in the construction project, such as the planning, construction, and closeout stages (Ahmed, 1995). The cost management system should include processes to manage each stage such as cost estimation, budget, and control (Horngren et al., 2002). Moreover, Oberlender and Oberlender (1993) mentioned that a cost management plan is a "project money plan" and it represents the financial forecast action for the project. They argue that the cost management plan requires the implementation of specific tasks to articulate this plan such as estimation, budgeting, control, payment processing, and change management. These tasks must be implemented in specific orders and stages to obtain a reliable cost management plan for the project (Kerzner, 2017).

Given that the cost plan is influenced by project decisions, the cost plan should be flexible to deal with all changes and should be able to manage the data in a proper way (Potts and Ankrah, 2014). The cost management system has been defined by Shank (1989) as the framework of the project data. Such a system involves tools and techniques to direct project stakeholders during the entirety of a project's stages, such as estimation tools to support different managerial decisions, as well as providing a generic plan for the investment.

The most important activities in the cost management process are (1) the cost plan for preparing the needed data (e.g. the price list), determining which estimation technique must be adopted in the project based on the availability of the data (Jorgensen and Shepperd, 2006); (2) the cost estimation for the project design elements based on the completed design, which can be extracted from tender documents (Niazi et al., 2006); (3) the cost control and accounting process that takes place throughout the construction stage, such as preparing a payment invoice for the completed work (Leu and Lin, 2008); and (4) the calculation of final accounts while considering the economic assessment. In addition to the cost management during the completion of the project, it should also consider economic efficiency in its process, such as measuring the life-cycle cost of the building (Szekeres, 2005).

After discussing the structure of the cost management process, the next section highlights the challenges involved in the three main tasks, namely cost estimation, cost budgeting, and cost control.

Challenges of the traditional practice of cost estimation

Shane et al. (2009) state that bias in estimation is one of the most significant reasons for underestimating a budget. This situation is called *optimistic estimation* and usually the estimator uses this approach to show the client that they are more competitive than others. The procurement approach plays a significant role in cost estimation escalation due to the lack of a risk-sharing system (Harbuck, 2004). Allocating some risks to a party who cannot manage them will lead to an increase in the project cost as the contingency cost will not be enough to cover the consequences of risks (Love et al., 2011). Moreover, the lack of experience in dealing properly with the procurement approach can lead to an increase in the cost. For example, incorrect schedule acceleration can cause a cost overrun that is more than expected (ECONorthwest, 2002, Weiss, 2000).

Callahan (1998) mentions that the unplanned changes in the schedule lead to changes in the project budget during the execution process. Thus, some companies adopt a strategy to review their budgets periodically to ensure that their projects remain within the company budget, and if there is any change, these companies apply a technique which is called *expenditure timing adjustments* (Touran and Lopez, 2006, Hufschmidt and Gerin, 1970).

Complexity is inherent in the construction industry due to factors such as the location of the project or design changes in projects. Such factors can cause difficulties in determining the properly planned cost value, as repetitive changes in the project plans make the level of uncertainty very high (Touran and Lopez, 2006, Callahan, 1998). Consequently, coordination problems exist between different disciplines and some information will be missing, and as a result, the accuracy of cost estimation might be affected (Shane et al., 2009, Kaliba et al., 2009).

Scope changes such as changes in the design components or the proposed function of some parts in the project may lead to changes in project cost and schedule (Hussain, 2012, Khan, 2006). If these changes are not managed properly by the owner, they can represent a major change in the project scope, and in many cases, projects are executed with cost and schedule overruns (Alinaitwe et al., 2013).

Akintoye and Fitzgerald (2000) state that poor estimation procedures can cause misunderstanding in terms of the formats used, which do not provide an easy way to check, verify, and correct the estimated elements. Therefore, the procedures, formats, and methods should be understandable and clear to enable the user to determine the cost (Reilly, 2005). Moreover, a poor estimation can lead to data being missed, giving unreliable results which cause underestimation (Azhar et al., 2008). This will affect other processes such as scheduling and result in a misleading inventory plan, which definitely will cause a cost overrun due to considerable variance from the planned value (Shane et al., 2009). Poor cost estimation can affect the delivery of construction project in different aspects.

Misunderstanding of the contractual agreement plays an important role in misleading cost estimation; in particular, the misallocation of responsibilities between the different participants can cause cost estimation issues (Zaghloul and Hartman, 2003, Ali and Kamaruzzaman, 2010, Le-Hoai et al., 2008). Moreover, ambiguity in contract provision can be a reason for poorly allocating responsibilities, such as who is responsible for implementing reworks or change orders (Touran and Lopez, 2006). Poor execution cannot be ignored as one of the most important reasons for cost overrun, along with bad site management and a lack of collaboration due to the inability of the participants' representatives to make decisions (Shane et al., 2009, Enshassi et al., 2009).

The aforementioned challenges in cost estimation lead to incorrect contingency costs (Moselhi and Salah, 2012), and hence, the overall agreed budget will be unreliable. Also, the project will be affected either by misusing the contingency cost or facing a shortage in the contingency costs to cover the carried risks (Schexnayder et al., 2011).

After discussing the challenges faced during the traditional cost management process, the next section includes an introduction in terms of the overview of BIM-based cost management. This section shows how BIM can be used to automate the quantification process.

BIM and cost management

In moving towards efficient project delivery, the ultimate goal is to have a database of information that is available to all project participants, with confidence in its accuracy, universal utility, and clarity (Ashcraft, 2014, Oraee et al., 2017). The main drive for adopting BIM is to manage all project documents and stages (i.e. design, planning, and costing) in a single/ dynamic context, to secure the proper exploitation of available information (Redmond et al., 2012, Merschbrock et al., 2018, Abrishami et al., 2015). BIM design elements must contain the required information in various categories, including design or management (Banihashemi et al., 2018), to acquire smartly designed elements rather than traditional 3D components (Fu et al., 2006, Pärn and Edwards, 2017). BIM users should be capable of acquiring all the required information from a single BIM element to make informed decisions (Motamedi and Hammad, 2009, Shen et al., 2012, Abrishami et al., 2014). Four-dimensional modelling (4D BIM) can embed progress data in 3D model objects by adjusting the task–object relationship (Hamledari et al., 2017). The application of 4D BIM leads to streamlined workflows, efficient on-site management, and assessing the proper construction methods (Hartmann et al., 2008). As for cost management, BIM is one of the most efficient architectural, engineering, and construction (AEC) tools for increasing productivity on construction projects (Wang et al., 2016, Aibinu and Venkatesh, 2013, Lee et al., 2014). Colloquially termed *5D BIM* (Aibinu and Venkatesh, 2013), this BIM capability is the preferred technique for extracting quantities from 3D models, allowing cost consultants to incorporate productivity allowances and pricing values (Eastman et al., 2011a, Lee et al., 2014). The cost estimating process starts with exporting data from 3D models to BIM-based cost estimating software (e.g. CostX®) to prepare quantity take-off. Afterwards, the bills of quantities (BoQ) are generated and exported to an external database (Aibinu and Venkatesh, 2013). Prices and productivity allowances can also be added to project schedule preparation (Eastman et al., 2011a, Lee et al., 2014). Such automated quantification will shorten the quantity take-off processing time, and will automatically consider any changes in design – which is likely in fast-track projects (Wang et al., 2016, Popov et al., 2010).

5D BIM is mainly a process of retrieving quantities from the 3D BIM model; therefore, the role of quantity surveyors/cost estimators has been changed to adapt to this new technology, as highlighted in the next section.

5D BIM and quantity surveying

Quantity surveying has been a vital part of the construction process for more than 170 years (Cartlidge, 2011). Since that time, the role of the quantity surveyor has been to manage cost estimation and control as well as to optimize contractual and financial tradeoffs, such as in the valuation and payment of construction projects (Ashworth et al., 2013). The role of the quantity surveyor has been developed to meet the requirements of value management approaches more than construction methods, and has been implemented by developing the tools and techniques used to capture cost management parameters, such as automated measurement (de Andrade et al., 2019). Moreover, the emergence of BIM has enabled the support and delivery of facilities management tasks as well as enhancing the holistic management process. Hence, the role of the quantity surveyor has changed in parallel with progress in the development of BIM (Stanley and Thurnell, 2014). The introduction of BIM has required a change in how the building is carried out in terms of design and procurement strategy, as well as all the other parameters necessary to achieve the necessary collaboration and integration in the AEC industry (Aranda-Mena et al., 2009, Qian, 2012). Consequently, cost management must change to be compatible with these approaches and an effective part of this process (Hanid et al., 2011).

Most definitions state that 5D BIM is the preferred method for extracting quantities from the BIM model to enable cost consultants to commence the costing process by inserting the productivity allowances and pricing values (Eastman et al., 2011a). Forgues et al. (2012) recommend that output data should be supported by another format to complete the measurement and pricing process. The cost estimation process begins by importing the BIM 3D model to any BIM-based cost estimation software, such as Exact COST-X or Visio office, to prepare the take-off of quantities (Mitchell, 2012). After that, the bill of quantities (BoQ) is generated and exported to an external database where prices and productivity allowances are added to prepare the project schedule (Eastman et al., 2011a, Forgues et al., 2012). Moreover, Hannon (2007) points out that such automated quantification will shorten what is typically a time-consuming process, and will automatically take into account any changes in the design development process.

In this section, the role of the quantity surveyor using 5D BIM was discussed, and the challenges that face quantity surveyors in using BIM are also presented. 5D BIM involves challenges and barriers like any other new process or technology. Therefore, this is highlighted in the next section.

5D BIM implementation: Barriers and challenges

Sylvester and Dietrich (2010) reported that 5D BIM has changed the role of the quantity surveyor, who once would have spent a lot of time extracting quantities from drawings to analyze and validate cost data to reach the optimal cost estimation value. Moreover, Shen and Issa (2010) asserted that using 5D BIM platforms reduces the number of errors generated by misleading manual calculations, leading to effectively estimated durations. In general, 5D BIM introduces a comprehensive process of allowing early decisions at early design stages by offering an automated quantity take-off for all designs, whether concepts or detailed designs (Forgues et al., 2012). Therefore, Smith (2014b) asserted that the BIM process introduces a holistic approach for all project functions such as design, management, construction, and sustainability matters simultaneously. McCuen (2008) reported that exploiting 4/5D BIM in the AEC industry leads to an increase in profitability. Moreover, Franco et al. (2015) claimed that the model has proven comprehensive and durable enough to assist in all phases of the project life cycle – from conception through design and construction to operations and maintenance.

Nassar (2011) argued that the estimation process is more than listing the design objects and the prices, and thus using BIM only is not adequate in the cost management process. Therefore, linking cost estimation programmes and BIM design platforms is important to complete the entire cost management process. Stanley and Thurnell (2014) reported that exploiting BIM estimation software produces an accurate cost estimation and gives the estimator reliable indicators to inform future projects. Nevertheless, McCuen et al. (2011) claimed that it is not necessary for the derived information from BIM model to be completely precise. On the other hand, data transfer between several platforms causes data wastage, which reduces the accuracy of the information (Azhar et al., 2012). Moreover, Sunil et al. (2017) reported that BIM leads to enhanced cost estimation and control of tasks, and this directly affects the role of cost managers and increases their capabilities and ways of making decisions. Moreover, BIM increases the involvement of quantity surveyors (QS) in different project tasks and eliminates the traditional isolated QS working environment, which reduces the availability of information. Nonetheless, the integration of and coordination between different models are not adequate, and thus the QS is still responsible for articulating the cost report semi-manually via linking several models such as a 3D design model and 5D platform to extract quantities and an Excel sheet to determine the prices by exporting the derived quantities (Smith, 2014a). On the other hand, the integration between cost estimation and schedule is processed manually by the QS, which makes this process complicated and time-consuming (Sunil et al., 2017). Moreover, Cho et al. (2012) also claimed that the project data is processed via a set of spreadsheets and estimating software, and that therefore, there is no single/dynamic platform to conduct the entire cost management process without any other supporting programmes.

There is no balance in the relationship between the amount of information required for cost estimation and the data added by designers (Kiviniemi et al., 2007). Moreover, the pricing format is not considered in BIM models, but it is required by quantity surveyors to modify the BoQ model for each project in terms of their breakdown structure (Wu et al., 2014).

A lacuna is created by the traditional approach of working in separate environments, whereby each discipline is implemented by using a different model. This results in confusion over what cost estimation model should be followed (Stanley and Thurnell, 2014). Consequently, the project core team member usually loses countless hours in adapting one model to meet the needs of the cost process (Meadati, 2009). Boon and Prigg (2012) contend that a balance of information between the different disciplines, such as the architectural and QS information, must be considered.

Stanley and Thurnell (2013) state that the nature of the construction industry is the reason for the delayed or less efficient implementation of 5D BIM. To address this, 5D BIM software companies should consider collaboration regarding the workflow of the cost data throughout the project stages, or between the different participants who lead, to make the cost management process effective and efficient (Olatunji et al., 2010).

The different challenges involved in 5D BIM implementation were discussed in this section. The next section includes an explanation of how BIM is employed to develop the cash-out plan (curve).

BIM and cash flow

Interdependencies between the cost (5D BIM) and schedule (4D BIM) are obvious because it is necessary to integrate the cost and schedule processes in a single system to establish appropriate control. However, in practice, the two parameters are still separate given that the schedule is represented by the work breakdown structure (WBS), whilst the costs are identified by the cost breakdown structure (CBS) (Fan et al., 2015). Hence, during the budgeting stage, the integration between the WBS and CBS becomes complex, leading to potential errors and mismatches (Jung and Woo, 2004).

Fan et al. (2015) state that the initial steps to integrate 4D BIM and 5D BIM are the creation of the project schedule and BIMs, and then the next step is the cost estimation of all BIMs. Subsequently, the generated cost items need to be linked with the project schedule (4D BIM), and the BIM element linked to the schedule. However, this process has some shortcomings when it comes to implementation, particularly in linking the BIM schedule to the generated cost. In sum, the BIM elements should be linked directly to the cost items to avoid the complicated process of integrating these elements with the schedule.

Kim (1989) developed a costing system model to manage cost estimation and budgeting control. However, the proposed model, called a *basic construction operation*, indicated the lowest level in construction operation, and this level has linked to three sources: WBS, CBS, and design files. The proposed system has been criticized by Rasdorf and Abudayyeh (1991) as it requires a high level of detail as well as the refinement of each operation to reach sub-task; that is, the system is not practical and applicable in the AEC industry.

Since the classification of construction works is vital to developing a reliable budget, Kang and Paulson (1998) developed a classification system based on four categories, namely, facilities, spaces, elements, and operations. Subsequently, the cost and schedule will be considered for each level in each category for a construction project; however, the proposed classification system is not suitable for quantity take-off in the cost estimation process (Wang et al., 2016). Therefore, the challenge of detailed cost estimation with a consistent WBS hierarchy persists in the AEC industry, particularly when using the work-packaging (WP) method, which relies on the cost/schedule control system criteria (C/SCSC) at the package level. However, this is not efficient due to the fact that construction operations involve long hierarchy levels to reach the sub-task level (Moder et al., 1983). Even though the method assigns the cost to WBS regardless of CBS, Rasdorf and Abudayyeh (1991) assert that it needs some improvements to make it applicable to complex projects.

Yang et al. (2007) developed a model to integrate the budget (override the resources) directly to the schedule in daily proportion to develop BCWS. In this model, each activity will be weighted daily as a ratio relative to the schedule, and this ratio will be used to measure progress. However, the shortcoming of this model is that a daily scale in construction projects may be impractical. Cho et al. (2012) developed a model which is entitled a 5W1H (What, When, Where, Who, Why, and How) to solve the challenge of integrating cost/schedule in a construction project. More specifically, the planner can follow the operation as multi-function within multi-level such as What (would be a column) and How (framework), and the answers to other questions give more details to enable the integration.

According to Eastman et al. (2011b), there is no fully functional BIM cost management software, and therefore the quantity surveyor should link between different platforms to carry out the main three tasks, namely estimation, budgeting, and control. Even though Lawrence et al. (2014) developed a model to update the estimated cost automatically based on design changes, the entire estimation would be unreliable due to plenty of missed information which is not embedded in the design. Moreover, Wang et al. (2016) developed a model to integrate cost/schedule-based BIM which creates links between the BIM design object, cost item, activity, and area (zone/floor). Even though the proposed model used BIM to formulate the project budget, the process does not support automation.

We discussed here how BIM is utilized to develop cost budgeting. Given that BIM supports completing tasks automatically, the next section will therefore consider existing attempts to integrate 4D and 5D to develop a cost plan.

4D/5D BIM automation

Integrating BIM into daily construction activities will facilitate the automatic updating of all site information, and as such can result in enhanced productivity, strengthened relationships amongst stakeholders, and increased trust in site-collected data (Omar and Dulaimi, 2015). As such, El-Omari and Moselhi (2011) asserted that using unsystematic procedures in collecting site data can lead to a huge loss of information, leading to unreliable results. 4D BIM automation will enhance the quality of the collected data and reduce human interference in the data collection process (Hartmann et al., 2008, Hamledari et al., 2017). Similarly, 5D BIM provides an effective methodology for cost data collection and analysis of construction projects (Wang et al., 2016, Aibinu and Venkatesh, 2013, Lee et al., 2014, Popov et al., 2010). Furthermore, Lee et al. (2014) recommended that BIM cost systems should participate in decision-making, rather than merely generating BoQs.

Automated data collection methods have intensively improved through various kinds of technology like barcoding, radio frequency identification, 3D laser scanning, photogrammetry, multimedia, and pen-based computers (El-Omari and Moselhi, 2011, Turkan et al., 2012, 2013). Eastman et al. (2011a), on the other hand, argue that there is no comprehensive BIM-based cost management platform that can perform all cost-related processes, namely estimation, budgeting, and control. Collected data is hence not effectively used across the construction industry, and research studies are shifting to explore the means towards analyzing data in efficient ways (Wang et al., 2016, Hosseini et al., 2018).

This section provided a clear view of current attempts to develop the cash-out plan by integrating 4D and 5D automatically.

Teaching map of 5D BIM

To teach 5D BIM to undergraduate or postgraduate students, the teaching process should be systematic to enable students to absorb the required theoretical knowledge before moving to practical and sophisticated applications. The following steps should be followed to prepare for/adopt 5D BIM in educational institutes:

- **Teaching cost management process:** Cost management comprises three main tasks (i.e. cost estimation, budget, and control), which should be fully comprehended by students before moving to learn 5D BIM. This is to enable students to understand how 5D BIM can contribute to conducting these tasks and what the limitations are.
- **Teaching traditional methods of cost estimation:** 5D BIM contributes mainly to automate extracting the BoQ from the 3D BIM model. Therefore, a student should be aware of how traditional methods are used to implement the same process. This should be accompanied by teaching different cost estimation methods, such as analogous and reserve estimation methods.
- **Teaching the 5D BIM implementation process:** After students understand all the essential theoretical knowledge, 5D BIM then can be introduced systematically through (1) teaching the process of retrieving cost information from the 3D BIM model, (2) teaching how the retrieved cost data (i.e. BoQ) can be used to estimate the whole project cost, and (3) teaching how the estimated cost can be used with other BIM documents such as 4D BIM to develop the project budget.
- **Teaching advanced 5D BIM integrations with other technologies:** Given that BIM can be integrated with several technologies such as immersive technologies and blockchain, it will be useful to show how 5D BIM can be enhanced by using these technologies, such as by automating all payments to enhance transparency among project parties.

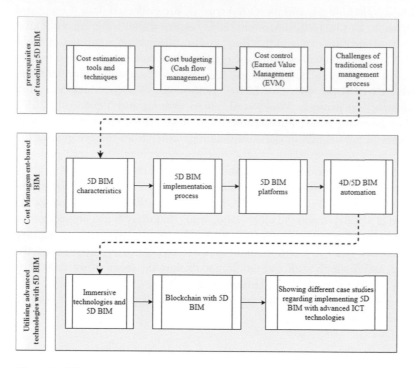

Figure 6.1 The process of preparing and teaching 5D BIM.

Figure 6.1 shows the process of preparing and teaching 5D BIM. The process is divided into three main stages, namely, the prerequisites of teaching 5D BIM, teaching 5D BIM, and using 5D BIM with other technologies.

Conclusion

The aim of this chapter was to review the application of 5D BIM for cost management and establish a teaching map for educators. The cost management system is a framework that comprises a set of tools and techniques for implementing specific tasks such as estimation, budgeting, control, payment processing, and change management. The most important activities in a cost management system are to prepare the cost plan, perform the cost estimation, perform the cost control, and provide the calculation of final accounts. In particular, cost estimation is an important element of construction project planning, and it is known to be an iterative process. Uncertainty, which is defined as controllable and uncontrollable factors that may occur during the project, is the major root cause of poor cost estimation. For example, changes in the scope of the project and bias may lead to inaccurate cost estimation. Also, ambiguous contractual agreements play an important role in inaccurate cost estimation. From its establishment, BIM has been perceived to contribute to the development of cost management systems and the role of quantity surveying. More specifically, 5D BIM offers the ability to extract quantities from 3D models. The cost estimating begins by exporting data from 3D models to BIM-based cost estimating software, and the preparation of BoQs. Such automated quantification reduces the time needed for the quantity take-off. The role of quantity surveying has changed in such a way that the focus is on analyzing and validating the cost data rather than spending time extracting quantities. Also, BIM helps quantity surveyors to be involved in different tasks of the project and eliminates the QS's traditional isolated working environment. However, the coordination of building information models developed by different trades is a challenge, and as a result, quantity surveyors need to articulate the cost report semi-manually by linking several models. The conventional approach of developing models in isolation makes BIM ineffective for use by quantity surveyors.

Recognizing the interdependencies between the cost (5D BIM) and schedule (4D BIM) is important to control cash flow. However, the integration of the work breakdown structure and the cost breakdown structure is a complex task. Several solutions have been proposed to tackle this challenge in BIM, such as creating the project schedule with BIM models and then performing the cost estimation of all design elements. It has been also proposed to automate the data collection methods to further integrate 5D BIM into daily construction activities and reduce human interfaces in data collection. However, the analysis of such data is a challenge and requires further study.

This chapter also proposes a teaching map for 5D BIM consisting of four steps. The first step is about teaching the three main tasks of the cost management process (i.e. cost estimation, budget, and control), teaching the traditional practices of cost estimation, teaching the 5D BIM implementation process (i.e. retrieving cost information from the 3D BIM model, how the retrieved BoQ can be used to estimate the whole project cost, and how the estimated cost can be used with other BIM documents) and teaching advanced 5D BIM integrations with other technologies.

Questions

1. What is the cost management process?
2. What are the characteristics of 5D BIM?
3. How has the role of quantity surveyor changed with 5D BIM?
4. What are the challenges of 4D/5D BIM automation?
5. How is BIM utilized to develop cost budgeting?

References

Abrishami, S., Goulding, J., Pour Rahimian, F. & Ganah, A. 2015. Virtual generative BIM workspace for maximising AEC conceptual design innovation: A paradigm of future opportunities. *Construction Innovation*, 15, 24–41.

Abrishami, S., Goulding, J., Rahimian, F. P. & Ganah, A. 2014. Integration of BIM and generative design to exploit AEC conceptual design innovation. *Information Technology in Construction*, 19, 350–359.

Ahmed, N. U. 1995. A design and implementation model for life cycle cost management system. *Information & Management*, 28, 261–269.

Aibinu, A. & Venkatesh, S. 2013. Status of BIM adoption and the BIM experience of cost consultants in Australia. *Journal of Professional Issues in Engineering Education and Practice*, 140, 04013021.

Akintoye, A. & Fitzgerald, E. 2000. A survey of current cost estimating practices in the UK. *Construction Management & Economics*, 18, 161–172.

Ali, A. & Kamaruzzaman, S. 2010. Cost performance for building construction projects in Klang Valley. *Journal of Building Performance*, 1, 110–118.

Alinaitwe, H., Apolot, R. & Tindiwensi, D. 2013. Investigation into the causes of delays and cost overruns in Uganda's public sector construction projects. *Journal of Construction in Developing Countries*, 18, 33.

Aranda-Mena, G., Crawford, J., Chevez, A. & Froese, T. 2009. Building information modelling demystified: Does it make business sense to adopt BIM? *International Journal of Managing Projects in Business*, 2, 419–434.

Ashcraft, H. W. 2014. The transformation of project delivery. *Construction Law*, 34, 35–58.

Ashworth, A., Hogg, K. & Higgs, C. 2013. *Willis's Practice and Procedure for the Quantity Surveyor*, Wiley.

Azhar, N., Farooqui, R. U. & Ahmed, S. M. 2008. Cost overrun factors in construction industry of Pakistan. *In*: First International Conference on Construction in Developing Countries (ICCIDC–I), Advancing and Integrating Construction Education, Research Practitioner, Karachi,, Pakistan. 499–508.

Azhar, S., Khalfan, M. & Maqsood, T. 2012. Building information modelling (BIM): Now and beyond. *Construction Economics and Building*, 12, 15–28.

Banihashemi, S., Tabadkani, A. & Hosseini, M. R. 2018. Integration of parametric design into modular coordination: A construction waste reduction workflow. *Automation in Construction*, 88, 1–12.

Boon, J. & Prigg, C. 2012. Evolution of quantity surveying practice in the use of BIM – the New Zealand experience. *In*: Proceedings of the CIB International Conference on Management and Innovation for a Sustainable Built Environment, Amsterdam, The Netherlands. 84–98.

Callahan, J. T. 1998. *Managing Transit Construction Contract Claims*, Transportation Research Board.

Cartlidge, D. 2011. *New Aspects of Quantity Surveying Practice*, Routledge.

Cho, D., Russell, J. S. & Choi, J. 2012. Database framework for cost, schedule, and performance data integration. *Journal of Computing in Civil Engineering*, 27, 719–731.

De Andrade, P. A., Martens, A. & Vanhoucke, M. 2019. Using real project schedule data to compare earned schedule and earned duration management project time forecasting capabilities. *Automation in Construction*, 99, 68–78.

Eastman, C., Teicholz, P., Sacks, R. & Liston, K. 2011a. *BIM Handbook: A Guide to Building Information Modeling for Owners, Managers, Designers, Engineers and Contractors*, Wiley.

Eastman, C. M., Eastman, C., Teicholz, P., Sacks, R. & Liston, K. 2011b. *BIM Handbook: A Guide to Building Information Modeling for Owners, Managers, Designers, Engineers and Contractors*, Wiley.

Econorthwest, P. B. Q. 2002. *Douglass, Inc. TCRP Report 78: Estimating the Benefits and Costs of Public Transit Projects: A Guidebook for Practitioners* [Retrieved June 21, 2014], Transportation Research Board.

El-Omari, S. & Moselhi, O. 2011. Integrating automated data acquisition technologies for progress reporting of construction projects. *Automation in Construction*, 20, 699–705.

Enshassi, A., Al-Najjar, J. & Kumaraswamy, M. 2009. Delays and cost overruns in the construction projects in the Gaza Strip. *Journal of Financial Management of Property and Construction*, 14, 126–151.

Fan, S.-L., Wu, C.-H. & Hun, C.-C. 2015. Integration of cost and schedule using BIM. 淡江理工學刊, 18, 223–232.

Forgues, D., Iordanova, I., Valdivesio, F. & Staub-French, S. 2012. Rethinking the cost estimating process through 5D BIM: A case study. *In*: Construction Research Congress 2012: Construction Challenges in a Flat World, West Lafayette, Indiana, United States. 778–786.

Franco, J., Mahdi, F. & Abaza, H. 2015. Using building information modeling (BIM) for estimating and scheduling, adoption barriers. *Universal Journal of Management*, 3, 376–384.

Fu, C., Aouad, G., Lee, A., Mashall-Ponting, A. & Wu, S. 2006. IFC model viewer to support nD model application. *Automation in Construction*, 15, 178–185.

Hamledari, H., Mccabe, B., Davari, S. & Shahi, A. 2017. Automated schedule and progress updating of IFC-based 4D BIMs. *Journal of Computing in Civil Engineering*, 31, 04017012.

Hanid, M., Siriwardena, M. & Koskela, L. 2011. What are the big issues in cost management? In RICS Construction and Property Conference, Lima, Peru. 738.

Hannon, J. J. 2007. Estimators' functional role change with BIM. *AACE International Transactions*, IT31, 1–8.

Harbuck, R. H. 2004. Competitive bidding for highway construction projects. *AACE International Transactions*, ES91, ES91.

Hartmann, T., Gao, J. & Fischer, M. 2008. Areas of application for 3D and 4D models on construction projects. *Journal of Construction Engineering and Management*, 134, 776–785.

Horngren, C. T., Bhimani, A., Datar, S. M., Foster, G. & Horngren, C. T. 2002. *Management and Cost Accounting*, Financial Times/Prentice Hall Harlow.

Hosseini, M. R., Maghrebi, M., Akbarnezhad, A., Martek, I. & Arashpour, M. 2018. Analysis of Citation Networks in Building Information Modeling Research. *Journal of Construction Engineering and Management*, 144, 04018064.

Hufschmidt, M. M. & Gerin, J. 1970. Systematic errors in cost estimates for public investment projects. *In: The Analysis of Public Output*, NBER.

Hussain, O. A. 2012. Direct cost of scope creep in governmental construction projects in Qatar. *Global Journal of Management and Business Research*, 12, 1–12.

Jorgensen, M. & Shepperd, M. 2006. A systematic review of software development cost estimation studies. *IEEE Transactions on Software Engineering*, 33, 33–53.

Jung, Y. & Woo, S. 2004. Flexible work breakdown structure for integrated cost and schedule control. *Journal of construction Engineering and Management*, 130, 616–625.

Kaliba, C., Muya, M. & Mumba, K. 2009. Cost escalation and schedule delays in road construction projects in Zambia. *International Journal of Project Management*, 27, 522–531.

Kang, L. S. & Paulson, B. C. 1998. Information management to integrate cost and schedule for civil engineering projects. *Journal of Construction Engineering and Management*, 124, 381–389.

Kerzner, H. 2017. *Project Management: A Systems Approach to Planning, Scheduling, and Controlling*, Wiley.

Khan, A. 2006. Project scope management. *Cost Engineering*, 48, 12–16.

Kim, J. 1989. *An Object-Oriented Database Management System Approach to Improve Construction Project Planning and Control*. Doctoral Dissertation, University of Illinois.

Kiviniemi, A., Rekola, M., Belloni, K., Kojima, J., Koppinen, T., Makelainen, T. & Heitanen, J. 2007. *Senate Properties: BIM Requirements 2007 Quantity Take-Off* (Vol. 7), Senate Properties.

Lawrence, M., Pottinger, R., Staub-French, S. & Nepal, M. P. 2014. Creating flexible mappings between Building Information Models and cost information. *Automation in Construction*, 45, 107–118.

Le-Hoai, L., Dai Lee, Y. & Lee, J. Y. 2008. Delay and cost overruns in Vietnam large construction projects: A comparison with other selected countries. *KSCE Journal of Civil Engineering*, 12, 367–377.

Lee, S.-K., Kim, K.-R. & Yu, J.-H. 2014. BIM and ontology-based approach for building cost estimation. *Automation in Construction*, 41, 96–105.

Leu, S.-S. & Lin, Y.-C. 2008. Project performance evaluation based on statistical process control techniques. *Journal of Construction Engineering and Management*, 134, 813–819.

Love, P. E., Davis, P. R., Chevis, R. & Edwards, D. J. 2011. Risk/reward compensation model for civil engineering infrastructure alliance projects. *Journal of Construction Engineering and Management*, 137, 127–136.

McCuen, T. L. 2008. Scheduling, estimating, and BIM: A profitable combination. *AACE International Transactions*, BIM11, 11–18.

McCuen, T. L., Suermann, P. C. & Krogulecki, M. J. 2011. Evaluating award-winning BIM projects using the national building information model standard capability maturity model. *Journal of Management in Engineering*, 28, 224–230.

Meadati, P. 2009. BIM extension into later stages of project life cycle. *In*: 45th Annual International Conference on Associated Schools of Construction, Gainesville, Florida, US. 121–129.

Merschbrock, C., Hosseini, M. R., Martek, I., Arashpour, M. & Mignone, G. 2018. Collaborative role of sociotechnical components in BIM-based construction networks in two hospitals. *Journal of Management in Engineering*, 34, 05018006.

Mitchell, D. 2012. 5D BIM: Creating cost certainty and better buildings. *In*: RICS Cobra Conference, Brisbane (Australia).

Moder, J. J., Phillips, C. R. & Davis, E. W. 1983. Project management with CPM, PERT, and precedence diagramming. ISBN-13: 978-0442254155, ISBN-10: 0442254156.

Moselhi, O. & Salah, A. 2012. Fuzzy set-based contingency estimating and management. *In*: Proceedings of the International Symposium on Automation and Robotics in Construction (ISARC). IAARC Publications, Eindhoven, Netherlands. 1.

Motamedi, A. & Hammad, A. 2009. Lifecycle management of facilities components using radio frequency identification and building information model. *Journal of Information Technology in Construction*, 14, 238–262.

Nassar, K. 2011. Assessing building information modeling estimating techniques using data from the classroom. *Journal of Professional Issues in Engineering Education and Practice*, 138, 171–180.

Niazi, A., Dai, J. S., Balabani, S. & Seneviratne, L. 2006. Product cost estimation: Technique classification and methodology review. *Journal of Manufacturing Science and Engineering*, 128, 563–575.

Oberlender, G. D. & Oberlender, G. D. 1993. *Project Management for Engineering and Construction*, McGraw-Hill.

Olatunji, O., Sher, W. & Ogunsemi, D. 2010. The impact of building information modelling on construction cost estimation. *In*: W055-Special Track 18th CIB World Building Congress, May 2010, Salford, UK. 193.

Omar, H. & Dulaimi, M. 2015. Using BIM to automate construction site activities. *Building Information Modelling (BIM) in Design, Construction and Operations*, 149, 45.

Oraee, M., Hosseini, M. R., Papadonikolaki, E., Palliyaguru, R. & Arashpour, M. 2017. Collaboration in BIM-based construction networks: A bibliometric-qualitative literature review. *International Journal of Project Management*, 35, 1288–1301.

Pärn, E. A. & Edwards, D. J. 2017. Conceptualising the FinDD API plug-in: A study of BIM-FM integration. *Automation in Construction*, 80, 11–21.

Popov, V., Juocevicius, V., Migilinskas, D., Ustinovichius, L. & Mikalauskas, S. 2010. The use of a virtual building design and construction model for developing an effective project concept in 5D environment. *Automation in Construction*, 19, 357–367.

Potts, K. & Ankrah, N. 2014. *Construction Cost Management : Learning from Case Studies*, Taylor and Francis. 2nd ed.

Qian, A. Y. 2012. *Benefits and ROI of BIM for Multi-Disciplinary Project Management*, National University of Singapore.

Rasdorf, W. J. & Abudayyeh, O. Y. 1991. Cost-and schedule-control integration: Issues and needs. *Journal of Construction Engineering and Management*, 117, 486–502.

Redmond, A., Hore, A., Alshawi, M. & West, R. 2012. Exploring how information exchanges can be enhanced through Cloud BIM. *Automation in Construction*, 24, 175–183.

Reilly, J. Cost estimating and risk management for underground projects. *In*: Proceedings of the International Tunneling Conference, 2005. Citeseer.

Schexnayder, C., Molenaar, K. & Shane, J. 2011. Estimating large complex projects. *Revista Ingeniería de Construcción*, 22, 91–98.

Shane, J. S., Molenaar, K. R., Anderson, S. & Schexnayder, C. 2009. Construction project cost escalation factors. *Journal of Management in Engineering*, 25, 221–229.

Shank, J. K. 1989. Strategic Cost Management: New. *Journal of Management Accounting Research*, 1, 47–65.

Shen, W., Hao, Q. & Xue, Y. 2012. A loosely coupled system integration approach for decision support in facility management and maintenance. *Automation in Construction*, 25, 41–48.

Shen, Z. & Issa, R. R. 2010. Quantitative evaluation of the BIM-assisted construction detailed cost estimates. *Journal of Information Technology in Construction (ITcon)*, 15, 234–257.

Smith, P. 2014a. BIM & the 5D project cost manager. *Procedia: Social and Behavioral Sciences*, 119, 475–484.

Smith, P. 2014b. BIM implementation: Global strategies. *Procedia Engineering*, 85, 482–492.

Stanley, R. & Thurnell, D. 2013. Current and anticipated future impacts of BIM on cost modelling in Auckland. In Proceedings of the 38th Australasian Universities Building Education Association Conference, 20–22 November 2013, Auckland, New Zealand.

Stanley, R. & Thurnell, D. 2014. The benefits of, and barriers to, implementation of 5D BIM for quantity surveying in New Zealand.

Sunil, K., Pathirage, C. & Underwood, J. 2017. Factors impacting Building Information Modelling (BIM) implementation in cost monitoring and control. *In:* 13th International Postgraduate Research Conference (IPGRC). University of Salford, University of Salford, UK. 210–224.

Sylvester, K. E. & Dietrich, C. April, 2010. Evaluation of building information modeling (BIM) estimating methods in construction education. *In*: Proceedings of 46th ASC Annual International Conference. Citeseer, The Associated Schools of Construction, Windsor, CO. 7–10.

Szekeres, V. 2005. Evaluation of factors behind the stagnation of Japan's economy. *In*: 3rd International Conference on Management, Enterprise and Benchmarking, Budapest. 171–185.

Touran, A. & Lopez, R. 2006. Modeling cost escalation in large infrastructure projects. *Journal of Construction Engineering and Management*, 132, 853–860.

Turkan, Y., Bosche, F., Haas, C. T. & Haas, R. 2012. Automated progress tracking using 4D schedule and 3D sensing technologies. *Automation in Construction*, 22, 414–421.

Turkan, Y., Bosché, F., Haas, C. T. & Haas, R. 2013. Toward automated earned value tracking using 3D imaging tools. *Journal of Construction Engineering and Management*, 139, 423–433.

Wang, K.-C., Wang, W.-C., Wang, H.-H., Hsu, P.-Y., Wu, W.-H. & Kung, C.-J. 2016. Applying building information modeling to integrate schedule and cost for establishing construction progress curves. *Automation in Construction*, 72, 397–410.

Weiss, L. L. 2000. *Design/Build: Lessons Learned to Date*, South Dakota Department of Transportation.

Welde, M. & Odeck, J. 2017. Cost escalations in the front-end of projects: Empirical evidence from Norwegian road projects. *Transport Reviews*, 37, 612–630.

Wu, S., Wood, G., Ginige, K. & Jong, S. W. 2014. A technical review of BIM based cost estimating in UK quantity surveying practice, standards and tools. *Journal of Information Technology in Construction*, 19, 534–562.

Yang, Y.-C., Park, C.-J., Kim, J.-H. & Kim, J.-J. 2007. Management of daily progress in a construction project of multiple apartment buildings. *Journal of Construction Engineering and Management*, 133, 242–253.

Zaghloul, R. & Hartman, F. 2003. Construction contracts: The cost of mistrust. *International Journal of Project Management*, 21, 419–424.

7 Building information modelling for facilities management

Skills, implementation, and teaching map

Sandra Matarneh and Faris Elghaish

Introduction

Facilities management (FM) involves a wide range of multi-disciplinary services with the overall purpose of maintaining and enhancing building assets to ensure occupants' wellbeing (Becerik-Gerber et al., 2010). The key challenge for FM teams is to have real-time, accurate, and comprehensive information to perform their day-to-day activities and to provide their senior management with accurate information for decision-making (Atkin and Brooks, 2009).

Currently, various technology platforms, data repositories, or database management systems, including computer-aided facility management (CAFM) systems, are used for information management in different facilities. In most FM practices today, the data required for the computerized maintenance management systems (CMMS) come from various sources. These data are collected and entered manually into these systems, manipulated several times during the project life cycle, and entered manually into each FM system several times, as these systems lack interoperability between each other, resulting in error-prone processes (Becerik-Gerber et al., 2010; Teicholz, 2013; Patacas et al., 2015).

Using building information management (BIM) in FM practice facilitates the information management of a building's components and systems during its life cycle (Teicholz, 2013). The more accurate and up-to-date information is available to the FM team, the greater the opportunity for the enhancement of processes throughout the operation and maintenance (O&M) phase. One of the key success factors for BIM implementation in FM is to identify the required data for day-to-day activities (Liu and Issa, 2015). Accordingly, the FM industry is beginning to acknowledge the importance of having a standardization for data format specification. Standards such as UK PAS1192-3:2014 have been issued which support data management and provide a specification for information management for the operational phase of facilities using BIM. It has been suggested that the Construction Operation Building Information Exchange (COBie) be used as a data exchange method (BSI, 2014a). COBie is a neutral spreadsheet format that organizes a facility's non-geometric data in a structured simple format for use by the owner/FM teams (Thabet and Lucas, 2017). However, even though different ways, processes, and tools have been developed to exchange information during a facility's life cycle, there is still a lack of understanding of what sort of information needs to be used by the FM team during the O&M phase and how to transfer this information seamlessly into existing FM systems. This chapter aims to examine how internal and external contexts shape the exploration and exploitation of learning when implementing BIM in facilities management. This is done by presenting a general overview of facilities information systems and highlights the applications and challenges of using building information modelling (BIM) in FM practice.

Facilities information management

Today's facilities are ever more sophisticated, and available and reliable information for O&M activities is vital (Jordani, 2010). The key challenge for facility managers is to have real-time accurate and complete information to perform their day-to-day activities and to provide their senior management with accurate information for the decision-making process (Atkin and Brooks, 2009). Currently, there are various technology platforms, data repositories, or databases such as computer-aided facility management (CAFM) and computerized maintenance management system (CMMS) that are used for these purposes in different facilities. In most current practices, data is extracted from paper construction documents and is re-entered manually in one of these computerized information systems (Teicholz, 2013).

Most of the facility information needed to support FM practice is often created and accumulated throughout the design and construction phases and is often handed over to the owner/FM teams when the construction is completed, in the form of papers and/or electronic copy. However, this late delivery of unstructured information causes a serious challenge for owners/FM teams to check and verify whether the delivered information includes the required information to perform FM activities during the O&M phase (Teicholz, 2013; Patacas et al. 2015). Moreover, the fragmentation in the construction industry and the consequent lack of communication between project stakeholders at different phases of the facility life cycle result in the handover of unstructured and incomplete information to the owner/FM teams. In the case of the delivered information being solely in the form of a paper copy, it remains in the owner/FM team's storage system until it becomes outdated or damaged. In the best-case scenario, FM teams start to scan the delivered hardcopy documents to transfer them to a digital format. However, scanning documents does not actually mean a digital format, since the FM teams cannot update them or conduct any query on this scanned database (East, 2007).

East and Brodt (2007) point out that the owner will have to pay at least three times for the handover of the construction information: the first time is when the cost of providing this information was embedded in the design and construction costs; the second time is when the FM firms are paid to survey the existing facility conditions to obtain as-built drawings; and the third time is when the delivered/collected information has to be re-entered manually into FM systems (East and Brodt, 2007).

The project information handed over to the owner/FM teams is bulky and includes many documents such as as-built drawings, specifications, operation and maintenance manuals, and warranties and guarantees of the installed systems and equipment. Processing a large amount of fragmented information to organize and re-enter them into FM systems is a costly and time-consuming process, resulting in a lengthy and error-prone process which can be extended for up to six months to finalize this task (Gallaher et al., 2004; Patacas et al., 2015). As an example, according to a study conducted by Penn State University, DoD & DoD Sandusky Laboratories, University of California, each maintenance work order required 7–120 minutes to collect related information to complete that work order. If the required information could be made available within 15 minutes to execute this work order, the cost savings for the total 6,356 work orders per year were estimated to be about 583,316.00 USD per year, considering the average cost of executing a work order is 50.00 USD per hour (Alevras and Arabia, 2014). The National Institute of Standards and Technology (NIST) pointed out in their study that two-thirds of the estimated $15.8 billion lost in the US capital facilities industry were associated with inadequate interoperability during the O&M phase to cover for expenses related to manual information re-entry, information verification, redundancy, and idle labour time spent in looking for unavailable information (Gallaher et al., 2004; Rundell, 2006; Jordani, 2010).

BIM has been developed to enable the project stakeholders during different phases to collect, manage, exchange, and share the facility's information during its life cycle (Isikdag et al., 2008). A BIM database includes information about the facility's geometry and its components. Thus, owners/FM teams can minimize their share of the cost related to inadequate interoperability by adopting BIM. BIM allows for managing the vast complexity and large amount of information generated during the facility's life cycle, making the information available during the O&M phase for the FM team's use (Rundell, 2006; Azhar, 2011; Becerik-Gerber et al., 2012; Kassem et al., 2015).

Although BIM is emerging as the main database for a building's life cycle, the use of BIM in the O&M phase is limited. Becerik-Gerber et al. (2012) conducted interviews with FM practitioners to outline the role of BIM in FM. Their study indicates that the existing FM information management is being done manually and that integrating BIM in FM could leverage FM practice.

Although BIM is currently recognized by academics and FM practitioners, it is still unclear how to efficiently integrate it into FM practice. Moreover, it is still unclear what information is needed for FM teams' use and how to transfer this information from BIM models to FM systems. Issues of information exchange and interoperability need to be addressed to facilitate transferring the relevant information to FM systems and to facilitate BIM implementation in FM (Becerik-Gerber et al., 2012; Kassem, et al., 2015).

Among many ambiguous issues that need to be cleared up to extend BIM implementation in FM, interoperability remains the main issue. Interoperability is the capability to exchange information among various applications to enable the automation of information exchange and access and to avoid manual data re-entry. Due to the wide variety of BIM and FM platforms, interoperability between these platforms remains one of the key challenges in using BIM in FM practice (Arayici, 2015; Ham and Golparvar-Fard, 2015; Kassem et al., 2015; Ibrahim et al., 2016).

Recently, there have been various attempts to solve the interoperability issue by introducing different universal data standards, such as the Industry Foundation Classes (IFC) and XML schemas, and structured specifications such as the Construction Operations Building Information Exchange (COBie) (Azhar et al., 2012). However, these attempts still have their inherent limitations. Pragmatic strategies for purposeful information exchange among BIM models and different FM information systems such as CMMS are required to overcome the interoperability challenge.

Moreover, most of the existing studies related to integrating BIM in FM practice are focusing only on the human and organizational issues, business and legal barriers, and avoiding the technical barrier (interoperability barrier). However, most of the proposed theoretical framework of BIM information exchange for FM use is based on the assumption that information can be transferred seamlessly between the various BIM and FM systems (Kensek, 2015). Successful integration of BIM in FM demands a proper approach to address the lack of a standardized information exchange process and the lack of interoperability between BIM and FM systems.

This chapter shows that the exploration of BIM is seen as essential for future viability, and facility managers, owners, and students can explore BIM through facility scenarios as to how BIM could provide a viable solution to resolve information gaps. Exploitation is based on previous experiences where apparent similar changes happened and inferences are imposed onto a new BIM context. The emphasis is on BIM leveraging information processes rather than changing facilities management practices. The chapter therefore shows that a balance between exploration and exploitation learning is essential in order for BIM to leverage facilities management practice. The following sections will discuss the existing facilities management systems and highlight the current challenges and obstacles hindering the integration of the various facilities management systems to achieve efficient information management.

Facilities information systems

Over the last few decades, the scope of FM has both evolved and become more complex. FM's role, which at one time entailed mainly operating and maintaining individual facilities, has now evolved to include other responsibilities such as health and safety, code compliance, and energy and sustainability management. As the scope and responsibilities of FM have increased, the supporting information technologies have too. Currently, there is a wide range of FM information systems available to support the day-to-day activities of FM (Whittaker, 2017).

Since the late 1980s, FM information systems have been established to automate FM information collection and to provide FM teams with the tools to track, plan, manage, and report on facilities information. These systems enable decision-makers to automate many of the data-intensive FM functions and accordingly results in continuous cost savings and improved utilization of facilities throughout their entire life cycle (NRC, 2008).

There is no ideal FM information system suitable for all conditions to meet the specific demands of any FM team. However, FM information systems continue to evolve at a rapid pace. Even the basic computerized maintenance management systems (CMMS) continue to add functional modules to enhance capabilities. Furthermore, the use of handheld technologies that seamlessly interface with FM information systems continues to expand (Whittaker, 2017).

FM systems consist of a variety of software applications and information sources that may include object-oriented database systems, computerized maintenance management systems (CMMS), integrated workplace management systems (IWMS), and also project delivery systems, computer-aided design (CAD) systems, Revit, and building information models (BIM), as well as interfaces to other systems such as building automation systems (BAS) and enterprise resource planning (ERP) applications (Whittaker, 2017). Today most systems are web-based and provide a host of features, including facilities-related scheduling and analysis capabilities. Data may be collected from a variety of sources through technology interfaces or human transfer processes and may be stored, retrieved, and analysed from a single data-store (Whittaker, 2017).

The generally accepted terminology commonly used to describe the various types of FM information technologies is presented as follows:

Computerized maintenance management systems (CMMS): CMMS is a conventional software that is used to support FM teams in scheduling and recording operations and preventive/planned maintenance activities associated with the facility's equipment. CMMS also supports the FM team in prioritizing work orders and in planning for periodic/preventive maintenance. Moreover, all historically recorded information related to work order execution is loaded into the CMMS database for future planning and control (Vanier, 2001). Although CMMS has the potential to increase the efficiency of the FM team and serve as a maintenance history database, more than 50% of CMMS implementation fails to achieve its purpose (Berger, 2009).

Computer-aided facility management (CAFM) systems: CAFM systems were traditionally software applications that included core CMMS functionality and incorporated CAD- or geographical information systems (GIS)-based spatial management capabilities. They were generally used to manage building space allocation and space planning, in addition to the basic work order (WO) processes. Today, this class of software has expanded to include more FM functionality and is now generally referred to as *integrated workplace management systems* (IWMS) (Lee et al., 2013).

Integrated workplace management systems (IWMS): The term IWMS refers to FM information systems with the broadest functionality to support real-estate and FM requirements. Effectively, IWMS have evolved from CAFM systems and can encompass the entire life cycle of the facility, from design to construction and operations. IWMS are enterprise-class software platforms that integrate five key functional domains within a single hosted database. The functional domains typically include maintenance management, space management and planning, real-estate and lease management, project portfolio management, and environmental sustainability (Clarke and D'arjuzon, 2019).

Building automation system (BAS): This is a software package used to automatically monitor and control mechanical equipment, including heating, ventilating, and air-conditioning (HVAC), lighting, and other systems through a building management system or building automation system (BAS) (Elmualim and Pelumi-Johnson, 2009). It is a computer-driven system programmed to control mechanical equipment. It is also called the *building control system* or *energy management system* (EMS) (Marinakis et al., 2013).

Enterprise resource planning (ERP): ERP is a software package used as a financial management system to manage organizations' business processes and automate other functions related to facility services, technology, financial management, and human resource management (Lee et al., 2013).

Currently, data is handed over to facility managers through a handover process which is integrated into one of the facility management systems mentioned above. Information from construction projects is established and then formatted to fit facilities management systems. When considering BIM, there is an implicit change with obvious obstacles such as interoperability issues, learning curves, user resistance, and disruption to business activities (Love et al., 2015). To facilities management, information is a key commodity and BIM may offer "added value" to leverage the existing facilities information management. The key is to understand the existing facilities management systems requirements.

This section introduced the existing facilities management systems. In the following section, the challenges facing facilities management organizations in reaping the full benefits of these existing systems will be discussed.

Challenges related to facilities management systems

FM information systems are developed to streamline facility workflow processes, to provide data for facilities decision-making, and to help measure FM performance. Although FM information systems can deliver significant benefits to businesses, the FM team faces several challenges in successfully utilizing these systems. Aziz et al. (2016) emphasize that the quality of information entered into FM information systems is a key factor in utilizing these systems successfully. According to Whittaker (2014), the implementations of most FM information systems either fail or lead to underutilized solutions. The most common reasons for the failure of these systems to meet the FM team's needs include the following:

- Lack of understanding of what the FM team wants to get out of the system prior to feeding these systems with facility information;
- Lack of clear expected outcomes of performance measures;
- Poor definition and application of data standards;

- Lack of clear system configuration standards to enable the reporting of the desired key performance indicators (KPIs);
- Poorly defined information exchanges or application interface requirements;
- Lack of understanding of the resources required to implement and maintain the technology and data;
- Lack of training of FM teams on workflow processes aligned with the software standard operating procedures;
- Lack of training of FM teams on managing critical data and consistency in workflow processes.

The challenge for FM teams is to overcome these failure points by focusing on these aims: identification of the system application goals and objectives, development of realistic outcomes for performance measures, establishment of consistent and holistic data standards, appropriate software configuration to allow analytics and reporting, and development and training on data capture and maintenance and consistent workflow processes (Whittaker, 2014).

In addition to the challenge of overcoming the above failure points, the main challenge that needs to be overcome is related to the lack of interoperability among the different FM information systems.

The integration of the component systems is an ongoing requirement for facilities management information system development. Software products with modules that can provide all or most of these capabilities or that have created interfaces to other industry-leading products are being developed in response to this need. Whenever possible, data is shared and is not duplicated. The best example of this is the location data (site, building, floor, room) that is managed in the CAFM system and is shared by the CMMS. This reduces data entry and allows all location-based information to be validated upon entry.

To date, little attention has been given to the ongoing issue of the interoperability between BIM technologies and current and legacy FM technologies (e.g. computer-aided facility management systems [CAFM]) during the handover of information and data to the operation stage. The existing systems need to be linked to BIM technologies to enhance, support, and leverage the existing information and process. It is essential that BIM data is transferred or linked to existing FM legacy systems and used to improve current methods of operation in order to support the business case for adopting BIM on existing assets.

During the life of the building, a facility should exist for information to be updated which is also responsive to change. There is a need for standardized data libraries and open systems that can be utilized by any CAFM or facilities management system. Without such non-proprietary formats, facility owners and managers must dictate which proprietary information systems to use, or re-input information into a CAFM system. Re-inputting information into relevant FM systems is inefficient, time-consuming, and costly for owners and facility managers.

This section addressed the challenges of the existing facilities management systems and highlighted the need to resolve the interoperability between the existing legacy of facilities management systems. The following sections will discuss in more depth the integration of BIM into facilities management practice.

BIM integration in facilities management practice

Facilities management encompasses a group of multi-disciplinary practitioners from independent disciplines who are working together to optimize the performance of a building's functions while ensuring it meets the end-users' needs (Atkin and Brooks, 2009; Becerik-Gerber et al., 2012). FM functions rely on an extensive range of data and information which are usually fragmented between various disciplines. Alvarez-Romero (2014) summarizes the traditional handover process in which the FM team have often been provided with hard-copy and electronic forms of O&M manuals. Usually, these documents are provided several months after completion of the facility's construction, and substantial effort and time may also be needed to integrate such information into FM information systems.

Traditionally, FM information is managed by dispersed information systems (e.g. CMMS, CAFM, BAS), in which data have to be re-entered many times for each FM information system individually, and are not synchronized between systems, resulting in error-prone, inconsistent data, as well as time and effort being spent in the process (Becerik-Gerber et al., 2012). BIM technologies and processes facilitate FM information management throughout the facility's life-cycle phases. BIM can be used as a single source of accurate and up-to-date FM information, which is an opportunity for the FM team to reduce the cumbersome and error-prone data-entry process, and accordingly minimize facility information loss during its life cycle (Eastman et al., 2011; Al-Shalabi and Turkan, 2015).

To summarize, BIM, with its capabilities, can act as a data pool during the facility life cycle, including the O&M phase. However, some research has contradicted this position and concluded that the value of integrating BIM in FM is considered to be marginal due to a lack of alignment between BIM embedded data and FM required data (Bosch et al., 2015). This view resonates with the conclusions of Kassem et al. (2015), who concede that BIM–FM integration represents a major challenge.

There are some actual case studies showing the tangible benefits of integrating BIM in FM. One of the earliest efforts to use BIM for FM was in the IFC-model-based Operations and Maintenance of Building project (Nisbet, 2008). In this project, a college building was designed using BIM and an IFC schema to capture the required FM information. The FM information was then transferred to the Maximo data structure (FM information system). The outcome of this project provided the basis for the development of COBie.

Another early exemplar was the Sydney Opera House case study, where integrating BIM in FM showed the different applications of BIM in FM and underlined the need to change the business processes and workflows. The project identified the key barrier to integrating BIM in FM, which was the lack of IFC standard support by FM tools (CRC, 2007; Eastman et al., 2011). A recent implementation of BIM for FM is in the Manchester Town Hall Complex project (Codinhoto et al., 2013). The project identified the lack of awareness of BIM potential in FM and the lack of guidelines for BIM implementation in FM as key challenges. Another case study of using BIM for FM is the existing Northumbria University campus buildings, where many challenges facing BIM in FM applications were identified.

Other examples by Dempsey (2009) revealed that there was a 98% reduction in time and effort when creating the FM database using BIM. A study by Ding et al. (2009) supports these findings and reveals that the integration of BIM in FM brought about a 98% reduction in the time required for updating FM databases. Moreover, the School of Cinematic Arts at the University of Southern California (USC) may well be the first project to implement BIM throughout the project's life cycle in the United States. (Smith and Tardif, 2009). This project used BIM to monitor the HVAC and electrical systems in the building (Becerik-Gerber and Kensek, 2010). Finally, the General Services Administration (GSA) and NASA are joining forces to integrate BIM for FM at the NASA Langley Research Center. The overall objective of that project was to test if the integration between BIM and CMMS is possible and valuable (Kasprzak and Dubler, 2012).

The overall aim of integrating BIM for FM was not to add an additional information system but to support the standardization of data delivery, define data ownership, and facilitate data accessibility (Sabol, 2013). However, developing technologies and processes to fully integrate BIM with FM applications and data repositories will be an ongoing challenge.

Organizations will continue to implement facilities information systems and building automation systems on a parallel basis. In the future, they will be seeking opportunities for further enhancements through a shared data repository to link between these different technologies and to overcome the interoperability issue. BIM is envisioned to bridge this gap and to link different facilities information and automation systems. Before exploring how BIM can do this, the next section will introduce where BIM can be implemented in facilities management.

BIM application areas in facilities management

BIM offers significant potential for supporting FM practice. The following is a summary of the main BIM applications in FM:

- **Locating building components:** Locating building components and equipment is a repetitive, laborious, and time-consuming task for FM teams. Usually, the FM team depends on hard copies of documents or on their experience to analyse the problem situation (East et al., 2013). A BIM model can be utilized to visualize the location of equipment and link the items to their related data. In addition, integrating BIM with the FM database helps in providing the equipment's maintenance history to better diagnose the problem. Moreover, safety, security, and productivity could be enhanced by the real-time components being located through the utilization of a BIM model and radio frequency identification (RFID) technology (Costin et al., 2012). Lin et al. (2014) proposed the use of a barcode-based system to locate different building components.
- **Facilitating real-time data access:** To perform accurate maintenance activities, FM teams need an accessible database. BIM, with its capabilities, can act as a unified digital database in which the collected data, along the facility life cycle, could be used to establish a knowledge management database. Motawa and Almarshad (2013) incorporated the case-based reasoning (CBR) technique with BIM and proposed a knowledge-based system to support O&M activities. For effective maintenance schedules, Motamedi et al. (2014) applied failure-cause detection patterns based on knowledge-assisted and BIM-enabled visual analytics.
- **Visualization and marketing:** BIM provides the FM team with a more reliable method of visualization, which enables them to conduct a what-if analysis and accordingly improve decision-making. In addition, the benefits of rendering tools and walk-through options have the potential to support marketing by creating images for the interior spaces and furniture, which can significantly influence the customers.
- **Checking maintainability:** BIM can support maintainability studies by addressing accessibility issues, such as examining the availability of enough access spaces for the removal/replacement of equipment. As suggested by Becerik-Gerber et al. (2012), BIM-based maintainability studies are related to the following areas: preventive maintenance, accessibility, and sustainability of materials.
- **Creating and updating digital assets:** Usually when a project is constructed and handed over to the client, digital assets are manually prepared and transferred to the FM systems, in which they tend to be error-prone. BIM offers the opportunity to capture, digitize, and automatically transfer the required assets' data efficiently. The digital assets

include equipment and systems such as manufacturer/vendor information, HVAC information, and documents including warranties and specifications.

- **Space management:** Space management includes assigning spaces, forecasting requirements, and streamlining the moving process. The types of required information for these activities are related to space descriptions, numbers, boundaries, areas, etc. Traditionally, CAD files were used to present this information, in which deficiencies occur. A BIM model can visualize space and host spatial attributes, which helps in recognizing underutilized spaces, estimating space requirements, and conducting space analysis.

- **Planning and feasibility studies for non-capital construction:** A building continuously changes based on the end-users' requirements and deteriorates due to many factors, including the weather; this ends with the need to renovate. A BIM model can be used to help with the renovation of a facility. In addition, the extracted historical data relating to the facility, such as material specifications and cost, could be used as a reference for the planned work.

- **Emergency management:** In the case of an emergency, the most important priority is the availability and accessibility of information. BIM can support emergency responders in identifying and finding possible emergency problems and locating hazards through its graphical interface. A BIM model can also help in simulating emergencies to develop a response plan.

- **Controlling and monitoring energy:** Usually, energy management systems are used to control and monitor a facility's energy consumption. These systems work individually and are not compatible with other FM systems. The BIM model's graphical interface and its linkage to building sensors, metering, and sub-metering information could allow automated control and real-time monitoring of data. Using BIM with an understanding of occupant behaviour can help apply "what-if" scenarios to analyse and simulate energy systems' performance.

- **Personnel training and development:** Traditionally, training is conducted through several methods, including presentations, site visits, hand-by-hand demonstration, and self-study, which is considered a time-consuming process. The BIM model can allow trainees to virtually walk through the model and, therefore, help them gain a better understanding of assigned zones.

There are several applications for BIM in facilities management. Among these several applications, the main driver for using BIM in FM is to improve the handover processes. Data and information collected through a BIM process during the building life cycle will reduce the cost and the time required to collect and build FM systems (even with current interoperability challenges). BIM will also eliminate the need to duplicate information in downstream FM systems.

To better capture the value of implementing BIM-enabled FM, the following two sections will highlight the benefits and challenges of BIM implementation in facilities management.

Benefits of BIM for facilities management

AEC/FM practitioners are constantly seeking new technologies and approaches to gain a competitive advantage in the current challenging economies and competitive markets. According to Baladhandayutham and Venkatesh (2012), AEC/FM needs to improve its performance efficiency and productivity and needs to be more client-centred. The use of BIM technologies is seen to be a promising tool for the AEC/FM industries to enhance their performance and competitiveness. The benefits of utilizing BIM in the facility life cycle are well acknowledged. According to Azhar et al. (2012), the essential benefit of utilizing BIM in FM is that it delivers accurate information about the facility's spaces, components, and systems, in which these types of information can enhance the efficiency of the FM practices.

Eastman et al. (2011) identified many benefits of utilizing BIM during the O&M phase, which are as follows: (1) improving the commissioning and handover process of FM data; (2) enhancing FM practices by providing an accessible bank of information that can be analysed, exchanged, and updated; and (3) integration with FM systems through different software packages, such as FM: Interacts and FM: Systems.

Langdon (2012) details a number of benefits of BIM in the FM field, which include (1) creating an FM database automatically using the BIM as-built model; (2) enabling FM costing and procurement; and (3) enabling real-time updated facility information to be made available by updating the BIM as-built model through the facility life cycle.

Kasprzak and Dubler (2012) believe that BIM has become a successful tool for handing over accurate information which supports FM teams in the decision-making process. However, Changyoon et al. (2013) recommend that for successful FM practices, effective information management of the facility's information needs to be implemented throughout each stage of the facility's life-cycle process. It can be challenging to collect and store facility data and make them available to FM systems. This is where BIM has an opportunity to improve FM practices, as it enables maintenance information to be accessed and linked to its related component, as well as enabling speedy identification of any problem area by using BIM's visualization capabilities. Codinhoto et al. (2013) point out that BIM enables the FM team to perform a what-if analysis technique, which saves effort and time spent looking for accurate relevant information. Moreover, Kelly et al. (2013) outline the main benefits of integrating BIM in FM, which include (1) improving the information handover by providing augmented manual processes; (2) improving the FM data accuracy; and (3) enhancing the efficiency of work order performance by providing an accessible bank of data which enables quick problem location and interventions.

Other valuable benefits of using BIM in FM are identified by Volk et al. (2014). These include data documentation through the as-built model, quality control, energy and space management, assessment and monitoring, emergency management, maintenance of warranty and service information, and the ability to continuously update facility information to reduce errors in the renovation process. Similarly, Kassem et al. (2015) summarize the benefits of BIM for FM as the following: improving current information handover processes; increasing the accuracy of FM data; facilitating the accessibility of FM data; and increasing the efficiency of work order performance.

Brinda and Parsanna (2014) list several BIM benefits related to various stakeholders:

1. Maintenance workers: the BIM model can reduce redundant field trips to locate problems by providing accurate field conditions and relevant maintenance information, which can reduce the cost by providing accessible information and accordingly prompt responses to work orders;
2. Building operators: the BIM model can identify and track facility equipment and accordingly provide accurate equipment inventories. It can also identify hidden facility components, maintain the facility maintenance history, and therefore enhance the facility's performance by enabling analysis and comparison between actual and predicted energy performance;

3. Building occupants: the use of the BIM model can increase building occupants' satisfaction by decreasing the time needed for a work order response.

Arayici et al. (2012) articulate similar benefits of using BIM in FM: space planning, accurate quantification of assets such as equipment and furniture, and avoiding interoperability inconsistencies among FM software. In addition, efficient space management, the existence of an accurate FM database, and accessible data for maintenance are three common reasons for BIM to have better integration with FM (Kensek, 2015).

The overall purpose of using BIM for FM is to leverage the facility's information system through its life cycle to provide effective and efficient, safe, and healthy work environments (Jordani, 2010).

This section presented the benefits of BIM utilization in FM. However, in the most current practice, stakeholders are not implementing BIM exclusively in facilities management practices. In the most current facilities management operations that use BIM, most functions are still done manually, even though facility managers know that adopting BIM during operational building can decrease the chance of errors and increase efficiency (Becerik-Gerber et al., 2012; Motamedi et al., 2014, Matarneh et al., 2019). There are several challenges yet to overcome to facilitate BIM implementation in facilities management practices. The following section sheds light on the main challenges hindering BIM implementation in facilities management.

Challenges of BIM for facilities management

The benefits of BIM in terms of productivity, cost and time reductions, and efficient information exchange are generally accepted in the AEC/FM industry. Yet BIM adoption in FM has been much slower than expected (Azhar et al. 2008). A survey conducted by Becerik-Gerber and Kensek (2010) found "BIM for FM" to be the least interesting topic for both practitioners and academics. Participants listed several challenges to account for this: organization-wide resistance to major change, a lack of interoperability among BIM and FM software packages, and the lack of real-case studies that support the value of BIM for FM.

The outcome of the survey agrees with the conclusions of Wong and Jay (2010), who found that even with the growing research interest in BIM, the adoption of BIM has been narrowed to specific phases of a facility's life cycle. These authors concluded after reviewing over 50 BIM-related case studies that there was a lack of interest in adopting BIM in post-construction phases. They considered that the limited available maintenance budgets and condition of assets are the main challenges.

According to Azhar (2011), there are two types of challenges facing BIM adoption, which can be divided into two broad categories: legal and technical. The first challenge is related to the lack of BIM data ownership and responsibility and liability for updating the BIM model data to ensure its accuracy, while the second challenge is related to interoperability. This claim was backed up by Gu and London (2010), who concluded that the lack of processes for updating BIM models with as-built information is the main challenge facing BIM adoption in FM.

Becerik-Gerber et al. (2012) list a number of technology and process-related challenges:

- Unclear roles and responsibilities for loading data into the model or databases and maintaining the model;
- Diversity in BIM and FM software tools, and interoperability issues;

- Lack of effective collaboration between project stakeholders for modelling and model utilization;
- Necessity of, yet difficulty relating to, software vendor's involvement, including "fragmentation among different vendors, competition, and lack of common interests".

In addition to technology and process-related challenges, there are organizational challenges, which include

- Cultural barriers toward adopting new technology;
- Organization-wide resistance need for investment in infrastructure, training, and new software tools;
- Undefined fee structures for additional scope;
- Lack of sufficient legal framework for integrating owners' view in design and construction;
- Lack of real-world cases and positive proof of return of investment.

Arayici et al. (2012) also highlight the lack of evidence as to how BIM can support the FM decision-making process. Aguilar and Ashcraft (2013) mentioned that BIM adoption in FM depends on the understanding of the information management system between different stakeholders and suggested that this should be clarified and legally controlled. Sabol (2013) points out that the overwhelming amount of information in BIM models is difficult for the FM team to manage, update, and maintain.

Kelly et al. (2013) also listed several challenges facing BIM adoption in FM:

- The lack of tangible benefits of BIM in FM despite agreement about the potential of BIM in FM;
- The interoperability between BIM and FM technologies;
- The lack of clear requirements for the implementation of BIM in FM;
- The lack of clear roles, responsibilities, contracts, and liability framework;
- A lack of real-world case studies of BIM applications in FM.

Liu and Issa (2012) identified another challenge which was related to the lack of understanding of the end user's requirements to improve the business processes. According to Volk et al. (2014), adopting BIM in FM has many challenges, particularly in existing facilities which include difficulties in capturing existing data, modelling uncertain data under changing environmental conditions, and objects' relations in existing buildings. Kiviniemi and Codinhoto (2014) found that, unlike its acknowledged benefits in the design and construction phases, there is little hard evidence of BIM's benefits in the operation and maintenance phase.

Love et al. (2015) discussed the limited application areas of BIM in FM, and indicated that this is due to the required financial investment, interoperability issues between systems, lack of standardized tools and processes, and lack of understanding of the required data for FM. Kassem et al. (2015) suggest that BIM adoption in FM has been too slow compared to other life-cycle phases, due to the lack of real-world case studies that show BIM's benefits in FM, clients' lack of awareness, and FM professionals' lack of skills and understanding.

Bosch et al. (2015) explain that the added value of using BIM in FM is marginal, due to the lack of alignment between people, processes, and technologies. One of the significant challenges mentioned by Ilter and Ergen (2015) is the level of understanding of the FM team, who should identify the required FM information. Similarly, Kassem et al. (2015) also stated that current BIM information for FM was insufficient and inaccurate due to the lack of a process that ensures that the model has been updated with any changes that have occurred after the design phase.

Shalabi and Turkan (2017) consider that the main challenges facing BIM adoption in FM practices are the "limited awareness of expected BIM benefits for FM among FM professionals, lack of data exchange standards, and unproven productivity gains illustrated by case studies". However, Nicał and Wodyński (2016) considered that the main barriers that are challenging to deal with are interoperability issues and cultural changes.

Ibrahim et al. (2016) summarized all the challenges identified in the literature that are slowing BIM adoption in FM practice and listed them in three main categories, which are business, legal, technical, human, and organizational.

Adopting BIM in mainstream FM encompasses multiple disciplines to ensure higher functionality of the built environment by integrating people, place, processes, and technology. Essentially, BIM is mostly used for the operations phase, and commercially available technologies focus on transferring information from the design and construction phase to the operations phase by enabling the creation and capturing of digital facility information throughout the facility life cycle (Akcamete et al., 2010; Volk et al., 2014). More attention is still needed to focus on solving the interoperability issue between BIM and existing facilities management systems to streamline facilities information transfer and provide efficient facilities information management.

This section discussed how BIM can be leveraged in FM practices and detailed the challenges and obstacles facing the utilization of BIM in FM. Some common challenges have been presented in most of the research literature, which are interoperability and data exchange between BIM and FM technologies, a lack of clear understanding of the required data for FM day-to-day activities, and a lack of real-world case studies. The next sections will focus on interoperability and data exchange issues.

Data exchange approaches

Research on how to link FM and BIM has been conducted to overcome the interoperability barrier (Kang and Hong, 2015). Accordingly, different approaches have been developed to link BIM and FM, which suggest using one or more of the following methods (Ibrahim et al., 2016):

- Design pattern and application programming interface (API);
- Web service;
- Extract, transform, and load (ETL) and data warehouse (DW);
- BIM-based neutral file format;
- Information delivery manual (IDM) and model view definition (MVD).

As mentioned above, there are different approaches currently being developed for linking BIM and FM to overcome the interoperability barrier. The four main approaches are:

1. **Manually and spreadsheets:** Where owners are using CAFM and CMMS systems, they have two options: either to enter information manually or to use customized digital spreadsheets that are compatible with the FM systems, including information from BIM data and hard copies of documents (Arayici, 2015). However, this approach is laborious and time-consuming.
2. **Industry Foundation Classes (IFC):** these were developed by the National BIM Standards (NBIMS) BuildingSMART as an open, vendor-neutral, and independent

BIM data repository (Building-SMART-Alliance, 2015). IFC defines building objects, including their geometry, properties, and relationships, to support multi-disciplinary coordination, information sharing, and exchange across IFC-compliant applications, and the handover of information for analysis and other tasks (Thein, 2011). IFC is an object-oriented database that enables data sharing (through ifcXML and aecXML) to support analysis for heat loss, cooling loads, etc. and/or information handover to an FM team.

IFC has gone through a number of evolutions, the latest version of which is IFC4, which was issued in 2013 as ISO 16739:2013 (Building-SMART-Alliance, 2015). Currently, several BIM software vendors have considered IFC importers/exporters within their applications. This allows models to be imported/exported from BIM authoring applications, i.e. Autodesk Revit into 5D estimating applications such as Exactal CostX (Dhillon, et al. 2014).

Redmon et al. (2012) claim that true interoperability will only be achieved when every software application being used on a project can read and write to and from a centralized web-hosted database, thus standardizing the process of passing information between stakeholders and representing the latest information on the project.

However, IFC does not completely solve the interoperability problem. Some researchers have found that there is a degree of data loss during information exchange using IFC between heterogeneous software applications (Eastman et al., 2011; Redmon et al., 2012). For example, Patacas et al. (2016) found geometry errors during the information exchange, while Motamedi et al. (2014) found that the exported IFC file of the model does not contain all the required logical relationships between the components, spaces, and distribution systems, and missing relationships have to be added manually. Other examples are presented by Yang and Ergan (2016), who revealed the limitations of IFC's representation for HVAC troubleshooting. Sampaio and Simões (2014) claim that the IFC format is not yet sufficiently developed to properly implement the data exchange between BIM and FM, as they faced a problem in retaining the colour added to the model.

On the other hand, Shalabi and Turkan (2017) provided a schema that integrates corrective maintenance data in a 3D-IFC-BIM environment, and they did not report any problems. Although the IFC schema has driven interoperability progress forward, the model does not provide adequate conditions for accurate interoperability (Sacks et al., 2010). Moreover, some applications are still not compatible, directly or indirectly, with IFC (Arayici, 2015).

3. **Construction Operation Building Information Exchange (COBie):** this system was recently developed to support the collection of structured information during the design and construction phases and handover to the FM team (East et al., 2013). Although COBie looks promising for resolving the interoperability barrier between the design and construction phases and with the O&M phase (Open-BIM-Network, 2012), COBie "does not provide details on what information is to be provided, when and by whom" (East and Carrasquillo-Mangual, 2013).

However, the main problem with COBie is that it is seen as a spreadsheet rather than an .xml-based information exchange (John et al., 2013). Apparently, an FM team may require more information than the information COBie can provide. Several studies have shown its capabilities if used during the early design and construction phases. For example, Lavy and Jawadekar (2014) conducted a study using three case studies where BIM and COBie were used. Their study concluded that BIM and COBie should be started earlier in the design phase. Although COBie outlines the required information specifications, it is static and should be extended by the professionals involved based on the project's requirements in the FM phase. Furthermore, COBie's capability related to spatial and system decomposition information is limited (Ilter and Ergen, 2015).

In addition to COBie, there are a number of different information exchange standards under development, such as Building Automation Modelling information exchange, HVAC information exchange, electrical system information exchange, life-cycle information exchange, and many others which are being developed (Building-SMART-Alliance, 2015).

4. **Proprietary middleware:** This is a customized computer software provided by a single software vendor. This approach enables two independent systems (BIM and FM) to interact by providing a single information source and updating information system-atically. This approach links BIM and FM systems (bi-directional link) using API, web services, design patterns, and a BIM-based neutral file format such as IFC and COBie. However, it is costly and complex, and its implementation is fixed during program-ming in the proprietary middleware approach (Kang and Choi, 2015). One of the most effective proprietary middleware packages for BIM integration in FM in the real-world market is "Ecodomus".

The inherent power of BIM for FM is mainly associated with streamlining information flow between the project stakeholders during the facility life cycle and facilitating information hando-ver to FM teams (Matarneha et al., 2019; Reza Hosseini et al., 2018). Yet information flow among project stakeholders is neither automated nor seamless. There are still technical issues to be overcome – mainly identifying the required FM information for data exchange purposes and boosting interoperability between BIM and FM systems (Gao and Pishdad-Bozorgi, 2019; Matarneha et al., 2019; Yalcinkaya and Singh, 2019). Although standard data formats are capa-ble of exchanging data between different platforms, particularly IFC and COBie schemas, the data exchange process between BIM and FM systems using open standard data formats is not a straightforward process. For example, the integration between BIM and the computer-aided facility management (CAFM) system has been actively criticized for inadequate data interoper-ability, particularly the inability to transfer semantic FM information properly (BIFM, 2013). The next section will discuss in depth the interoperability issue between BIM and FM systems.

Interoperability

Throughout a facility's life cycle, there are different stakeholders, resulting in various interac-tions, which makes the need for interoperability essential (Grilo and Jardim-Goncalves, 2010). Interoperability is defined as the ability of two or more systems or components to manage and exchange information to enable automation in information flow and to avoid system errors resulting from manual data re-entry (Gallaher et al., 2004; Grilo and Jardim-Goncalves, 2010). In short, interoperability enables different vendors to share data within a heterogeneous envi-ronment with independent parties who share a common data model (Arayici et al., 2011).

Gallaher et al. (2004) point out that when interoperability problems occur, they form a frag-mented organizational structure and business process. The estimated cost of inadequate inter-operability in the US capital facilities supply chain in 2002 was $15.8 billion, where two-thirds of this cost was borne by owners and operators. According to the US Census Bureau Report 2004b, the value of the capital facilities in the United States in 2002 was over $374 billion. This means that even a small improvement in efficiency could achieve substantial economic benefits.

Comparing the associated costs of the O&M phase, it has higher costs than other life-cycle phases. This is due to the fact that information management and accessibility hinder efficient facil-ities operation. The major costs of inadequate interoperability for owners and operators are at the operation and maintenance stage, and these are mitigation costs, which are usually incurred by owners and operators. Mitigation costs are the costs associated with the manual re-entry of information, information verification, and rework due to incorrect information, while avoidance costs are usually incurred by general contractors, which are the costs associated with the use and maintenance of redundant information technology systems (i.e. training, IT support, data transfer, and sharing). Quantified delay costs are mainly incurred by owners and operators. Inadequate interoperability issues in the capital facilities industry arise from the fragmented nature of the construction industry; traditional practices, which continue to be paper-based; a lack of standardization; and inconsistent technology adoption among stakeholders (Gallaher et al., 2004).

Interoperability is the key underpinning of BIM, since it enables participants from different disciplines who are using different software applications to share information and work collaboratively (Cheung et al. 2012). This collaboration cannot be achieved without having accurate and complete information, including specification, geometric shape, parametric properties, assembly data, and overall design intent (Eastman et al., 2011).

The AEC/FM industry has recently started looking for solutions to embrace software systems that support interoperability; nevertheless, they are lagging behind other industries (Shalabi and Turkan, 2017). Furthermore, interoperability between BIM and existing FM technologies is one of the main challenges in implementing BIM in FM practice, and this is due to the diversity between the BIM platforms and the FM platforms (Ibrahim et al., 2016; Shalabi and Turkan, 2017).

The area of interoperability for BIM and FM is an ever-growing research domain. A number of researchers, FM practitioners, professional organizations, and software vendors are focusing their efforts on overcoming the interoperability challenge by introducing a set of universal open data standards, such as the IFC and XML schemata and COBie and its subsets, to facilitate data exchange (Sabol, 2013).

Although there are many solutions currently available which have made the BIM for FM process a reality, there are still some ongoing challenges related to extracting information from BIM directly to CAFM systems, a lack of interoperability between software applications, and information overload.

At present, there is no "one size fits all", and such a solution seems distant. Seamless interoperability does not yet exist (Grilo and Jardim-Goncalves, 2010; Ibrahim et al., 2016). Accordingly, well-established concrete strategies for successful data exchange, interoperability, and integration of the FM required information are needed to overcome the interoperability problem and provide a seamless data exchange process (Kensek, 2015; Kassem et al., 2015; Ibrahim et al., 2016).

This section discussed the interoperability between different BIM and FM platforms, showing that the industry still needs to provide more practical solutions to streamline the information exchange process between the various BIM and FM software applications. The next section will present a teaching map for BIM-based facility management for undergraduate or postgraduate students.

Teaching map for BIM-based facility management

In order to teach BIM-based facility management to undergraduate or postgraduate students, the teaching process should be systematic to enable students to absorb the required theoretical knowledge before moving to the practical and sophisticated applications. The following steps should be followed to prepare/adapt BIM-based facility management in educational institutes:

*   **Teaching whole life-cycle costing (WLCC):** WLCC comprises of three main terms, which are capital cost, operating costs, and income; students should be aware of

the traditional methods used to calculate capital and operating costs before using BIM to implement the same process.

- **Teaching 5D BIM implementation process to calculate the capital cost:** Once students understand all the essential theoretical knowledge, 5D BIM can then be introduced systematically through (1) teaching the process of retrieving cost information from the 3D BIM model, (2) teaching how the retrieved cost data (i.e. BoQ) can be used to estimate the whole project cost, and (3) teaching how the estimated cost can be used with other BIM documents such as 4D BIM to develop the project budget.
- **Teaching the utilization of BIM to estimate the operating cost:** 6D BIM refers to facility management and sustainability using BIM tools; students should be introduced to a set of tools such as Revit and CostX.
- **Teaching advanced BIM tools and other advanced technologies to automate the calculation process:** Since BIM can be integrated with several technologies such as blockchain and Internet of Things (IoT), it will be useful to show how BIM can be enhanced by using these technologies such as tracking the performance of an asset using the IoT.

Figure 7.1 shows the steps of preparing and teaching facility management–based BIM in different stages.

Summary

This chapter has presented a general overview of current FM information systems and highlighted the need for BIM integration in FM practices. It is believed that BIM can leverage FM performance and reduce the costs of the longest and most expensive phase – the operation and maintenance phase. To achieve this, information management should be considered as a priority to provide facility managers with the required information to enable them to work efficiently.

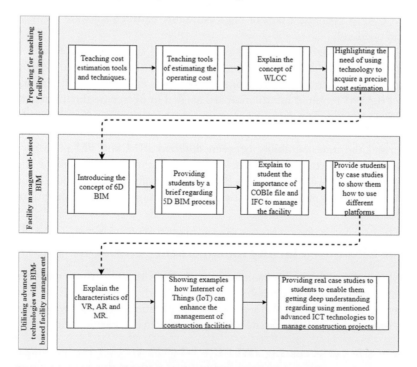

Figure 7.1 The teaching map of facility management–based BIM.

BIM as a data pool is a promising tool for facilities management teams which can provide them with all the required information for their day-to-day activities. However, to achieve the successful integration of BIM into FM practice, several challenges have to be solved, the main ones being the issues of information exchange and interoperability between BIM and FM systems.

Questions

1. How BIM can enhance FM practice?
2. Name three facilities management systems.
3. What is the overall aim of integrating BIM in FM practice?
4. Name three BIM applications in FM practice.
5. What are some of the benefits of BIM integration in FM practice?
6. What is the main challenge facing BIM implementation in FM?
7. Name three approaches to transferring information from BIM to FM systems.

References

Aguilar, K., & Ashcraft, H. (2013). *Legal Issues When Considering BIM for Facilities Management* (P. Teicholz, Ed.). Wiley.

Akcamete, A., Akinci, B., & GarrettJr., J. (2010). Potential utilization of building information models for planning maintenance activities. In W. Tizani (Ed.). Proceedings of the International Conference on Computing in Civil and Building Engineering. Nottingham, UK: Nottingham University Press.

Alevras, G., & Arabia, S. (2014). Incorporating BIM into facilities management. In IFMA's World Workplace, The Facility Conference & Expo. New Orleans, LA.

Al-Shalabi, F., & Turkan, Y. (2015). A novel framework for BIM enabled facility energy management: A concept paper. In 5th International/11th Construction Specialty Conference. Vancouver, British Columbia.

Alvarez-Romero, S. (2014). *Use of Building Information Modeling Technology in the Integration of the Handover Process and Facilities Management.* The Faculty of the Worcester Polytechnic Institute.

Arayici, Y. (2015). *Building Information Modelling* (1st ed.). Bookboon Publisher.

Arayici, Y., Coates, P., Koskela, L., Kagioglou, M., Usher, C., & O'Reilly, K. (2011). Technology adoption in the BIM implementation for lean architectural practice. *Automation in Construction*, 20(2), 189–195. Retrieved from http://dx.doi.org/10.1016/j.autcon.2010.09.016

Arayici, Y., Onyenobi, T., & Egbu, C. (2012). Building information modelling (BIM) for facilities management (FM): The mediacity case study approach. *International Journal of 3-D Information Modeling*, 1(1), 55–73.

Atkin, B., & Brooks, A. (2009). *Total Facilities Management.* Wiley-Blackwell.

Azhar, S. (2011). Building information modeling (BIM): Trends, benefits, risks, and challenges for the AEC industry. *ASCE Journal of Leadership and Management in Engineering*, 11(3), 241–252.

Azhar, S., Hein, M., & Sketo, B. (2008). Building information modeling: Benefits, risks and challenges. In 44th Associated Schools of Construction National Conference. Auburn, AL.

Azhar, S., Khalfan, M., & Maqsood, T. (2012). Building information modelling (BIM): Now and beyond. *Construction Economics and Building*, 12(4), 15–28.

Aziz, N. D., Nawawi, A. H., & Ariff, N. R. (2016). Building information modelling (BIM) in facilities management: Opportunities to be considered by facility managers. *Procedia - Social and Behavioral Sciences*, 234, 353–362.

Baladhandayutham T., Venkatesh S., (2012) An analysis on application of lean supply chain concept for construction projects. *Synergy*, X(I).

Becerik-Gerber, B., & Kensek, K. (2010). Building information modeling in architecture, engineering, and construction: Emerging research directions and trends. *Journal of Professional Issues in Engineering Education and Practice*, 136(3), 139–147.

Becerik-Gerber, B., & Rice, S. (2010). The perceived value of building information modeling in the U.S. building industry. *Journal of Information Technology in Construction*, 15, 185–201.

Berger, D. (2009, Apr. 13). 2009 CMMS/EAM Review: Power up a winner – How to find the right asset management system for your plant. Retrieved March 05, 2018, from https://www.plantser vices.com/articles/2009/066/.

Bosch, A., Volker, L., & Koutamanis, A. (2015). BIM in the operations stage: Bottlenecks and implications for owners. *Built Environment Project and Asset Management, 5*, 331–343.

Brinda, T. N., & Parsanna, E. (2014). Developments of facility management using building information modelling. *International Journal of Innovative Research in Science, Engineering and Technology, 3*(4), 11379–11386. Retrieved from HYPERLINK "http://www.ijirset.com" www.ijirset.com

British Institute of Facilities Management (BIFM). (2013). *Sustainability in Facilities Management Report*. Published by the British Institute of Facilities Management. Available on: https://www.sustainabilit yexchange.ac.uk/files/sustainability_in_facilities_management_report_2013.pdf

Building-SMART-Alliance. (2015). *BuildingSMART Alliance: National Institute of Building Sciences Website*. Retrieved from https://www.architectmagazine.com/technology/nibs-buildingsmart-alliance-re leases-version-3-of-the-us-national-bim-standard_o.

Changyoon, K., Hynsu, L., Hongjo, K., & Hyoungkwan, K. (2013). *BIM-Based Mobile System for Facility Management*. Seoul: Yonsei University.

Cheung, E., Chan, A., & Kajewski, S. (2012). Factors contributing to successful public private partnership projects: Comparing Hong Kong with Australia and the United Kingdom. *Journal of Facilities Management, 10*(1), 45–58.

Clarke, S., & D'arjuzon, R. (2019, March). Green quadrant integrated workplace management systems 2019. Retrieved May 2019, from https://www.ibm.com/downloads/cas/9LV6ML7B.

Codinhoto, R., Kiviniemi, A., Kemmer, S., Martin-Essiet, U., Donato, V., & Guerle-Tonso, L. (2013). *BIM-FM. Manchester Town Hall Complex*. University of Salford.

Costin, A, Pradhananga, N, & Teizer, J. (2012b). Leveraging passive RFID technology for construction resource field mobility and status monitoring in a high-rise renovation project. *Automation in Construction*, Elsevier, 24, 1–15.

Council, N. R. (2008). *Core Competencies for Federal Facilities Asset Management Through 2020: Transformational Strategies*. Washington, DC: The National Academies Press. doi: 10.17226/12049. Available on: https://www.nap.edu/catalog/12049/core-competencies-for-federal-facilities-asset-management -through-2020-transformational

Dempsey, J. (2009). A coast guard pilot to make better facility decisions. *Journal of Building Information Modeling*, 26.

Ding, L., Drogemulle, R., Akhurst, P., Hough, R., Bull, S., & Linning, C. (2009). Towards sustainable facilities management. In K. H. P. Newton (Ed.), *Technology, design and process innovation in the built environment* (pp. 373–392). Oxon: Taylor & Francis.

Dhillon, R., Jethwa, M., & Rai, H. (2014). Extracting building data from BIM with IFC. *International Journal on Recent Trends in Engineering and Technology, 11*, 202–210.

East, W. (2007). 'Construction Operations Building Information Exchange (COBie) requirements definition and pilot implementation standard.' Washington, DC: Engineering Research and Development Center, US Army Corps of Engineers.

East, W., & Brodt, W. (2007). BIM for construction handover. *Journal of Building Information Modeling*, 28–35.

East, W., & Carrasquillo Mangual, M. (2013). The COBie Guide: A commentary to the NBIMS-US COBie standard. Engineer Research and Development Center.

Eastman, C., Teicholz, P., Sacks, R., & Liston, K. (2011). *BIM Handbook: A Guide to Building Information Modeling for Owners, Managers, Designers, Engineers and Contractors*. New York: Wiley.

Elmualim, A., & Pelumi-Johnson, A. (2009). Application of computer-aided facilities management (CAFM) for intelligent buildings operation. *Facilities, 27*(11/12), 421–428.

Gallaher, M. P., O'Connor, A. C., Dettbarn, J. L., & Gilday, L. T. (2004). Cost analysis for inadequate interoperability in the U.S. capital facilities industry. National Institute of Standards and Technology. Retrieved January 18, 2017, from http://fire.nist.gov/bfrlpubs/build04/art022.html

Gao, X., & Pishdad-Bozorgi, P. (2019). BIM-enabled facilities operation and maintenance: A review. *Advanced Engineering Informatics*, 39, 227–247, ISSN 1474–0346, doi: 10.1016/j.aei.2019.01.005.

Grilo, A. C., & Jardim-Goncalves, R. (2010). Value proposition on interoperability of BIM and collaborative working environments. *Automation in Construction* , *19*(5), 522–530.

Gu, N., & London, K. (2010). Understanding and facilitating BIM adoption in the AEC industry. *Automation in Construction*, 19(8), 988–999.

Ham, Y., & Golparvar-Fard, M. (2015). Mapping actual thermal properties to building elements in gbXML-based BIM for reliable building energy. *Automation in Construction*, 49, 214–224.

Hosseini, M. Reza, Roelvink, Rogier, Papadonikolaki, Eleni, Edwards, David John, & Pärn, Erika (2018), Integrating BIM into facility management: Typology matrix of information handover requirements. *International Journal of Building Pathology and Adaptation*, 36(1), 2–14, doi: 10.1108/IJBPA-08-2017-0034.

Ibrahim, K. F., Abanda, F. H., Vidalakis, C., & Woods, G. (2016). BIM for FM: Input versus output data. *Proceedings of the 33rd CIB W78 Conference* 2016. Brisbane, Australia.

Ilter , D., & Ergen, E. (2015). BIM for building refurbishment and maintenance: current status and research directions. *Structural Survey*, 33(3), 228–256.

Isikdag, U., Underwood, J., & Aouad, G. (2008). An investigation into the applicability of building information models in geospatial environment in support of site selection and fire response management processes. *Advanced Engineering Informatics*, *22*(4), 504–519. doi:http://dx.doi.org/10.1016/j.aei.2008.06.001

John, G., Tebbit, J., Wiggett, D., & Mordue, S. (2013). *BIM for the Terrified: A Guide for Manufacturers*. Construction Products Association and NBS.

Jordani, D. A. (2010). BIM and FM: The portal to lifecycle facility management. *Journal of Building Information Modeling*, Spring 2010, 13–16.

Kang, T.-W., & Choi, H.-S. (2015). BIM perspective definition metadata for interworking facility management data. *Advanced Engineering Informatics*, *294*, 958–970.

Kang, T. W., & Hong, C. (2015). A study on software architecture for effective BIM/GIS-based facility management data integration. *Automation in Construction*, *54*, 25–38.

Kasprzak, C., & Dubler, C. (2012). Aligning BIM with FM: streamlining the process for future projects. *Australasian Journal of Construction Economics and Building*, *12*(4), 68–77.

Kassem, M., Kelly, G., Dawood, N., Serginson, M., & Lockley, S. (2015). BIM in facilities management applications: A case study of a large university complex. *Built Environment Project and Asset Management*, 5(3), 261–277. doi: 10.1108/BEPAM-02-2014-0011

Kelly, G., Serginson, M., Lockley, S., Dawood, N., & Kassem, M. (2013). BIM for facility management: A review and a case study investigating the value and challenges. In Proceedings of the 13th International Conference on Construction Applications of Virtual Reality (pp. 191–199). London.

Kensek, K. (2015). BIM guidelines inform facilities management databases: A case study over time. *Buildings*, 5(3), 899–916. doi: 10.3390/buildings5030899

Kiviniemi, A., & Codinhoto, R. (2014). Challenges in the Implementation of BIM for FM: Case Manchester town hall complex. *Computing in Civil and Building Engineering*, 665–672.

Langdon, D. (2012). *Getting the most out of BIM: A guide for clients*. An AECOM.

Lavy, S. and Jawadekar, S. (2014), A case study of using BIM and COBie for facility management 3. *International Journal of Facility Management*, 5(2), 13–27.

Lee, J., Jeong, Y., Oh, Y., Lee, J., Ahn, N., Lee, J., & Yoon, S. (2013). An integrated approach to intelligent urban facilities management for real-time emergency response. *Automation in Construction*, 30, 256–264.

Lin, Y., Su, Y., & Chen, Y. (2014). Developing mobile BIM/2D barcode-based automated facility management system. *The Scientific World Journal*, 2014, 16 pages. doi: 10.1155/2014/374735

Liu , R., & Issa, R. R. (2012). Automatically updating maintenance information from a BIM database. In ASCE TCCIT International Conference on Computing in Civil Engineering. doi:https://doi.org/10.1061/9780784412343.0047

Liu, R. & Issa, R. R. (2015), Survey: Common knowledge in BIM for facility maintenance. *Journal of Performance of Constructed Facilities*, 30(3).

Love, P., Zhou, J., Matthews, J., Chun, P., & Carey, B. (2015). Systems information model for managing electrical, control, and instrumentation assets. *Built Environment Project and Asset Management*, 5, 278–289.

Marinakis, V., Karakosta, C., H.Doukas, Androulaki, S., & Psarras, J. (2013). ICT evolution in facilities management (FM): Building information modelling (BIM) as the latest technology. *Sustainable Cities and Society*, *6*(1), 11–15.

Matarneh S.T., Danso-Amoako, M., Al-Bizri S., Gaterell, M., & Matarneh R. (2019). Building information modeling for facilities management: A literature review and future research directions. *Journal of Building Engineering*, 24 (2019), DOI: 100755, 10.1016/J.JOBE.2019.100755.

Motamedi, A., Hammad, A., & Asen, Y. (2014). Knowledge-assisted BIM-based visual analytics for failure root cause detection in facilities management. *Automation in Construction*, 73–83. doi: 10.1016/j.autcon.2014.03.012

Motawa, I., & Almarshad, A. (2013). A knowledge-based BIM system for building maintenance. *Automation in Construction*, 29, 173–182. doi: 10.1016/j.autcon.2012.09.008

Nicał, A. K., & Wodyński, W. (2016). Enhancing facility management through BIM 6D. In Creative Construction Conference 2016, 25–28.

Nisbet, N. (2008). *COBIE Data Import/Export Interoperability with the MAXIMO Computerized Maintenance Management System*. Construction Engineering Research Laboratory.

Open-BIM-Network. (2012). Open BIM network. Retrieved from http://www.openbimnetwork.com/assets/applets/OPEN_BIM_Focus_-_Issue_4-_October_2012.pdf.

Patacas, J., Dawood, N., Vukovic, V., & Kassem, M. (2015). BIM for facilities management: evaluating bim standards in asset register creation and service life planning. *Journal of Information Technology in Construction - ITcon*, 313–329.

Patacas, J., Dawood, N., Greenwood, D., & Kassem, M. (2016). Building owners and facility managers in the validation and visualisation of asset information models (AIM) through open standards and open technologies. *Journal of Information Technology in Construction*, 21, 434–455.

Redmon, A., Hore, A., Alshawi, M., & West, R. (2012). Exploring how information exchanges can be enhanced through Cloud BIM. *Journal of Automation in Construction*, 24, 175–183.

Rundell, R. (2006). How can BIM benefit facilities management? Retrieved January 13, 2017, from http://www.cadalyst.com/cad/building-design/1-2-3-revit-bim-and-fm-3432

Sabol, L. (2013). Chapter 2 – BIM technology for FM, IFMA (Ed.). In *BIM for Facility Managers* (pp. 17–45). Hoboken, NJ: Wiley.

Sacks, R., Koskela, L., Dave, B., & Owen, R. (2010). The interaction of lean and building information modeling in construction. *Journal of Construction Engineering and Management*, *136*(9), 968–980.

Sampaio, A., & Simões, D. (2014). Maintenance of buildings using BIM methodology. *The Open Construction and Building Technology Journal*, 8, 337–342.

Shalabi, F., & Turkan, Y. (2017). IFC BIM-based facility management approach to optimize data collection for corrective maintenance. *Journal of Performance of Constructed Facilities*, 31(1). doi:10.1061/(ASCE)CF.1943-5509.0000941

Smith, D., & Tardif, M. (2009). *Building information modelling: A strategic implementation guide for architects, engineers, contractors and real estate asset management*. Hoboken, NJ.

Teicholz. (2013). *BIM for Facility Managers*. New York: Wiley.

Thabet, W., & Lucas, D. J. (2017). A 6-step systematic process for model-based facility data delivery. *Journal of Information Technology in Construction* (ITcon), 22, 104–131. http://www.itcon.org/2017/6

Thein, V. (2011). Industry foundation classes (IFC), BIM interoperability through a vendor-independent file format. USA: Bentley Sustaining Infrastructure. Available on: https://docplayer.net/16482970-Industry-foundation-classes-ifc.html

Vanier, D. (2001). Asset management: "A" TO "Z". In *Innovations in Urban Infrastructure* (pp. 1–16). APWA International Public Works Congress.

Volk, R., Stengel, J., and Schultmann, F. (2014). Building information modeling (BIM) for existing buildings: Literature review and future needs. *Automation in Construction* 38, 109–127.

Yang, X., & Ergan, S. (2016). Leveraging BIM to provide automated support for efficient troubleshooting of HVAC-related problems. *Journal of Computing in Civil Engineering*, 30(2). doi:10.1061/(ASCE)CP.1943-5487.0000492

8 BIM, sustainability, and energy optimization

Zeynep Işık, Yusuf Arayici, Hande Aladağ, Gökhan Demirdöğen, and Farzad Khosrowshahi

Introduction

Sustainable development aims to satisfy the needs of the present without risking that future generations will not be able to meet their own needs (Brundtland Report, 1987). Economic (profits), environmental (planet), and social (people) are considered to be the main pillars of sustainability. In the context of the environmental pillar, being aware of your resource consumption and reducing unnecessary waste are among the key issues. In addition to the effective resource use – in the context of a circular economy – reduction in energy consumption, emissions, and waste production should be gathered by increasing energy and material cycles. Therefore, important acts in reducing energy can be defined as energy conservation and optimization.

With the parallel rise of sustainable construction, building information modelling (BIM) has come into prominence in the AEC industry since it provides the necessary technology to improve energy efficiency in buildings (Gholami et al., 2013; Habibi, 2017) and accordingly allows for sustainable, energy-efficient development. BIM provides environmental benefits by helping to reduce the carbon footprint of a building and to optimize the energy consumption of a building. It also facilitates predicting and reducing the amount of energy a building consumes in the construction, operation, and maintenance phases. It follows that there would be an improvement in the economics of building due to the reduction in energy consumption (Yoon et al., 2009; Kim et al., 2018). Thus, it leads BIM to be considered a new model for energy-efficient buildings. For instance, Eleftheriadis et al. (2017) indicated that BIM could offer engineers the essential decision-making procedures to leverage increasing demand for sustainable structural designs, whereas Kim et al. (2018) suggested that environmental friendliness and economic feasibility of a building (optimization between energy efficiency and investment costs) can be achieved by engineers by implementing BIM. Habibi (2017) also stated BIM's role in the exploration of alternative approaches to enhancing energy efficiency in buildings. Considering that engineers have to participate in many decision-making stages in business life by combining their academic knowledge with economic, environmental, and social aspects to create an output, engineering education should offer intellectual and technical mastery along with practical wisdom and awareness for engineers (Sheppard et al., 2007). With this background, this chapter aims to present the importance of BIM's role in improving energy efficiency in buildings by covering the integration of sustainability, energy optimization, and BIM for new graduates and educators. In this context, this chapter explains:

- The significant benefits and barriers of BIM use in sustainable building design, practical strategies and techniques for applying BIM to sustainable design, Green BIM, and the current state and benefits of BIM-based sustainability analyses under the heading "Integration of BIM and sustainability";

- How BIM is linked to energy saving and sustainable development, the relationship between BIM and life-cycle assessment, and BIM-based building energy performance analysis (energy simulations tools) under the heading "Integration of BIM and energy optimization";
- How BIM is used for energy optimization in facility management (FM) under the heading "Integration of BIM and facility management (FM)".

Integration of BIM and Sustainability

Contributions of BIM to sustainable building design

BIM is one of the most current and emerging movements in the AEC industry due to its contribution to sustainability. Hence, this part provides readers an overview of the significant benefits of and barriers to the use of BIM for sustainable building design. A literature review reveals a wide range of areas that project teams can benefit from BIM usage to achieve sustainable building design (Autodesk, 2009; Dowsett and Harty, 2013; Wong and Fan, 2013; Tiemeyer, 2016; Ayman et al., 2018). The prominent areas are listed below.

- **Integrated project delivery (IPD):** The traditional design and construction processes in the AEC industry are defined as cyclical, where the gathered ideas and information provided from all project stakeholders are coordinated and necessary improvements are made by each project stakeholder. Within the scope of these cyclical processes, there is both information and document sharing among project stakeholders. However, this information chain might be open to miscommunication between stakeholders, especially as the number of stakeholders and the size and complexity of the project increase. At this point, adopting BIM-based tools will reduce the time needed to establish healthy communication between stakeholders as well as contributing to the improvement of the design and acceleration of construction. In addition, with successful BIM implementation, the extra cost of design changes can be eliminated. Therefore, a shift in the project delivery culture can be achieved with the help of BIM.
- **Design optimization:** Design teams can benefit from BIM-based tools via visualizing designs, conducting energy optimization analyses, and then developing various design alternatives. Thus, the use of BIM supports the fulfilment of sustainability goals with the selection of the most optimal design among the alternatives developed by the design team.
- **Energy and resource consumption:** BIM solutions and integrated analysis tools can contribute to the selection of best solutions to reduce the consumption of resources such as energy, water, and materials.
- **Energy analysis:** Energy analysis tools in BIM can provide insights for effective daylighting practices, water usage predictions and costs, and HVAC and ventilation settings for designers and contractors. Thus, BIM use assists designers and contractors in financial and design decision-making by showing how these systems will function in the integrated process initially as well as allowing an understanding of the energy costs of the project throughout its life cycle.
- **Material management:** Material selection is an important factor in determining the environmental impact of construction as well as determining the project's total cost.

BIM-based tools that allow the consistent integration of all building information provide accurate material quantity calculations. In addition, the integration of unit prices with the material quantities that can be extracted directly from BIM model also enables budget tracking and cost analysis for designers and contractors.

- **Reducing waste and inefficiency:** With the use of BIM-based tools, project documents such as model views, drawing sheets, schedules etc. that frequently change as a result of stakeholders' inputs can become more consistent. This acquisition avoids the waste of energy and material resources to harmonize uncoordinated, inconsistent, or missing construction documents.

- **Increase on-site renewable opportunities:** BIM-based tools can be beneficial for designing systems that minimize water use, protect existing wetlands, and focus on net-zero water usage. As a natural consequence of efficiency in water usage, costs and the impact on water and wastewater systems can be minimized.

- **Improved facility management:** It is possible for facility managers to access the necessary data in managing the O&M activities of the facility via BIM-based software. With the access and use of relevant data in BIM models, facility managers can assist multi-disciplinary collaboration among stakeholders (Singh et al., 2011), optimize data collection for corrective maintenance (Shalabi and Turkan, 2017), manage the maintenance operations of building equipment (Fargnoli et al., 2019), support safe maintenance and repair practices (Wetzel and Thabet, 2015), and even ensure end-users' integration and coordination (Mirarchi et al., 2018). Moreover, including facility managers in the early design process through BIM could reduce maintenance work during the operational phase of facilities (Wang et al., 2013). In addition, the data stored in BIM-based software can be used to train maintenance workers during the design and construction process.

- **Using BIM in retrofitting, retro-commissioning, and renovations:** While BIM is mostly used for new construction projects, retrofitting, retro-commissioning, and renovation projects can benefit from promising aspects of BIM. For example, Ilter and Ergen (2015) indicated that data from BIM models can be transferred to cost-estimating software to calculate the cost of retrofits, whilst Hammond et al. (2014) stated that BIM's functions are also practical for integrating sustainable design principles into the retrofit and/or renovation of existing structures. However, BIM adaptation in maintenance, refurbishment, or deconstruction activities does not yet take place in the majority of existing buildings yet due to the high modelling/conversion effort needed to integrate captured building data into semantic BIM objects, and the challenges of handling uncertain data in BIM occurring in existing buildings (Volk et al., 2014). Despite these challenges, BIM models generated from original and retrofit design plans, coupled with laser scanning in some cases, will let designers, managers, and owners monitor, benchmark, and suggest retrofit opportunities for operational efficiency and cost savings.

- **Build consensus:** Simulations prepared with BIM models allow users to visualize their projects in real-world situations. Also, with the help of augmented reality and virtual reality, it is possible for all stakeholders to see early designs or proposed modifications before starting construction. For instance, with virtual walkthroughs or 3D model site plans, it would be possible for stakeholders to visualize all the details and solutions related to the project through BIM models. As a result, the possibility of customers finding designs insufficient will be reduced and common understanding related to the project will be improved.

- **Increase investor confidence:** The ability to develop more cost-effective sustainable design alternatives through BIM analysis tools increases investor confidence that funds will be used appropriately to support the optimum performance of the building.
- **Meet demands for sustainability and energy efficiency:** Building owners and tenants are recognizing the financial rewards from green building retrofit investments. Therefore, the interest of owners and tenants in sustainable buildings is growing since it is recognized that efficiency in terms of sustainability increases the value of a real-estate property. One of BIM's contributions in terms of sustainable building production is the ability to making appropriate decisions about a building's form and material selection that will later create a valuable impact on the building's life cycle. The ability to conduct in-depth sustainability analyses via BIM-based tools can assist project teams to fulfil the prerequisites for the applied green building rating system (Azhar et al., 2011; Jalaei and Jrade, 2015), which indicates the provision of the appropriate conditions for sustainability and energy efficiency.

Even though recognition of the contribution of BIM is growing in the AEC industry, there are also some barriers to adopting BIM for sustainable design (İlhan and Yaman, 2015). Olawumi et al. (2018) conducted a study with the aim of determining the barriers to the integration of BIM and sustainability practices in construction projects using a Delphi survey of international experts. According to Olawumi et al. (2018)'s study, a total of 38 barriers were identified and their order of significance was determined. The results of the study show that "industry's resistance to change from traditional working practices", "longer time in adapting to new technologies (steep learning curve)", and "lack of understanding of the processes and workflows required for BIM and sustainability" are the salient ones. The results also show that barriers with a high significance are clustered under the categories of "education, knowledge and learning", "attitude and market", and "organizational and project related issues" in descending order. Meanwhile, the study's findings are in line with Ayman et al. (2018)'s study, which accentuated the common and additional barriers of BIM-based sustainability. Authors listed "investment cost", "AEC market deficiency", "rigidity to change", "client influence", and "immature, inconsistent and growing state" as the common barriers to BIM-based sustainability, whereas additional barriers were categorized under the categories "technological", "legal", and "extra initial cost". Their study highlighted the level of accuracy and questionable reliability of simulations of energy and building performance as prominent barriers, along with interoperability deficiency, issues regarding the accessibility and security of data, legal issues around ownership, and additional financial support for hiring a BIM coordinator.

Recommendations on how to apply BIM to sustainable design

In a previous section, the barriers to adopting BIM for sustainable design were presented. It can be stated that barriers with a high significance are clustered according to "education, knowledge and learning", "attitude and market", and "organizational and project related issues". Hence, this section aims to provide recommendations to obviate these barriers in engaging BIM in sustainable construction practices in terms of those clusters.

Even though simulating the energy performance of a building with BIM-based tools creates significant savings in terms of cost and time, stakeholders in the AEC industry are less aware of and knowledgeable about the use of this software (Anton and Diaz, 2014; Ahn et al., 2014). In the context of the "education, knowledge and learning" cluster, professions and organizations in the AEC industry should adopt recent trends and advancements into their practice. Hence, improvements in human resource capacity in the AEC industry can be accomplished in terms of BIM-based sustainability.

"Resistance to change", "lack of commitment by clients", and "the top echelon of construction firms" are the salient barriers to implementing innovative development in the AEC industry (Abubakar et al., 2014; Olawumi et al., 2018). Stakeholders should adopt BIM and sustainability principles in their projects as a way of taking a forward-thinking and proactive approach, since demands and mandates for green buildings and sustainable smart cities are rapidly increasing. Thus, the obstructive effects of barriers under "attitude and market" can be overcome (Olawumi et al., 2018).

To overcome the problems under the cluster of "organizational and project related issues", project organization and team collaboration should become more visible in complex or labour-intensive projects which might affect the implementation of innovative concepts like BIM and sustainability (Boktor et al., 2014). Since the investment and training costs of BIM-based software are found to be prohibitive by construction companies (Hanna et al., 2013), technology investment subsidies for construction firms provided by governments can help organizations to overcome their hesitation in engaging BIM and sustainability in their activities (Olawumi et al., 2018).

Green BIM: Practical strategies and techniques

Being inclusive of the intelligent data required by energy analysis and simulations to make the most optimum design decisions in terms of energy efficiency, BIM contributes to sustainability by designing iterative modelling. This energy-efficient design (green building design) is relevant to green building strategies, the importance of which is increasingly recognized in the construction industry. The term *green BIM* refers to BIM implementations that have the objective of providing required data for energy performance evaluation and energy optimization in acquiring energy efficiency during a building's life cycle (Ebrahim & Wayal 2020). Hence, it enhances design and construction efficiency. Stored data related to sustainable materials in BIM can further be utilized to quantify the environmental impacts of systems and materials to support decisions as well as to conduct life-cycle assessments in sustainable building production.

In recent years, growing mandates and regulations that aim to minimize the environmental impact and energy consumption of buildings and increase emission mitigation have prompted individuals and international organizations to initiate rating systems for green and sustainable construction. In this vein, different rating systems are being used in various countries. Prominent among these are "Leadership in Energy and Environmental Design (LEED)" (in the United States), "Building Research Establishment Environmental Assessment Methodology (BREEAM)" (in the United Kingdom), "Green Star" (in Australia), "the Comprehensive Assessment System for Building Environmental Efficiency" (in Japan), and "Building Environmental Assessment Method (BEAM) Plus" (in Hong Kong) (Wong and Zhou, 2015). In the field of green and sustainable buildings, buildings can also benefit from green BIM to go green and meet the requirements of a green building rating system (Azhar et al., 2011). These rating systems are beneficial in evaluating and benchmarking sustainability under categories such as sustainable sites, water efficiency, energy and atmosphere, materials and resources, indoor environmental quality, innovation in operations, and regional priority.

Green BIM, which aims to engage BIM implementations with sustainable design and construction techniques (Bonenberg and Wei, 2015), can be defined with three main pillars that interact with each other. As shown in Figure 8.1, these pillars are "BIM attributes", "green attributes", and "project phases" (Lu et al., 2017). The interaction between BIM attributes and project phases focuses on "how BIM could support the different phases and the whole lifecycle of green buildings", whereas the interaction between BIM attributes and green attributes focuses on "how BIM could support the various sustainability aspects of green buildings" (Lu et al., 2017). In this sense, the interaction between BIM attributes and green attributes reflects the potential of BIM applications to support green building assessment frameworks (Lu et al., 2017).

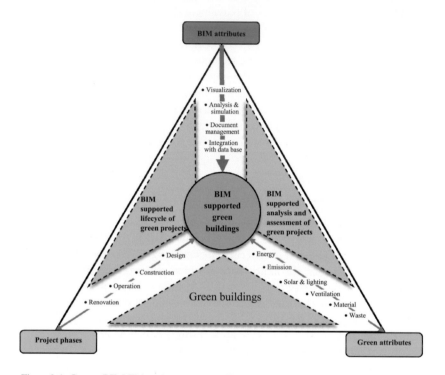

Figure 8.1 Green BIM Triangle taxonomy (Lu et al., 2017).

Since green BIM allows the energy analysis of different design options for a greener building at the early phase of the design, green BIM offers a roadmap that leads to the adoption of sustainable building rating systems. This roadmap can be summarized as follows:

- **Green BIM with the planning site location and analysis:** In order to develop an efficient site location plan and determine the spatial orientation and facade of the buildings, data about site topography, vegetation, and weather conditions are essential. However, considering the deficiencies of traditional site analysis, the capability of BIM and GIS enables the simulation of spatial data and model sites and building layout (Bonenberg and Wei, 2015).
- **Green BIM with planning building design and analysis:** This consists of building design and analysis related to the use of natural ventilation, natural lighting, solar energy, rainwater recycling, waste recycling, green materials, and the application of software featuring energy-efficient computing (Bonenberg and Wei, 2015).

- **Green BIM in the building construction process:** Since construction activities are ultimately responsible for the problem of on-site carbon emissions, the integration of BIM and GIS can help stakeholders to better visualize and analyse how greenhouse gas emissions that result from construction activities affect the environment (Hajibabai et al., 2011).

- **Green BIM in building operation:** The energy consumption rate of a building during the operation phase constitutes the major portion of the total energy consumption, as the carbon emission rate of the building in the operation phase is equal to one-third of the building's total carbon emission through its life cycle (Wong and Zhou, 2015). In this sense, BIM can be considered as an invaluable tool for monitoring the sustainability performance of buildings in the operation phase (Lu et al., 2017). Potential application areas for BIM-based analysis tools for managing environmental performance during the operation phase predominately consist of evaluations of heating and cooling, daylighting, the subsequent heat and energy loads, and appropriate building equipment selection (Wong and Zhou, 2015). Moreover, using BIM as an enabling technology for a cloud-based building data service in the operational phase is another potential application areas in terms of energy management (Curry et al., 2013).

- **Green BIM in building renovations and retrofit:** It is believed that retrofitting/renovation projects can benefit from BIM applications as much as new construction projects (Berstein et al., 2015), such that there are a significant amount of studies in the literature showing how BIM can be used in retrofitting/renovation projects. Assessing deconstruction strategies to enable the retrieving of energy and capital invested in building components (Akbarnezhad et al., 2014), supporting energy rehabilitation processes ranging from energy usage diagnoses to retrofitting decision-making (Lagüela et al., 2013), creating a virtual retrofit model to support streamlined decision-making in building retrofit projects (Woo and Menassa, 2014), and analysing the energy performance of retrofit projects via BIM and 3D laser scanning (Larsen et al., 2011) are some examples in studies that aim to boost the use of BIM in retrofitting/renovation projects.

- **Green BIM in building repair and maintenance:** Retrofitting existing buildings in the maintenance phase can significantly reduce the energy consumption of the building by protecting natural resources (Hammond et al., 2014). Hence, facility managers' interest in incorporating sustainable design attributes, reducing operation costs, limiting environmental impacts, and increasing building resilience will become more prevalent (Wong and Zhou, 2015).

- **Green BIM in building demolition:** In the literature, there are studies on the impact of BIM-based system development for the estimation and planning of demolition waste. For example, one BIM-based system that is mentioned gives users the ability to extract material and volume information through the BIM model for demolition and renovation waste estimation and planning, while users consequently can predict the number of truck delivery journeys and the amount of statutory waste disposal charges (Cheng and Ma, 2012, 2013). Likewise, another developed model gives users the ability to assess the impacts of various building deconstruction options in terms of their economic costs and environmental benefits (Akbarnezhad et al., 2014).

Determination of current state and benefits of BIM-based sustainability analyses

BIM-based tools are used in various sustainability analyses. There is no single BIM-based software that can do all types of green analyses. Generally, software tools are useful only in particular areas. The main areas in which current BIM-based tools are used for green issues are extrapolated below.

- **Generative design:** Generative design software allows designers to benchmark all possible design alternatives within their design goals by using spatial, material, methodological, and economic parameters. CAD tools such as Rhinoceros and Revit are among parametric and generative design software tools that enable users to generate design alternatives in terms of kinetic facades, daylight aspects, and innovative solar shading solutions (Loonen et al., 2017).
- **Energy performance analysis and evaluations:** Users can benefit from current BIM-based tools in energy performance analysis and evaluations for performing a whole-building energy analysis, for analysing different energy conservation measures, for generating feasibility evaluations of renewable energy, and for detecting energy faults (Lu et al., 2017).
- **Carbon emissions analysis and evaluations:** Current BIM-based software can be used to calculate carbon emissions and to specify optimum design alternatives for carbon emission reduction to achieve carbon neutrality. Building-system components and the external environment are incorporated into carbon emissions analysis by using information about local electricity emissions, hydrocarbon production on the construction site, and energy conversion approaches in order to assess the effect of those factors on carbon emissions throughout a building's life cycle (Lu et al., 2017).
- **Natural ventilation system analyses and optimization:** Users can benefit from current BIM-based software in natural ventilation system analysis and evaluations for estimating natural ventilation capacity as well as in forming ventilation strategies for a better thermal comfort level. This software generally takes into account the heating and cooling loads of buildings to predict the natural ventilation capacity of the building. In terms of ventilation strategies, alternatives such as single-sided ventilation, cross-ventilation, whole-building ventilation, or chimneys are offered, which will then allow the user to select the right ventilation system for their project (Lu et al., 2017).
- **Solar radiation and lighting analysis:** Current BIM-based software is applicable for solar radiation for the exterior of buildings with the aim of optimizing the impact of the sun on a building, developing a lighting-condition analysis for both the exterior and interior of buildings, and creating a point-by-point lighting simulation by comparing natural and artificial light. By allowing the evaluation of solar gain and temperature changes on building surfaces, BIM-based software enables users to optimize a building's position and orientation, which will then allow the user to select the right shading systems for their project. Within lighting analyses, BIM-based software allows users to make the most of natural daylight. As a side note, BIM-based software programmes can be used to visualize how shading systems and lighting designs will perform in terms of solar and lighting conditions (Lu et al., 2017).
- **Water usage analysis:** Users can benefit from current BIM-based software for the optimization of water usage estimation and water distribution systems. These software programmes can calculate water consumption by taking into account factors such as building type and the number of users. They also have the ability to convert predicted results into water cost reports (Ham & Golparvar-Fard, 2015; Lu et al., 2017).
- **Acoustics analysis:** Current BIM-based software can be used to simulate acoustic performance and to enhance the visual and audial effects of simulated acoustics. For example, visualization maps can be used to show various acoustic effects of the building that also reflect generated simulation results (Lu et al., 2017).

- **Thermal comfort analyses:** Users can benefit from current BIM-based software in thermal comfort analyses with the aim of simulation and optimization at the design stage along with monitoring comfort levels by integrating sensors.

Evaluation of various building performance analyses software tools used in the AEC industry

Autodesk Revit is one of the commercial BIM-based software programmes available on the market which allows users to design a building and structure and its components in a 3D/4D model. Based on a survey of 91 design and construction firms in the United States, Azhar et al. (2009) found that there are three commonly used BIM-based sustainability analysis software programmes: Autodesk Ecotect™, Autodesk Green Building Studio (GBS)™, Integrated Environmental Solutions (IES)®, and Virtual Environment (VE)™.

The above-mentioned BIM-based sustainability analyses are carried out using BIM-based software. BIM-based software is designed in such a way that additional information such as sustainability and maintenance information can be inserted into a BIM framework (Wong and Zhou, 2015). Some authors review current BIM-based software used for green analysis in terms of their applications areas and capabilities (Lu et al., 2017; Loonen et al., 2017; Al Ka'bi, 2019). According to the findings of Lu et al. (2017) and Loonen et al. (2017), there are several prominent BIM-based software specifically designed and developed for green analysis, which are listed below along with their areas of application (note that there is no single BIM-based software tool that can do all types of green analyses).

- Autodesk® Green Building Studio (GBS): used for energy, carbon emissions, natural ventilation, solar and daylight, and water analyses in the design, operation, and maintenance phases.
- Integrated Environmental Solutions® Virtual Environment (IES-VE): can be used for energy, carbon emissions, natural ventilation, solar and daylight, and water analyses in the design phase. It can test internal and external solar shading effects and compare the simulated results with the expected design (Loonen et al., 2017). IES-VE compares simulated data and comfort requirements to the standards and enables the evaluation of indoor thermal comfort (Habibi, 2016).
- Bentley Hevacomp: can be used for energy, carbon emissions, natural ventilation analyses in the design phase.
- AECOsim: can be used for energy, carbon emissions, solar and daylight analyses in the design phase.
- EnergyPlus: can be used for energy, carbon emissions, solar and daylight, and water analyses in the design phase. It supports the simulation of green walls and roofs. It can also simulate the performance of building envelopes with moveable insulation.
- HEED: can be used for energy and carbon emissions analyses in the design phase.
- DesignBuilder: can be used for energy, carbon emissions, natural ventilation, and solar and daylight analyses in the design phase.
- eQUEST: can be used for energy, natural ventilation, and solar and daylight analyses in the design, construction, and operation and maintenance phases. It enables energy cost estimations, the development of dynamic models showing how natural lighting affects the amount of electricity required for lighting, and the automatic implementation of energy efficiency measures.
- DOE2: can be used for energy, natural ventilation, and solar and daylight analyses in the design phase.

- FloVENT: can be used for natural ventilation analyses in the design phase.
- ODEON Room Acoustics Software: can be used for acoustic analyses in the design phase.
- TRNSYS: can be used for energy, natural ventilation, and solar and daylight analyses in the design phase. It supports the simulation of green walls and roofs.
- ESP-r: can be used for modelling adaptive facade technologies, and for thermo-physical performance predictions. It supports the simulation of green walls and roofs.
- IDA ICE v4.7: can be used for modelling thermotropic/chromic windows. It can also simulate the performance of building envelopes with moveable insulation.
- RIUSKA: is commonly used for indoor air temperature simulations, HVAC-system simulations, and energy consumption simulation of a whole building or a single space (Al Ka'bi, 2019).

Azhar et al. (2009) conducted a survey with 91 design and construction firms in the United States and found that Autodesk Green Building Studio (GBS) and Integrated Environmental Solutions and Virtual Environment (IES-VE) take the lead as the most preferred BIM-based sustainability analysis software tools.

Integration of BIM and energy optimization

Energy efficiency improvements in the AEC industry via BIM technology

BIM-able energy analysis tools can provide insights for designers and contractors in terms of effective daylighting practices, water usage predictions, HVAC, and ventilation settings. What-if tests performed through BIM tools can specify the correct energy and water requirements for a building. Moreover the use of BIM throughout a building's life cycle reduces material waste and helps minimize waste costs as well as overall building costs in the long term. According to a study performed by The Boston Consulting Group (2016), using BIM to conduct a building-wide energy analysis can save up to 20% of energy costs. This ratio would show an increase in building renovations and retrofit due to the energy-inefficient nature of existing buildings.

BIM and life-cycle assessment

BIM-based tools are also eligible for performing life-cycle assessments by analysing the materials to boost cost effectiveness in the long term. BIM-integrated activities in life-cycle assessment can be examined under three main categories:

- **Reducing energy consumption with design (design for energy):** This refers to the concept of energy-efficient generative design with the integration of BIM. Generative design software allows designers to benchmark all possible design alternatives within their design goals with the help of using spatial, material, methodological, and economic parameters as a part of the design exploration process (Autodesk, 2009). In addition, the generative design approach with the integration of BIM allows the creation of sophisticated geometry by using algorithms, which redounds to the quality of the design as a result.

  A generative BIM workspace, created by combining BIM technology with computational methods, will help designers to generate more energy-efficient and optimized construction practices. With the help of BIM-based tools, it is possible to generate and compare a permutation of models considering design constraints and the design context determined by client requirements. Using land types, land and building relationships, climate/meteorological data, wind analysis data, material data, etc. as input generative design parameters, the best decisions can be generated by considering mass interaction, facade configuration, waste reduction, and energy-use issues to achieve a socially, economically, and environmentally regenerative built environment.

- **Energy monitoring in the facility:** This refers to the concept of integrating BIM and Big Data analytics for tracking energy consumption and generating alternative renewable energy use.
- **Energy optimization and cost effectiveness:** The energy performance of a building can be discussed in terms of five different aspects: the building envelope, the air conditioning and ventilation, the water heating system, the dynamic equipment, and daylighting. Among those categories, the building envelope is the most crucial by virtue of the fact that any improvement in the building envelope would lower energy costs during operation and reduce energy waste and carbon dioxide emissions as the consequence of a good energy-saving design (Egwunatum et al., 2016).

BIM-based building energy performance analysis: Energy simulation tools

The importance of energy analysis during the design phase becomes apparent throughout a building's life cycle in terms of its economic and environmental aspects. In this context, there are many energy simulation software tools which are designed to perform energy analysis for various purposes. The most commonly used simulation software tools are GBS (Autodesk Green Building Studio), Energy Plus (Energy Simulation Software tool), the IDA ICE (Indoor Climate Energy), IES-VE (Integrated Environmental Solutions – Virtual Environment), EnergyBuild, OpenStudio, Building Energy Analysis Model (BEAM), eQuest, TRNSYS, and RIUSKA (Jaric et al., 2013; Kamel and Memari, 2019; Al Ka'bi, 2019). Among those simulation software tools, some are most suitable for simplifying the creation of energy simulation models (i.e. Ecotect, eQuest, IES-VE, DesignBuilder, OpenStudio, Riuska) whilst others are better in terms of utilizing the generated databases to decrease the simulation time of the energy models (i.e. EnergyPlus and DOE2) (Al Ka'bi, 2019). According to studies by Jaric et al. (2013), Loonen et al. (2017), and Al Ka'bi (2019), the features of the main energy simulations tools are the following:

- Designers can utilize ARCHICAD's built-in building energy modelling capabilities by using EcoDesigner with the aim of providing building energy modelling (BEM) workflow. It does not require any special preparations and takes all the necessary data from the existing model. Before evaluation is started, some additional information about the project like location, activity, MEP system, energy type, and the availability of green energy systems must be provided. Using the provided structural and opening lists, designers can re-evaluate all the parameters of the construction elements and openings, gain a better understanding of the energy model, and modify data if needed. The built-in U-value calculator enables users to achieve proper heat transmission coefficient values. Based on those data, EcoDesigner calculates yearly energy consumption, CO_2 emissions, and monthly energy balance.
- The Revit Conceptual Energy Analysis Tool enables the analysis of annual carbon emissions, monthly heating loads, monthly cooling loads, and annual energy use. The tool

works with a massing model, enabling designers to acquire knowledge about building performance in the early design phases.

- With the graphical display of information through Ecotect, designers have the ability to use real-time data to measure how the environment may impact the building's performance over time.

- Autodesk Green Building Studio (GBS) generates estimations related to energy consumption, carbon emissions due to using fossil fuel, simulation parameters, performance metrics, and expected costs. It can present results in tabular report form. The understandable summary of information gathered in the seedling stages of the design process via GBS allows designers to choose the most appropriate design alternative.

- OpenStudio is a graphic energy modelling tool that consists of software tools to support whole-building energy modelling using EnergyPlus and advanced daylight analysis using Radiance. Users can provide high-level visualization about schedules, loads constructions and materials, and HVAC systems.

- IES-VE is an integrated analysis tool which can be used to investigate the performance of a building either retrospectively or during the design stages of a construction project. The interaction between the user and the software is done through a graphical user interface (GUI). The software presents graphical results via specific inputs given by the user. On the other hand, its limited flexibility for modelling adaptive facades can be considered an unfavourable feature of the software.

- eQUEST is a freeware energy analysis tool which performs whole-building energy performance simulation analysis in the design process. With features such as interactive graphics, combining a building creation wizard with energy efficiency measure wizards and fast processing, it is one of the salient energy simulation tools. Another important feature of eQUEST is that it allows users to see energy simulations of the building on an hourly basis.

- EnergyPlus is a whole-building energy simulation tool that leads to more accurate predictions about thermal comfort metrics and visual comfort metrics, including lighting and shading. This energy simulation tool takes into account detailed information on air flow, moisture, and heat transfer to create a comprehensive energy performance model.

- IDA Indoor Climate and Energy (IDA ICE) is a flexible, whole-building performance simulation tool for simulation of building's energy consumption, indoor air quality, and thermal comfort. The ability to define custom control macros enabling simulation users to control the operation is a particularly useful feature in the context of adaptive facades.

- TRNSYS is an energy simulation tool that presents connections between the building and various other sub-systems/components. The integration of a multi-zone building model with other components such as weather data, HVAC systems, occupancy schedules, thermal energy storage, and solar energy systems provides convenience in terms of managing complexity in the built environment.

Among the reviewed energy simulation tools, EcoDesigner, Green Building Studio (GBS), and Ecotect are the software tools designed to work together with BIM applications and provide seamless data exchange. Ecotect is the application that provides the user with the largest set of building simulations, but it requires professional knowledge. On the other hand, for users who do not have a nodding acquaintance with energy efficiency simulation, EcoDesigner and Green Building Studio can be more practical since they provide quick and easy simulation results (Jaric et al., 2013).

Integration of BIM and facility management (FM) in the context of energy controlling and monitoring

BIM dimensions refer to levels of information and data entered into a 3D model using BIM-based software. In fact, according to the BIM fundamentals, there are seven dimensions generally accepted by academics and practitioners. There is a consensus that 2D/3D BIM depict digital modelling, 4D BIM is linked with time (planning or scheduling), and 5D BIM is related to cost (cost estimation, budgets, and quantity take-offs). Despite this consensus, there is no clear agreement on the implication of dimensions after 5D (Charef et al., 2018). For instance, some authors argue that 6D BIM can embed O&M manuals, plans and technical support and link 6D BIM with the representation of the as-built model for facilities management. Some authors also use 6D BIM for health and safety. 7D BIM has also a similar level of unclarity. It is evident that 7D BIM can be used for referring to energy and facility management. However, there is an agreement (but not a consensus) that practitioners generally use 6D BIM to refer to sustainability (energy analysis, efficiency studies) whilst 7D BIM is commonly used for facility management activities (Charef et al., 2018).

Since optimizing asset management from design to demolition is targeted in the operation and maintenance phase, the BIM model should feature a comprehensive set of well-structured information regarding the building assets (Nicał and Wodyński, 2016). Thus, facility managers tend much more to use 7D BIM throughout the operation and maintenance phase of a building's life cycle.

The operation and maintenance phase is considered to be the most cost-consuming period in a building's life cycle. For this reason, facility managers are in need of BIM implementations to provide knowledge management with more control over data and documents relating to the building. This acquisition of BIM in knowledge management will enable facility managers to make appropriate decisions with a chance to evaluate all possible costs that will occur during the life cycle and reduce costs to the minimum level (Heralova, 2014). It is for this same reason that some public administrations keen to implement BIM for cost minimization. For example, the UK government has mandated BIM adoption by all centrally procured projects to reduce capital costs.

FM practices can exploit the use of BIM in various aspects. Some benefits of BIM implementation in FM include improving the handover process, localizing facility components, enhancing fault detection, allowing real-time data accessibility, facilitating emergency management, preparing the rental contracts, collaboration and visualization of data, and personnel training (Lee and Akin, 2011; Kivits and Furneaux, 2013; Love et al., 2013; Koch et al., 2014; Kensek, 2015; Love et al., 2015a, b; Williams et al., 2015; Gheisari and Irizarry, 2016; Gurevich et al., 2017; Carbonari et al., 2018; Miettinen et al., 2018; Tucker and Masuri, 2018; Matarneh et. al, 2019). Hereby, BIM use in FM activities optimizes building performance and occupant value.

The main advantage of linking BIM to facility management is the efficient use of energy acquired as a result of energy optimization and energy monitoring. The ability to integrate data and documents produced by different project stakeholders at different stages with information gathered from energy management systems through BIM-based tools provides significant contributions in further energy analysis. Moreover, this ability of BIM-based tools is particularly useful for comparing the actual energy performance of a facility with performance parameters in terms of monitoring the energy performance of facilities during the operation and maintenance phase (Matarneh et al., 2019). During the execution of BIM-enabled FM implementations, various approaches such as sensor data analytics, metering data analytics, mapping algorithms, and visual programming allow facility managers to provide real-time monitoring and automated control (Becerik-Gerber et al., 2012; Ignatov and Gade, n.d.). In addition, using a colour-coding scheme in as-built BIM model to compare energy consumption values in 2D and 3D views and what-if scenarios to simulate how energy systems will work under different configurations would assist facility managers to make the most energy-saving management decisions. Linking BIM-based tools with occupancy sensors is also conducive to tracking historical data of energy usage for each room/zone/occupant. Therefore, energy consumption behaviour could be analysed and predicted, and energy-related budgeting and conservation activities could be supported (Becerik-Gerber et al., 2012).

BIM can also be a leverage for facility managers to benchmark various energy alternatives in terms of the operational cost and environmental impact of each alternative. Within this benchmarking process, BIM-based tools allow facility managers to analyse the savings of various facility improvements and building-system retrofits to achieve optimization in the energy performance of the building throughout its lifespan.

To sum up, the integration of BIM and FM in the context of energy controlling and monitoring can improve the efficiency of FM. To enhance the value of this integration, identifying the required energy data from BIM models, integrating different energy information streams and utilizing actual facility energy data in BIM-based simulations to support energy retrofit decisions should be considered as future agendas (Matarneh et al., 2019).

The BIM FM model, which is derived from the BIM as-built model and then linked with the facility management system for ongoing management, would require integration with other systems to enhance the benefits of BIM in FM: the computer-aided facility management (CAFM) system, computerized maintenance management system (CMMS), and integrated workplace management system (IWMS). CAFM systems are deployed on a departmental basis to track space and maintenance, whereas IWMSs are deployed on an enterprise basis to manage the use of workplace resources including the management of a company's real-estate portfolio and facilities assets. On the other hand, CMMS are useful for tracking remedial and scheduled maintenance.

Conclusion

While BIM is defined as a necessary technology to reach the full potential of built environment, sustainable construction involves building environmentally responsible and resource-efficient constructions throughout their life cycle. Although the topics of BIM and sustainability are among the emerging trends in the AEC industry, there are not enough attempts to adapt these trends in construction practices during the life cycle of a building in practice. In other words, BIM is not sufficiently adopted as a basic part of design and construction processes to achieve sustainability goals in the AEC industry. Since buildings are responsible for substantial electricity consumption, energy use, and carbon dioxide emissions, the adaptation and implementation of BIM and sustainability integration throughout the whole life cycle of construction practices will increase energy efficiency in the built environment. Moreover, simulating the energy performance of a building using BIM-based tools will result in significant savings in cost and time that will eventuate in significant economic contributions as an outcome.

In this context, all professions and organizations in the AEC industry have a responsibility to ensure that BIM-enabled sustainability practices are disseminated and the number of best practices is increased. Therefore, professions in the AEC industry should be trained and educated on these topics to turn them into pioneers who encourage the implementation and dissemination of BIM-enabled sustainability practices. In this way it would be possible to overcome the barriers to adopting and applying BIM in sustainability practices such as industry resistance and a lack of understanding of the required workflows.

Questions

1. What are the contributions of BIM to sustainable building?
2. Please explain your thoughts about how to overcome the barriers to adopting and applying BIM in sustainability practices in the AEC industry.
3. Please explain how AEC practitioners can use green BIM applications throughout the life cycle of a building.
4. Please explain how BIM technology can boost energy efficiency in construction.
5. Please name the prominent BIM-based building energy performance analysis tools.
6. Please explain in which areas AEC practitioners can benefit from BIM-based energy simulation tools.
7. Please explain the benefits of BIM use in facility management.

References

Abubakar, M., Ibrahim, Y. M., Kado, D., & Bala, K. (2014). Contractors' perception of the factors affecting building information modelling (BIM) adoption in the Nigerian construction Industry. *Computing in Civil and Building Engineering 2014*, 167–178.

Ahn, K. U., Kim, Y. J., Park, C. S., Kim, I., & Lee, K. (2014). BIM interface for full vs. semi-automated building energy simulation. *Energy and Buildings, 68*, 671–678.

Akbarnezhad, A., Ong, K. C. G., & Chandra, L. R. (2014). Economic and environmental assessment of deconstruction strategies using building information modeling. *Automation in Construction, 37*, 131–144.

Al Ka'bi, A. H. (2019). Comparison of simulation applications used for energy consumption in green building. In 11th International Conference on Computational Intelligence and Communication Networks (CICN) (pp. 65–68). Honolulu, HI, USA, IEEE.

Antón, L. Á., & Díaz, J. (2014). Integration of LCA and BIM for sustainable construction. *International Journal of Social, Behavioral, Educational, Economic, Business and Industrial Engineering, 8*(5), 1378–1382.

Autodesk (2009). Generative design. https://www.autodesk.com/solutions/generative-design.

Autodesk (2009). *The Advantages of BIM-Enabled Sustainable Design for Improving Commercial Building Performance: AUTOCAD White paper.* http://images.autodesk.com/adsk/files/advantages_of_bim_enabled_sustainable_design.pdf.

Ayman, R., Alwan, Z., & McIntyre, L. (2018). Factors motivating the adoption of BIM-based sustainability analysis. In The International Seeds Conference, Dublin, Eire.

Azhar, S., Brown, J., & Farooqui, R. (2009). BIM-based sustainability analysis: An evaluation of building performance analysis software. In Proceedings of the 45th ASC Annual Conference (Vol. 1, No. 4, pp. 276–292), Gainesville, Florida.

Azhar, S., Carlton, W. A., Olsen, D., & Ahmad, I. (2011). Building information modeling for sustainable design and LEED® rating analysis. *Automation in Construction, 20*(2), 217–224.

Becerik-Gerber, B., Jazizadeh, F., Li, N., & Calis, G. (2012). Application areas and data requirements for BIM-enabled facilities management. *Journal of Construction Engineering and Management, 138*(3), 431–442.

Bernstein, H., Jones, S., & Russo, M. (2015). Green BIM: How building information modeling is contributing to green design and construction. *Journal of Information Technology in Civil Engineering and Architecture, 2*, 20–36.

Boktor, J., Hanna, A., & Menassa, C. C. (2014). State of practice of building information modeling in the mechanical construction industry. *Journal of Management in Engineering, 30*(1), 78–85.

Bonenberg, W., & Wei, X. (2015). Green BIM in sustainable infrastructure. *Procedia Manufacturing, 3,* 1654–1659.

Brundtland Report (1987). Our Common Future. World Commission on the Environment and Development. http://www.un-documents.net/our-common-future.pdf.

Carbonari, G., Stravoravdis, S., & Gausden, C. (2018). Improving FM task efficiency through BIM: A proposal for BIM implementation. *Journal of Corporate Real Estate. 20*(1), 4–15.

Charef, R., Alaka, H., & Emmitt, S. (2018). Beyond the third dimension of BIM: A systematic review of literature and assessment of professional views. *Journal of Building Engineering, 19*, 242–257.

Cheng, J., & Ma, L. (2012). A BIM-based system for demolition and renovation waste quantification and planning. In Proceedings of the 14th International Conference on Computing in Civil and Building Engineering (ICCCBE 2012), Moscow.

Cheng, J. C., & Ma, L. Y. (2013). A BIM-based system for demolition and renovation waste estimation and planning. *Waste Management, 33*(6), 1539–1551.

Curry, E., O'Donnell, J., Corry, E., Hasan, S., Keane, M., & O'Riain, S. (2013). Linking building data in the cloud: Integrating cross-domain building data using linked data. *Advanced Engineering Informatics, 27*(2), 206–219.

Dowsett, R. M., & Harty, C. F. (2013). Evaluating the benefits of BIM for sustainable design: A review. In Proceedings of the 29th Annual ARCOM Conference (pp. 13–23). Association of Researchers in Construction Management, Reading, UK.

Ebrahim A., Wayal A. S. (2020) Green BIM for sustainable design of buildings. In Gunjan V., Singh S., Duc-Tan T., Rincon Aponte G., Kumar A. (eds*) ICRRM 2019 – System Reliability, Quality Control, Safety, Maintenance and Management*. ICRRM 2019. Springer, Singapore.

Egwunatum, S., Joseph-Akwara, E., & Akaigwe, R. (2016). Optimizing energy consumption in building designs using building information model (BIM). *Slovak Journal of Civil Engineering, 24*(3), 19–28.

Eleftheriadis, S., Mumovic, D., & Greening, P. (2017). Life cycle energy efficiency in building structures: A review of current developments and future outlooks based on BIM capabilities. *Renewable and Sustainable Energy Reviews, 67*, 811–825.

Fargnoli, M., Lleshaj, A., Lombardi, M., Sciarretta, N., & Di Gravio, G. (2019). A BIM-based PSS approach for the management of maintenance operations of building equipment. *Buildings, 9*(6), 139.

Gheisari, M., & Irizarry, J. (2016). Investigating human and technological requirements for successful implementation of a BIM-based mobile augmented reality environment in facility management practices. *Facilities. 34*(1/2), 69–84.

Gholami, E., Sharples, S., Shokooh, J. A., & Kocaturk, T. (2013). Exploiting BIM in energy efficient refurbishment. In PLEA: 29th Conference, Sustainable Architecture for a Renewable Future, Munich, Germany, 10–12 September 2013.

Gurevich, U., Sacks, R., & Shrestha, P. (2017). BIM adoption by public facility agencies: Impacts on occupant value. *Building Research & Information, 45*(6), 610–630.

Habibi, S. (2016). Smart innovation systems for indoor environmental quality (IEQ). *Journal of Building Engineering, 8*, 1–13.

Habibi, S. (2017). The promise of BIM for improving building performance. *Energy and Buildings, 153*, 525–548.

Hajibabai, L., Aziz, Z., & Peña-Mora, F. (2011), Visualizing greenhouse gas emissions from construction activities. *Construction Innovation, 11*(3), 356–370.

Ham, Y., & Golparvar-Fard, M. (2015). Mapping actual thermal properties to building elements in gbXML-based BIM for reliable building energy performance modeling. *Automation in Construction, 49*, 214–224.

Hammond, R., Nawari, N. O., & Walters, B. (2014). BIM in sustainable design: Strategies for retrofitting/renovation. In *Computing in Civil and Building Engineering*, 1969–1977.

Hanna, A., Boodai, F., & El Asmar, M. (2013). State of practice of building information modeling in mechanical and electrical construction industries. *Journal of Construction Engineering and Management, 139*(10), 04013009.

Heralova, R. S. (2014). Life cycle cost optimization within decision making on alternative designs of public buildings. *Procedia Engineering, 85*, 454–463.

Ignatov, I. I., & Gade, P. N. (2019). Data formating and visualization of BIM and sensor data in building management systems. In *19th International Conference on Construction Applications of Virtual Reality: Enabling digital technologies to sustain construction growth and efficiency*, Bangkok, Thailand (pp. 141–151).

İlhan, B., & Yaman, H. (2015). BIM and sustainable construction integration: An IFC-based model. *Megaron, 10*(3), 440–448.

lter, D., & Ergen, E. (2015). BIM for building refurbishment and maintenance: Current status and research directions. *Structural Survey, 33* (3), 228–256.

Jalaei, F., & Jrade, A. (2015). Integrating building information modeling (BIM) and LEED system at the conceptual design stage of sustainable buildings. *Sustainable Cities and Society, 18*, 95–107.

Jarić, M., Budimir, N., Pejanović, M., & Svetel, I. (2013). A review of energy analysis simulation tools. In TQM 2013, Proceedings of 7th International Working Conference of Total Quality Management – Advanced and Intelligent Approaches (pp. 103–110). Belgrade: Mechanical Engineering Faculty.

Kamel, E., & Memari, A. M. (2019). Review of BIM's application in energy simulation: Tools, issues, and solutions. *Automation in Construction, 97*, 164–180.

Kensek, K. (2015). BIM guidelines inform facilities management databases: a case study over time. *Buildings, 5*(3), 899–916.

Kim, J. U., Hadadi, O. A., Kim, H., & Kim, J. (2018). Development of a BIM-based maintenance decision-making framework for the optimization between energy efficiency and investment costs. *Sustainability, 10*(7), 2480.

Kivits, R. A., & Furneaux, C. (2013). BIM: Enabling sustainability and asset management through knowledge management. *Scientific World Journal, 2013*, 1–14.

Koch, C., Neges, M., König, M., & Abramovici, M. (2014). Natural markers for augmented reality-based indoor navigation and facility maintenance. *Automation in Construction, 48*, 18–30.

Lagüela, S., Díaz-Vilariño, L., Martínez, J., & Armesto, J. (2013). Automatic thermographic and RGB texture of as-built BIM for energy rehabilitation purposes. *Automation in Construction, 31*, 230–240.

Larsen, K. E., Lattke, F., Ott, S., & Winter, S. (2011). Surveying and digital workflow in energy performance retrofit projects using prefabricated elements. *Automation in Construction, 20*(8), 999–1011.

Lee, S., & Akin, Ö. (2011). Augmented reality-based computational fieldwork support for equipment operations and maintenance. *Automation in Construction, 20*(4), 338–352.

Loonen, R. C., Favoino, F., Hensen, J. L., & Overend, M. (2017). Review of current status, requirements and opportunities for building performance simulation of adaptive facades. *Journal of Building Performance Simulation, 10*(2), 205–223.

Love, P. E., Matthews, J., Lockley, S., Bosch, A., Volker, L., & Koutamanis, A. (2015a). BIM in the operations stage: Bottlenecks and implications for owners. *Built Environment Project and Asset Management, 5*(3), 331–343.

Love, P. E., Matthews, J., Lockley, S., Kassem, M., Kelly, G., Dawood, N., & Serginson, M. (2015b). BIM in facilities management applications: A case study of a large university complex. *Built Environment Project and Asset Management, 5*(3), 261–277.

Love, P. E., Simpson, I., Hill, A., & Standing, C. (2013). From justification to evaluation: Building information modeling for asset owners. *Automation in Construction, 35*, 208–216.

Lu, Y., Wu, Z., Chang, R., & Li, Y. (2017). Building information modeling (BIM) for green buildings: A critical review and future directions. *Automation in CONSTRUCTION, 83*, 134–148.

Matarneh, S. T., Danso-Amoako, M., Al-Bizri, S., Gaterell, M., & Matarneh, R. (2019). Building information modeling for facilities management: A literature review and future research directions. *Journal of Building Engineering, 24*, 100755.

Miettinen, R., Kerosuo, H., Metsälä, T., & Paavola, S. (2018). Bridging the life cycle: A case study on facility management infrastructures and uses of BIM. *Journal of Facilities Management, 16*, 2–16.

Mirarchi, C., Pavan, A., De Marco, F., Wang, X., & Song, Y. (2018). Supporting facility management processes through end-users' integration and coordinated BIM-GIS technologies. *ISPRS International Journal of Geo-Information, 7*(5), 191.

Nical, A. K., & Wodyński, W. (2016). Enhancing facility management through BIM 6D. *Procedia Engineering, 164*, 299–306.

Olawumi, T. O., Chan, D. W., Wong, J. K., & Chan, A. P. (2018). Barriers to the integration of BIM and sustainability practices in construction projects: A Delphi survey of international experts. *Journal of Building Engineering, 20*, 60–71.

Shalabi, F., & Turkan, Y. (2017). IFC BIM-based facility management approach to optimize data collection for corrective maintenance. *Journal of Performance of Constructed Facilities, 31*(1), 04016081.

Sheppard, S., Colby, A., Macatangay, K., & Sullivan, W. (2007). What is engineering practice? *International Journal of Engineering Education, 22*(3), 429.

Singh, V., Gu, N., & Wang, X. (2011). A theoretical framework of a BIM-based multi-disciplinary collaboration platform. *Automation in Construction, 20*(2), 134–144.

The Boston Consulting Group (2016). The transformative power of building information modelling. https://www.bcg.com/publications/2016/engineered-products-infrastructure-digital-transformative-power-building-information-modeling.aspx.

Tiemeyer, D., (2016). 7 sustainable design benefits of BIM services, https://blog.peterbassoassociates.com/blog/building-information-modeling.

Tucker, M., & Masuri, M. R. A. (2018). The development of facilities management-development process (FM-DP) integration framework. *Journal of Building Engineering, 18*, 377–385.

Volk, R., Stengel, J., & Schultmann, F. (2014). Building information modeling (BIM) for existing buildings: Literature review and future needs. *Automation in Construction, 38*, 109–127.

Wang, Y., Wang, X., Wang, J., Yung, P., & Jun, G. (2013). Engagement of facilities management in design stage through BIM: Framework and a case study. *Advances in Civil Engineering, 2013*, 1–8.

Wetzel, E. M., & Thabet, W. Y. (2015). The use of a BIM-based framework to support safe facility management processes. *Automation in Construction, 60*, 12–24.

Williams, G., Gheisari, M., Chen, P. J., & Irizarry, J. (2015). BIM2MAR: An efficient BIM translation to mobile augmented reality applications. *Journal of Management in Engineering, 31*(1), A4014009.

Wong, J. K. W., & Zhou, J. (2015). Enhancing environmental sustainability over building life cycles through green BIM: A review. *Automation in Construction, 57*, 156–165.

Wong, K. and Fan, Q. (2013), Building information modelling (BIM) for sustainable building design. *Facilities*, 31(3/4), 138–157.

Woo, J. H., & Menassa, C. (2014). Virtual retrofit model for aging commercial buildings in a smart grid environment. *Energy and Buildings, 80*, 424–435.

Yoon, S., Park, N., & Choi, J. (2009). A BIM-based design method for energy-efficient building. In Fifth International Joint Conference on INC, IMS and IDC (pp. 376–381). Seoul, South Korea, IEEE.

9 BIM for safety planning and management

Sambo Zulu, Allen Wan, Farzad Khosrowshahi, and Mark Swallow

Introduction

Digital technologies offer the construction industry opportunities to improve the delivery of projects. The integration of technology has proven vital during the global COVID pandemic, with many suggesting that technology will accelerate significantly in the construction sector from 2020 (NBS, 2020; BIM, 2020). The use of technology has been widely regarded for its opportunities to improve health and safety performance (Rolim et al., 2020; Martínez-aires et al., 2018). Health and safety is considered a core component of project performance (Mohammadi et al., 2018). The construction sector has, unfortunately, generated an adverse health and safety reputation and continues to report poor performance when compared to other major industries (Li et al., 2018). In 2019, the construction industry accounted for 30 of 147 fatalities across all industries in the United Kingdom (HSE, 2019). With such improvements still to be made, building information modelling (BIM) and its applications have been shown to have great potential to improve health and safety on construction projects (Wan et al., 2018; Swallow & Zulu, 2019; Zairi et al., 2016; Saeedfar, 2017). The demonstrable benefits of BIM, including visualization and sequencing (Mordue & Finch, 2014), logistic planning (Sulankivi et al., 2010), project time reduction (Hardin & McCool, 2015), waste reduction (Akinade et al. (2018), improved communication (Sulankivi et al., 2010), schedule progress and monitoring (Mordue & Finch, 2014), and accuracy of scheduling and logic (Mordue & Finch, 2014) have been shown to influence health and safety positively.

Considering health and safety in design, planning, and management is critical to ensuring adequate health and safety performance. In particular, planning focuses on the attention of health and safety managers and their teams to put in place measures that will aid the eventual management of project tasks. For example, the Construction Design Management 2015 regulations place emphasis on effective planning and management of health and safety throughout the construction process (HSE, 2015). Tools and techniques that can aid the planning and management of health and safety should, therefore, be encouraged. Studies have shown the potential for BIM to be the catalytic platform to help improve the design, planning, and management of health and safety. This has been achieved using tools such as virtual reality (Sacks et al., 2013; Swallow & Zulu, 2020), 4D simulation (Jin et al., 2019), drones (Tatum & Liu, 2017), sensor technologies, and clash detection, among others.

Health and safety is a vital aspect of construction management. Therefore, knowledge in this field is essential to those within the industry and graduates alike. Significant developments in BIM and digital technologies over recent years are providing innovative insights into safer design, planning, and site management. This chapter focuses on the opportunities for BIM applications to improve site health and safety planning and management.

BIM for safety

Although there is a need for improvement, the construction industry continues to make efforts to improve its health and safety performance (Forsythe, 2014). Technological innovations such as BIM provide opportunities for better health and safety management. For example, in a survey conducted by the NBS (2020), results showed that 70% of respondents agreed that BIM and digitization of the construction sector could improve health and safety. The importance of collaborative sharing of health and safety information and the effective use of digital technology has recently gained traction through the release of the UK BIM standard PAS 1192-6:2018. This standard provides guidance on the production of health and safety information used throughout the building project and asset life cycle, establishing a framework known as the "risk information cycle" (BSI, 2018). In addition to this standard, research studies have demonstrated the mechanisms through which BIM can be used as a platform for health and safety management on construction projects. Mordue and Finch (2014) explored this link and suggested that health and safety information and data can be incorporated into a BIM model, which would be useful for education and training and even to assist in altering behaviour within the architecture, engineering, and construction (AEC) sector.

The ability to remotely monitor construction-related activities is a plus for further improvement of productivity where material categorization is an important area. Combining BIM and image technologies from illumination, viewpoint, resolution, and scale can enhance health and safety planning and management. Dimitrov and Golparvar-Fard (2013) examined and evaluated the visual automation material detection system and found a high average accuracy of 97% for all materials. This is another step forward for better quality control of construction materials and structural safety. Zhou et al. (2012) have shown that information technologies or computer-based tools can assist in the management of site safety. This can be done in different ways, including at the design or construction stage, and can also cover various aspects such as construction activity or process and for product and project. BIM-based immersive virtual reality (Getuli et al., 2020), BIM-based performance monitoring for smart building management (Edirisinghe & Woo, 2020), and BIM-based live sensor data visualization using virtual reality (Natephra & Motamedi, 2019) are some other developing applications that can help improve health and safety practices on construction projects. A review of the literature shows that existing technological tools may be grouped under online databases, virtual simulation tools, geographic information systems, entity-based tools, 4D CAD tools, and sensor and warning tools. There may be overlaps between these categories. There is potential for the integration of these tools to enhance health and safety performance on construction projects.

BIM can enhance site safety through its functionality to perform compliance checking (New York City, 2013), BIM-driven prefabrication with spatial and dimension details, and hazard prediction and prevention involving a combination of BIM-enhance design and risk management. In addition, its integration with other technical tools such as radio frequency identification technology (RFID), global positioning systems, wireless sensor networks, accident investigation from 3D models, animations and scenario replay, simulation of the construction work process and site-safety planning (Kiviniemi et al., 2011), worker safety training (Balfour Beatty Construction, 2015), post-construction, and facility management (Mordue & Finch, 2014) provides a platform for further enhancement of health and safety planning and management.

Zhou et al. (2013) conducted a literature review and concluded that the application of advanced technology for construction safety management is becoming more evident. They found that after 2010, there were significantly more research papers reporting on how to adopt technologies for safety on construction sites. These were mostly from the United States of America, China, Japan, Hong Kong, Korea, the United Kingdom, and Taiwan. The studies ranged from the project (mostly from building projects) to the industry and process levels. Out of some 30 different types of technologies used for construction safety, there were relatively more papers discussing the following: virtual reality, sensor, database, and 4D technologies. These technologies can be applied to safety monitoring, safety assessment, technology application, hazard identification, and safety training. Ultimately, researchers identified research gaps in existing studies involving the pre-construction or repair stage, the impact of cost and technology adoption, the practical implementation of technology, and legal implications. According to NBS (2020), the use of technology within construction is increasing. For example, survey data showed that 78% of respondents are either already using technology with immersive capabilities or plan to within the next five years.

Stanford University's Center for Integrated Facility Engineering recognized the advantages of BIM to include seven categories: communication, facility performance, cost, schedule, project delivery, knowledge management, and safety (Kam et al., 2014). Despite similar experience in the aerospace and manufacturing sectors and BIM vendors claiming that BIM will "enhance quality and reduce human errors", Love et al. (2011) cautioned that there are few studies to support the idea that "BIM can reduce human errors related to mistakes, lapses of attention and omission". As the industry matures in the use of BIM, its benefits, such as a reduction in errors, safer designs, and better site planning and management, are being realized.

Contractors responsible for site works have more direct legal safety responsibilities than other parties, and thus they need to consider whether BIM can assist in improving safety. Ku and Taiebat (2011) investigated BIM implementation among contractors in the United States. The study reported that higher percentages of contractors use BIM for constructability and visualization, but only about 11% of them adopted BIM for safety. Swallow and Zulu (2019), on the other hand, found that only 31% of participants in their study used 4D planning. They compared the perception of 4D users and non-users of the benefits of 4D planning for health and safety and found that 4D users had a higher positive perception of the benefits of 4D planning for health and safety than did the non-4D users. It is vital, therefore, that the industry should be encouraged to adopt such technologies to improve health and safety awareness and performance.

Planning and virtual simulation

A review of the literature shows a consensus on the application of BIM to improve planning through the use of virtual simulation capabilities, including 4D planning. It is seen as a useful tool for improving the understanding of health and safety issues (Alomari et al., 2017). 4D planning requires synchronizing the information model components with schedule information, creating visual construction sequencing (Hardin & Mccool, 2015). 4D planning allows for the visualization and communication of project information (Romigh et al., 2017), which the project team can use to assess the logic and sequence of the proposed plan (Whitlock et al., 2018). Such an approach allows for opportunities to evaluate risk and explore the available options before project execution (HSE, 2018). It can also allow the project team to rehearse activities in a virtual world before commencement, thereby enhancing health and safety management (Gledson, 2016). The planning and subsequent simulation of site activities can help the project team to collaborate and identify in advance any potential hazards and put in place appropriate mitigating measures (Abed et al., 2019). The use of 4D models was demonstrated in a complex infrastructure project: Crossrail Moorgate Shaft, London, England (illustrated in Figure 9.1). In this project, 4D models were created in the early stages of pre-construction and used for collaborative review involving various stakeholders of the project, including the clients, the designers, and the site team. 4D planning allowed the project team to review the build sequences, enhancing the ability to identify safety hazards and co-ordinate access and logistics as well as temporary works in great detail. This collaborative process utilizing 4D planning resulted in a reduction of temporary works and reduced safety risk, with improved team co-ordination and effective feedback (Freeform, 2013).

Working towards Zero Harm and in line with Design for Safety, Balfour Beatty in the United Kingdom constructed 3D safety models for excavation processes which can be integrated with time variations for 4D presentations, showing typical excavation concerns and displaying relevant safety risks and control measures, for either training or communication purposes (Balfour Beatty Construction, 2015). In terms of safety planning, communication, and management, Kiviniemi et al. (2011) conducted a BIM-based pilot study in Finland addressing site-safety arrangements, demolition works, fall protection, and crane overlapping zones. They suggested that although the use of BIM for safety management is still in its infancy, the potential of this tool for improving site-safety planning is evident, promoting not only better integration of different construction activities but also communication among different project stakeholders. Anumba and Wang (2012) announced the development of a BIM-based iHelmet device to access and retrieve construction data and drawings from an off-site network to allow easier and quicker information flow.

In addition to permanent works, Kim and Teizer (2014) constructed a BIM-ready temporary scaffolding formula system enabling automation of design, schedule, quality, procurement, and costing. They reported that this tool has is limited by the "data availability of the model, changes to project variables and non-typical soil conditions". Once the setup of a "BIM for site safety" can be standardized, then the BIM system can be equipped with scaffold installation and dismantling details for better safety planning and monitoring. Li et al. (2012a) reported the success of using interactive training for hazard identification,

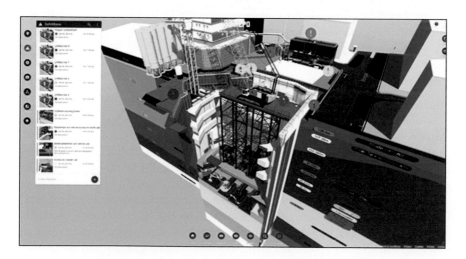

Figure 9.1 Crossrail moorgate shaft (Image: Freeform, 2013)

visualization, and training in the aircraft and automobile industry. The interactive hardware had a game engine titled Virtual Safety Assessment (VSA), which contained cases of proper usage of personal protective equipment (PPE) from typical construction accidents. VSA, as a form of BIM application, can help staff to develop safety-relevant know-how from 3D data, as the system can point out any weakness during PPE implementation.

Cheng and Teizer (2013) argued that there are benefits to adopting real-time data sensing and visualization systems for faster and more accurate information flow, especially for those long and larger capital projects involving complex construction tasks, different locations, and multiple stakeholders and contractors. The researchers developed virtual realities such as workers working towards live lifting plant and confirmed the potential for using the system in site monitoring, safety training, and accident prevention. In another study, Abed et al. (2019) utilized BIM technology to assist in identifying hazards related to falls. The study concluded that the BIM model improved safety managers' understanding of the works and resulted in greater accuracy of hazard recognition and selection of safety equipment in addition to providing these models for future safety training purposes.

The use of BIM for site safety can be extended beyond the construction stage. To mitigate security, explosion, smoke, fire, or other emergency scenarios, Ruppel and Schatz (2011) introduced a BIM-driven Rescue Game that provides designers, users, and emergency services with the ability to find the shortest time route to the exit. Working with a similar concept, Wang et al. (2014) developed a mobile BIM-based simulation platform for evacuation situations.

Given the importance of safety planning and the high risk of falling in the construction trades, Melzner et al. (2013) developed a fall checking algorithm and tested its potential using both German and US requirements within BIM settings. In addition to obtaining a better understanding of site safety, it was concluded that through BIM, site-safety planning workflow could be improved. In terms of a specified project, the Crossrail project in the United Kingdom reported two key BIM advantages, namely reducing risks and better safety training arrangements, as well as an improvement in overall safety performance (Crossrail, 2016).

Visualization of lifting: An example application area

Lifting operations present one of the main areas prone to accidents on construction sites. *Lifting* refers to the use of any mechanical equipment to perform raising and lowering of a load and for transporting the load while suspended. Lifting has been identified as one of the relatively high-risk construction tasks (Fang et al., 2016). For example, about 30% of Hong Kong's severe or fatal accidents involve lifting operations. Thus, lifting appliances and gears are subject to legislative control in many parts of the world, including Hong Kong and the United Kingdom. In Hong Kong, duty-holders such as clients (for the plant), competent persons (for the operating plant), and competent examiners (for maintenance purposes of inspection, examination, and testing of the plant) have a critical role in managing health and safety. In the United Kingdom, plant and lifting equipment are governed under key regulations such as Provision and Use of Work Equipment Regulations 1998 (PUWER) and Lifting Operations and Lifting Equipment Regulations 1998 (LOLER). In addition to these regulations, the approved code of practice produced by the Health and Safety Executive (HSE) specifies the safe use of cranes and lifting accessories. This pays particular attention to planning lifts, including the assessment of ground conditions, intended load, weather, and communication systems. The maintenance of cranes and their accessories, as well as the training of those involved in planning, supervising, and operating, are also critical considerations of their erection, use, dismantling, and transportation (HSE, 2014).

Both mobile and tower cranes are common types of construction plant found in the United States, Europe, and the Far East (Shapira & Lyachin, 2009). Crane lifting can be dangerous, yet it is critical for the completion of almost all construction work. Häkkinen (1993) reported that the most common types of crane accidents happen at the commencement of lifting and the loosening of the load, with slingers having a higher chance of getting injured than crane operators. The researcher suggested adopting proper risk management to address the risk, even though human errors are challenging to address. According to the Census of Fatal Occupational Injuries (CFOI), a total of 220 crane-related deaths were reported from 2011 to 2015, averaging 44 deaths per year over these five years.

To reduce the chances of operators being struck during crane lifting at blind spots, Li et al. (2013) utilized the information technology of global positioning systems and radio frequency identification devices in tower crane operation. The test trial was successful (without a false alarm) with an accuracy of 30cm within a 3m danger zone, activating warnings during unlawful entering of the danger zone. However, there are concerns that it is "difficult to set up" and about the "number of available tags", "source of battery power", and "workers feeling privacy being violated". Nevertheless, the system has the potential to serve as an input for a BIM model for other safety improvements or used for other hazards such as working at height.

Also using sensor technology with BIM integration for improving blind-spot lifting of a tower crane, Lee et al. (2012) verified the perception of user-friendliness of the system in Korea; they concluded that crane operators find the system very useful. However, one limitation was that the BIM model had to be manually updated; otherwise, blind-spot location tracking would become outdated.

In relation to rehearsing high-risk tower crane dismantlement tasks, Li et al. (2012b) evaluated the use of a close-to-reality virtual safety training system. The process allowed multi-users to co-ordinate and collaborate in a risk-free virtual environment in order to practice safe dismantling methods. Feedback from the study suggested an improvement in learning interest due to the visualization and ease of understanding offered by the virtual environment.

Given that heavy plant are major causes of construction site accidents in Hong Kong, Guo et al. (2012) have advocated that the industry should follow others (from the robot, laboratory, food and surgical trades) to utilize advanced computer information technology for safety training. This training tool has the advantages of being interactive, multi-user, flexible, and automatic. Using the game technology of 3DVIA Virtools, the dismantling process of the tower crane process was simulated (for example, displaying connecting lifting gears to the crane hook, releasing the screws of the crane structure, and lifting crane components to other locations); when comparing traditional and BIM-based training, participants commented that the top three benefits of this BIM-based training are the ease of recognizing plant operation, the ability to improve the operation procedure, and the higher potential for collaboration.

Table 9.1 below is an extract from a focus group study of the perceptions of operatives working for a lifting specialist company in Hong Kong. Their views demonstrate the viability of BIM to improve health and safety performance.

Sensor and devices

Construction is a labour- and information-intensive sector. Safety information and warnings, or the lack thereof, can be a matter of life or death. Hu and Zhang (2011) have explained that the aim of BIM has been around for a long time, and one potential BIM application is for structure collision analysis. They conducted a BIM-4D structure safety-based analysis on

Table 9.1 Views from a lifting specialist

Views from a lifting specialist

Below is a focus group discussion from a lifting specialist company in Hong Kong (Wang, 2018).

Participants were asked to discuss the benefits of using visual simulations to improve safety performance before and after watching the lifting visualization video. The participants had no experience in using BIM. However, after watching the lifting visualization video, the majority realized that a BIM model could provide relevant data for their lifting operation, leading to the improvement of lifting safety. The participants worked for a major lifting specialist, often involved in lifting plans and tasks. The most apparent reason for supporting BIM for safety was related to its ability to assist the lifting plan. Below are example comments:

Participant 1: "Very good if these (lifting relevant data) are available. We do have some data but not as scientific or systematic as those in the model. We send people to the site to observe site conditions in order to confirm what crane needs to be sent and under what conditions".

Participant 2: "The way I look at it is what this 3D BIM model can give you; maybe the data about the road width and the height of the loading area. In theory and assuming there is no change at site, this should help you to arrange what cranes should be assigned for this job as cranes have different lifting capacities … No site can guarantee there will be no change to the site conditions … When we plan indoor or night time operation, we often assume there will be no change in site conditions … The discussion here is whether or not this model can be beneficial to our trade. The facilitator is saying that there may be a downloadable "App" for the model. Remember this is not only for our company. We may have a way to retrieve relevant data but our competitors may not. This 3D tool is good for sites with blocked final positions, for example when you unload to the final position of some distance away".

Participant 3: "We may be using similar data. This is digitizing the data. We gather relevant data but under paper or graph format". (Participant 4)

The participants indicated that before lifting, they manually estimate the lifting capacity for each task at site; a BIM model can assist in determining the actual lifting capacity which is a significant advancement as opposed to guessing a load.

a stadium, tower block, and bridge project in China, and their results indicated that BIM-4D structure safety-based analysis offers the functional advantage of method statement review and site and resources planning as well as schedule and cost analysis (Hu & Zhang, 2011). In conclusion, the researchers remarked that BIM tools have the potential to be expanded into other aspects of site safety such as working at height, machine guarding, and the use of PPE.

To attempt to reduce the risk of workers being struck by moving plant or an object, Marks and Teizer (2013) studied various detection technologies including global positioning systems, a radio frequency identification device (RFID) and ultra-wideband for detecting and alerting site workers of approaching danger. The framework of RFID within BIM was proposed by Sattineni et al. (2013). Using RFID and sensors attached to helmets, the scholars tested the system at sites with sound detection and activation of visual and alert warnings, but weather conditions and the mounting or orientation of various devices limited the accuracy of the application.

The use of BIM to reduce paper and information checklists has been reported. Lorenzo et al. (2014) examined the case of using BIM and quick response (QR) code to enhance the information flow and even improve site management for safety in Italy. The theory was to set up individual workers' particular information data sheets, including safety-related items such as competency and training records, specified job task operating procedures, and even schematic assembly parts. This information had a corresponding QR code attached to either the worker (on the helmet, for example) or at the specified site location for verification and monitoring. From the evaluation questionnaire, the majority regarded the system as useful, particularly for larger sites, but there were concerns about its cost-effectiveness.

Falling from unprotected floor spaces or excavation openings represents a significant construction safety hazard. Instead of using physical fences or guard rails, Zhang et al. (2012) investigated the use of task and space databases from BIM and a real-time location system to alert workers when approaching dangerous site areas. The idea was to attach sets of ultrawideband tags at areas requiring fence protections, forming a part of the spatial BIM model. If workers (wearing tags) were to enter any unfenced locations, they would receive some form of system alert. Similarly, nearby plant operators could receive the same audio or vibration warning, either from a smartphone or portable device. There is evidence that construction safety can be improved by the use of emerging radio frequency (RF) remote sensing and actuating technology as it can warn workers in a real-time mode when equipment is failing (Teizer et al., 2010)

In an attempt to monitor another safety hazard at construction sites, Riaz et al. (2014) built a BIM and WiFi monitoring prototype for an oxygen gas and temperature detection system. The setup was evaluated through the use of a focus group; cited benefits were related to "produce health and safety data" and "future tool for monitoring safety". In terms of limitations, the concerns of "cost", "maintenance and repair", "reluctance to change", and "lack of BIM knowledge" were documented. The information part of a BIM can be site-safety relevant, and the above indicates how it can be used to detect dangers and warn relevant workers or systems.

The integration of the use of drones and BIM is another opportunity for the improvement of health and safety management on construction projects. The use of unmanned aerial vehicles (UAVs), or drones, for health and safety on construction projects is becoming popular (Boucher, 2015). Drones have the advantage of being able to capture information in hard-to-reach areas. In addition, they can be a supplement to the construction teams' efforts by tracking the building process, workforce monitoring, monitoring deliveries, and for quantification purposes. Ham et al. (2016) and Eschmann et al. (2012) observed that UAVs have the ability to photogrammetrically monitor a range of complicated infrastructures, such as bridges, dams, and cooling towers, subsequently allowing obtained data to be processed through building information modelling (BIM) software, generating interactive plans and simulations (Freimuth & Konig, 2015). Virtual reality has also proven to be a useful tool, particularly in the pursuit of improving safety training (Perlman et al., 2014; Sacks et al., 2013). These tools have often been explored within educational settings. For example, Swallow and Zulu (2020) used virtual reality with higher education students for safety training. They concluded that the use of virtual reality proved to be an effective training technique in providing a risk-free environment for inexperienced professionals, allowing them to select appropriate access equipment, and gave a further appreciation of costs associated with safe working practices.

Safety management

Safety management systems are designed to aid the management and reduction of site-safety risks. It is essential, therefore, that a holistic approach to safety management needs to be in place from the design stage to the completion of the project. Houssin & Coulibaly (2011) pointed out that the root cause of up to 60% of accidents was "design-related". Safety consideration in the downstream cycle tends to add sensor and system modification, causing unnecessary costs and generating trade complications to product or process development. Instead, the researchers proposed to integrate safety dimensions into the upstream design process, namely safety standard and specification, risk identification and analysis, design review, and the final step of inspection/testing. It is evident that although each stakeholder in the AEC sector has a role to play in contributing to safety management. The impact is most effective at the design stage.

Zhang et al. (2015) showed that there was still a gap between safety management and information technology. The researchers worked towards BIM ontology, enabling smoother automation and simulation for safety rule checking and hazard identification. From the interviews, the safety professionals agreed that the BIM semantic flow was representative and covered the main concepts of job hazard analysis in construction safety, including identification of potential hazard, mitigation options, safety benchmarks, resource implications, geometry information, site conditions, and visualization. The researchers pointed out that the major limitation of the system was the constant updating of site conditions such as that from plant location, terrain, and site layout conditions, as well as safety risk factors or the fact that the auditing system was not yet part of the ontology model.

Mordue and Finch (2014) suggested that designers are now partly responsible for site safety during construction stages, but some are not sure how to engage in site safety. Commonly, the construction industry is regarded as one of the most dangerous sectors, with workers exposed to different types of hazards, including falls from height. Zhang et al. (2013) have argued that there is a lack of active tools for designers to incorporate safety; they tested a rule-based BIM checking mechanism to automatically check falling hazards from holes or edges of walls or slabs, involving rule, model, execution, and report phases. The study concluded that by using BIM (even at the design stage), dangerous conditions can be determined via the automation of safety code checking. This tool should reduce the time needed to check safety requirements. However, the constantly changing site conditions will limit this detection setup; the selection of relevant safety control measures to prevent different types of fall hazards is still manually conducted. From a similar study in the United States and Germany, Melzner et al. (2013) outlined the benefits of the BIM fall-protection platform, including enhancement of workflow, improvement of turn-around time, forecasting of materials and bills of quantities, automation of risk analysis, implementation of safety measures, and visualization of simulation potential. However, the implementation of this BIM platform can be limited due to compatibility issues among different information formats, the frequent updates made to the BIM model, changes in safety rules, and systems not being able to predict workers' poor safety culture.

Park and Kim (2013) revealed that a streamlined BIM safety management system should consist of site planning, safety training, and relevant inspection. This BIM module was able to integrate construction activities/schedules/contractors to identify safety-sensitive locations and to implement virtual reality and gaming for safety training. The scholars tested the BIM setup at a Korean construction site for formwork and reinforcement activities. Specifically, the BIM safety model started with a safety manager, who created a virtual BIM safety site where the project team could identify and add relevant safety risks or measures to the BIM safety model. Then this new BIM safety system can be viewed and confirmed by others as the safe construction method to be implemented for both education and inspection purposes. One advantage of this system was that using a mobile device, other stakeholders would be able to monitor and update any inspection and training status. In terms of effectiveness, safety managers generally agreed that the BIM safety system is effective in providing relevant richer information and in improving real-time communication. The drawback of this BIM safety prototype was that there was a need to improve interoperability between site 3D models, images, checklists, and videos, as well as the accuracy of site sensor networks.

Summary

Similar to many other industries such as manufacturing and aerospace, the AEC industry is looking for ways to improve productivity, while keeping cost, time, constructability, and environmental concerns, as well as site safety, in check. This need to adapt is evident as the world faces changes due to the impact of a global pandemic. As BIM is becoming trendy in the construction industry with BIM mandates from countries such as the United States, the United Kingdom, Singapore, and Hong Kong, its ability to impact all aspects of the construction process should be welcome. The discussion above has demonstrated this ability. The discussion above demonstrates the potential for BIM to act as a driver for revisiting and enhancing health and safety planning and management. Studies with empirical evidence demonstrate different BIM applications to enhance health and safety management on construction projects. In addition to the incorporation of health and safety information and data into a BIM model, combining BIM and image technologies can also be a platform for improved health and safety performance on construction projects. Other areas of application include the ability to carry out compliance checking, BIM-driven prefabrication with spatial and dimension details, and hazard prediction and prevention involving a combination of BIM-enhance-design-out-risk, among others. The application of BIM to improve planning through the use of virtual simulation capabilities, including 4D simulation and virtual reality, is also acknowledged in the literature as a potential catalyst for enhanced health and safety performance. This is particularly important at the planning stage, as an opportunity is afforded to scrutinize and explore different options before commencement in order to show how the plan would play out in a "virtual world". The need for construction professionals to have the necessary health and safety knowledge and skills through training is essential. It has been demonstrated that embedding BIM and digital technologies into the education of those in the construction industry is key to both digital driving transformation and improving safety training. With technology that can significantly improve safety training and project planning readily available, the industry and educators alike should take this opportunity to embrace and use this as a platform to innovate.

Questions

1. Describe the importance of health and safety within the construction sector.
2. Describe the significance of the UK BIM standard PAS 1192-6:2018.
3. Describe technology explored for health and safety management within the construction industry.
4. Explain the term 4D. What are the applications of 4D within the construction sector?
5. Describe the use of virtual reality when used for safety training.
6. Explain the key considerations for lifting planning under LOLER 1998.

Bibliography

Abed, H. R., Hatem, W. & Jasim, N. (2019) Adopting BIM technology in fall prevention plans. *Civil Engineering Journal*, 5 (10), pp. 2270–228.

Akinade, O.O., Oyedele, L.O., Ajayi, S.O., Bilal, M., Alaka, H.A., Owolabi, H.A., & Arawomo, O.O. (2018) Designing out construction waste using BIM technology: Stakeholders' expectations for industry deployment. *Journal of Cleaner Production*, 180, 375–385.

AlBahnassi, H., & Hammad, A. (2012) Near real-time motion planning and simulation of cranes in construction: Framework and system architecture. *Journal of Computing in Civil Engineering*, (Jan/Feb), 54–63.

Alomari, K., Gambatese, J. & Anderson, J. (2017) Opportunities for using building information modeling to improve worker safety performance. *Safety*, 3 (1), p. 7.

Anumba, C. J., & Wang, X. (2012) *Mobile and Pervasive Computing in Construction*: Wiley.

Balfour Beatty Construction. (2015) *BIM for Zero Harm*. Balfour Beatty Construction.

The BIM (2020) *How to Build in 2030*. [Online]. Available from: <https://www.youtube.com/watch?v=PhRG1Q779lc>.

Boucher, P. (2015) Domesticating the drone: The demilitarisation of unmanned aircraft for civil markets. *Science and Engineering Ethics*, 21 (6), pp.1393–1412.

British Standard Institution (BSI) (2018) *PAS 1192–6:2018 Specification for the Collaborative Sharing and Use of Structured Health and Safety Information Using BIM*: BSI.

Cheng, T., & Teizer, J. (2013) Real-time resource location data collection and visualization technology for construction safety and activity monitoring applications. *Automation in Construction*, 34, 3–15.

Crossrail. (2016) *Building Information Modelling (BIM)*. Crossrail.

Dimitrov, A., & Golparvar-Fard, M. (2013) Vision-based material recognition for automated monitoring of construction progress and generating building information modeling from unordered site image collections. *Advanced Engineering Informatics*, 28 (1), 37–49. DOI: 10.1016/j.aei.2013.11.002

Edirisinghe, R., & Woo, J. (2020) bim-based performance monitoring for smart building management. *Facilities*. 39(1/2), 2021, 19–35.

Eschmann, C., Kuo, C.M., Kuo, C.H. and Boller, C. (2012) Unmanned aircraft systems for remote building inspection and monitoring. In Proceedings of the 6th European Workshop on Structural Health Monitoring, Dresden, Germany, Vol. 36.

Fang, Y., Cho, Y. K., & Chen, J. (2016) A framework for real-time pro-active safety assistance for mobile crane lifting operations. *Automation in Construction*, 72, pp. 367–379.

Forsythe, P. (2014) Proactive construction safety systems and the human factor. *Proceedings of the Institution of Civil Engineers: Management, Procurement and Law*, 167. pp. 242–252.

Freeform (2013) *Crossrail Moorgate Shaft Project (BAM Nutall, Kier and Crossrail)*: 4D Construction Group.

Freimuth, H., & König, M. (2015) Generation of waypoints for UAV-assisted progress monitoring and acceptance of construction work. In (Proc)15th International Conference on Construction Applications of Virtual Reality. October 5–7, 2015, Banff, Alberta, Canada. pp. 77–86.

Getuli, V., Capone, P., Bruttini, A., & Isaac, S. (2020) BIM-based immersive virtual reality for construction workspace planning: A safety-oriented approach. *Automation in Construction*, 114, p. 103160.

Gledson, B. (2016) Exploring the consequences of 4D BIM innovation adoption. In Proceedings of the 32nd Annual ARCOM Conference, 5-7 September 2016, Manchester, UK (Vol. 1. pp. 73–82).

Guo, H., Li, H., Chan, G., & Skitmore, M. (2012) Using game technologies to improve the safety of construction plant operations. *Accident Analysis & Prevention*, 48, 204–213. DOI: 10.1016/j.aap.2011.06.002

Häkkinen, K. (1993) Crane accidents and their prevention revisited. *Safety Science*, 16 (3), pp. 267–277. DOI: 10.1016/0925-7535(93)90049-J

Ham, Y., Han, K.K., Lin, J.J., & Golparvar-Fard, M. (2016) Visual monitoring of civil infrastructure systems via camera-equipped unmanned aerial vehicles (UAVs): A review of related works. *Visualization in Engineering*, 4 (1), p.1.

Hardin, B., & McCool, D. (2015) *BIM and Construction Management: Proven Tools, Methods, and Workflows*: Wiley.

Health and Safety Executive (HSE) (2014) *Lifting Operations and Lifting Equipment Regulations 1998: Approved Code of Practice and Guidance*: HSE

Houssin, R., & Coulibaly, A. (2011) An approach to solve contradiction problems for the safety integration in innovative design process. *Computers in Industry*, 62, 398–406. 334

HSE (2015) *Managing Health and Safety in Construction*: HSE.

HSE (2018) Improving health and safety outcomes in construction. Making the case for building information modelling (BIM). Health and Safety Executive [Online], RR 235. Available from: <http://www.hse.gov.uk/research/rrpdf/rr235.pdf>.

HSE (2019) *Workplace Fatal Injuries in Great Britain*: HSE.

Hu, Z., & Zhang, J.. (2011) BIM- and 4D-based integrated solution of analysis and management for conflicts and structural safety problems during construction: 2. Development and site trials. *Automation in Construction*, 20 (2), 167–180. DOI: 10.1016/j.autcon.2010.09.014

Jin, Z., Gambatese, J., Liu, D. & Dharmapalan, V. (2019) Using 4D BIM to assess construction Risks during the design phase. *Engineering, Construction and Architectural Management* [Online]. Available from: <https://www.emerald.com/insight/content/doi/10.1108/ECAM-09-2018-0379/full/html?skipTracking=true>

Kam, C., Fischer, M., Rinella, T., Mak, D., & Oldfield, J. (2014) Realising the promise of BIM in Hong Kong's construction industry. *Journal of Hong Kong's Construction Industry*, (May) 29–33.

Kim, K., & Teizer, J. (2014) Automatic design and planning of scaffolding systems using building information modeling. *Advanced Engineering Informatics*, 28 (1), 66–80. DOI: 10.1016/j.aei.2013.12.002

Kiviniemi, M., Sulankivi, K., Kahkonen, K., Makela, T., & Merivirta, M. (2011) *BIM-based Safety Management and Communication for Building Construction*. VTT Technical Research Centre of Finland, Vuorimiehentie, Finland- Research Notes 2597 [Online] vttresearch.com

Ku, K., & Taiebat, M. (2011) BIM experiences and expectations: The constructors' perspective. *International Journal of Construction Education and Research*, 7, pp. 175–197.

Lee, G., Cho, J., Ham, S., Lee, T., Lee, G., Yun, S. H., & Yang, H. J. (2012) A BIM- and sensor-based tower crane navigation system for blind lifts. *Automation in Construction*, 26, 1–10.

Li, H., Chan, G. & Skitmore, M. (2012a) Multiuser virtual safety training system for tower crane dismantlement. *Journal of Computing in Civil Engineering*, 26 (5), pp. 638–647.

Li, H., Chan, G., & Skitmore, M. (2012b) Visualizing safety assessment by integrating the use of game technology. *Automation in Construction*, 22, 498–505. DOI: 10.1016/j.autcon.2011.11.009

Li, H., Chan, G., & Skitmore, M. (2013) Integrating real time positioning systems to improve blind lifting and loading of crane operations. *Construction Management & Economics*, 31 (6), 596–605.

Li, X., Yi, W., Chi, H., Wang, X. & Chan, A. (2018) A critical review of virtual and augmented reality (VR/AR) applications in construction safety. *Automation in Construction* [Online], 86, pp. 150–162. Available from: <https://www.sciencedirect.com/science/article/pii/S0926580517309962>.

Lorenzo, T., Benedetta, B., Manuela, C., & Davide, T. (2014) BIM and QR-code. A synergic application in construction site management. *Procedia Engineering*, 85, 520–528.

Love, P.E.D., Edwards, D.J., Han, S., & Goh, Y.M.. (2011) Design error reduction: Toward the effective utilization of building information modeling. *Research In Engineering Design*, 22(3), 173–187. DOI: 10.1007/s00163-011-0105-x

Marks, E. D., & Teizer, J.(2013) Method for testing proximity detection and alert technology for safe construction equipment operation. *Construction Management and Economics*, 31 (6), 636–646. DOI: 10.1080/01446193.2013.783705

Martínez-aires, M. D., López-alonso, M. & Martínez-rojas, M. (2018) Building information modeling and safety management : A systematic review. *Safety Science Journal*, 101 (October 2015), pp. 2017–2019.

Melzner, J., Zhang, S., Teizer, J., & Bargstädt, H.-J. (2013) A case study on automated safety compliance checking to assist fall protection design and planning in building information models. *Construction Management and Economics*, 31 (6), 661–674. DOI: 10.1080/01446193.2013.780662

Mohammadi, A., Tavakolan, M. & Khosravi, Y. (2018) Factors influencing safety performance on construction projects : A review. *Safety Science*, 109 (October 2016), pp. 382–397.

Mordue, S. & Finch, R. (2014) *BIM for Construction Health and Safety*: RIBA Publishing.

Natephra, W., & Motamedi, A. (2019) *BIM-based Live Sensor Data Visualization Using Virtual Reality for Monitoring Indoor Conditions*. In: Haeusler, M., Schnabel, M.A., and Fukuda, T. (eds.), Intelligent & Informed - Proceedings of the 24th CAADRIA Conference - Volume 2, Victoria University of Wellington, Wellington, New Zealand, 15-18 April 2019, pp. 191–200.

National Building Specification (2020) *10th Annual BIM Report 2020* [Online]. Available from: <https://www.thenbs.com/knowledge/national-bim-report-2020>.

New York City, U. (2013) *Building Information Modeling Site Safety Submission Guidelines and Standards (BIM MANUAL)*: NYC Buildings.

Park, C.-S., & Kim, H.-J. (2013) A framework for construction safety management and visualization system. *Automation in Construction, 33*, 95–103. DOI: 10.1016/j.autcon.2012.09.012

Perlman, A., Sacks, R. & Barak, R. (2014) Hazard recognition and risk perception in construction. *Safety Science* [Online], 64, pp. 22–31. Available from: <https://www.sciencedirect.com/science/article/pii/S0925753513002877>

Riaz, Z., Arslan, M., Kiani, A., & Azhar, S. (2014) CoSMoS: A BIM and wireless sensor-based integrated solution for worker safety in confined spaces. *Automation in Construction*, 45, 96–106.

Rolim, A., Valente, G. & Keskin, B. (2020) Improving construction risk assessment via integrating building information modelling (BIM) with virtual reality. In Proceedings of 20th International Conference on Construction Applications of Virtual Reality, 2020. Middlesbrough: Teeside University, pp. 64–72.

Romigh, A., Kim, K. & Sattineni, A. (2017) 4D Scheduling: A visualization tool for construction field operations [Online]. In 53rd Associated Schools of Construction Annual International Conference Proceedings. pp. 395–404. Available from: <http://ascpro0.ascweb.org/archives/cd/2017/paper/CPRT130002017.pdf>.

Ruppel, U., & Schatz, K. (2011) Designing a BIM-based serious game for fire safety evacuation simulations. *Advanced Engineering Informatics*, 25 (4), 600–611.

Saeedfar, A. (2017) *Blog: Role of Safety in BIM. [Online] Assemble Systems*. Available online at: https://assemblesystems.com/blog/role-of-safety-in-bim/ (Accessed February 25, 2018).

Sacks, R., Perlman, A. & Barak, R. (2013) Construction safety training using immersive virtual reality. *Construction Management and Economics* [Online], 31 (9), pp. 1005–1017. Available from: <https://www.tandfonline.com/doi/full/10.1080/01446193.2013.828844?scroll=top&needAccess=true>.

Sattineni, A., Underwood, J., & Khosrowshahi, F. (2013) Conceptual framework for site safety monitoring using RFID and BIM. Paper presented at the Creative Construction, Budapest, Hungary.

Shapira, A., & Lyachin, B. (2009) Identification and analysis of factors affecting safety on construction sites with tower cranes. *Journal of Construction Engineering and Management*, 135 (1), pp. 24–33. DOI: 10.1061/(ASCE)0733-9364(2009)135:1(24)

Sulankivi, K., Kähkönen, K., Mäkelä, T., & Kiviniemi, M. (2010, May) 4D-BIM for construction safety planning. In Proceedings of W099-Special Track 18th CIB World Building Congress (Vol. 2010, pp. 117–128).

Swallow, M. & Zulu, S. (2019) Perception of the benefits and barriers of 4D modelling for site health and safety management. In Kumar, B., Rahimian, F., Greenwood, D. & Hartmann, T. ed., Proceedings of the 36th CIB W78 2019 AECO Conference. Newcastle: CIB, pp. 540–554.

Swallow, M. & Zulu, S. (2020) Virtual reality in construction health and safety education: Students experience through video elicitation: A pilot study. In Proceedings of 20th International Conference on Construction Applications of Virtual Reality. Middlesbrough, pp. 86–94.

Tatum, M. C. & Liu, J. (2017) Unmanned aerial vehicles in the construction industry. In 53rd ASC Annual International Conference Proceedings (vol. 2012. pp. 383–393).

Teizer, J., Allread, B.S., Fullerton, C.E., & Hinze, J., (2010) Autonomous pro-active real-time construction worker and equipment operator proximity safety alert system. *Automation in Construction*, 19 (5), pp.630–640.

Wan, A., Zulu, S. and Khosrowshahi, F (2018) Potential of using BIM for improving Hong Kong's construction industry. In *International Journal of 3-D Information Modelling*. DOI: 10.4018/IJ3DIM.2018070104

Wang, B., Li, H., Rezgui, Y., Bradley, A., & Ong, H. N. (2014) BIM-based virtual environment for fire emergency evacuation. *Scientific World Journal*, 2014, 589016. DOI: 10.1155/2014/589016

Whitlock, K., Abanda, F. H., Manjia, M. B., Pettang, C. & Nkeng, G. E. (2018) BIM for construction site logistics management. *Journal of Engineering, Project, and Production Management*, 8 (1), pp. 47–55.

Zairi, A., Ahmad, Z. & Mohd, Z. (2016) The application of technology in enhancing safety and health aspects on Malaysian construction projects. *Journal of Engineering and Applied Sciences*, 11 (11), pp. 7209–7213.

Zhang, S., Boukamp, F., & Teizer, J. (2015) Ontology-based semantic modeling of construction safety knowledge: Towards automated safety planning for job hazard analysis (JHA). *Automation in Construction*, 52, 29–41. DOI: 10.1016/j.autcon.2015.02.005

Zhang, C., Soltani, M., Setayeshgar, S., & Motamedi, A. (2012) Dynamic virtual fences for improving workers safety using BIM and RTLS. In 14th International Conference on Computing in Civil and Building Engineering. June 2012, Moscow, Russia, pp. 24–29.

Zhang, S., Teizer, J., Lee, J.K., Eastman, C.M., & Venugopal, M. (2013) Building information modeling (BIM) and safety: Automatic safety checking of construction models and schedules. *Automation in Construction*, 29, 183–195. DOI: 10.1016/j.autcon.2012.05.006

Zhou, Z., Irizarry, J., & Li, Q. (2013) Applying advanced technology to improve safety management in the construction industry: A literature review. *Construction Management and Economics*, 31 (6), 606–622. DOI: 10.1080/01446193.2013.798423

Zhou, W., Whyte, J., & Sacks, R. (2012) Construction safety and digital design: A review. *Automation in Construction*, 22, 102.

Section 1–3

Advanced discussions

10 Understanding BIM information management processes through international BIM standards

Mohammad Alhusban

Introduction

Several challenges are associated with the traditional process of construction projects. Traditionally, 2D drawings and documents were relied on for design development and project information management. This practice led to miscommunication and human error due to misinterpretations in the design and construction documents (Cohen, 2010). Moreover, a major consideration nowadays in the construction industry is the push towards sustainability in construction projects to achieve high energy performance and low environmental impact. This adds an extra layer of specialized construction information requirements, which increases the complexity of the design and delivery process. Moreover, fragmented management practices currently used in the construction industry lead to frequent reworking and/or redesigning of construction projects over their life cycles (Smith & Tardif, 2009). Reworking and redesigning affect the project performance in terms of cost, quality, and time. Reworking in construction projects is estimated to cost 11% of the original contractual costs (Forcada et al., 2017; Love et al., 2009). Additionally, project redesign results in quality defects and schedule delays (Lopez et al., 2010; Goodrum et al., 2008; Sun & Meng, 2009). Therefore, managing, coordinating, integrating, and updating the substantial amount of information from the construction project stakeholders over the life cycle of a project becomes crucially important (Hooper & Ekholm, 2010; Clough et al., 2008; Kim, 2014). Construction projects are faced with increased complexity in information management, and this has a significant impact on project outcomes such as cost, quality, and schedule performance.

Building information modelling (BIM) has been introduced as a response to the above-stated issues, and it can be considered one of the most effective technological and organizational innovations in the architecture, engineering, and construction industry (AEC) (Succar & Kassem, 2015). Technological innovation plays a key role in both short-term and long-term economic, societal, and environmental sustainability. BIM has been classified as an innovative (Davies & Harty, 2013; Brewer & Gajendran, 2012) and a disruptive piece of technology (Eastman et al., 2008). Disruptive innovation has been defined as:

> The extent to which it departs from industry norms [...] renders existing business models obsolete, changes the basis of competition in an industry and produces sustainable competitive advantage by changing the way a whole industry works.
>
> (Loosemore, 2013)

Therefore, BIM is considered to represent a major paradigm shift in the construction industry because it requires a change to the culture and the processes for achieving a more integrated approach (Succar, 2009; Ibrahim et al., 2004; HM Government, 2012; Hannele et al., 2012). A standardization in BIM processes is needed. A process is only successfully standardized if it is executed each time in a predefined (optimal) way by processing the same activities in the same order and producing exactly the same specified output. The standardization of work-flows is desirable within manufacturing and prefabrication industries where the same prod-ucts are generated repetitively; however, there is less clarity as to whether this definition is applicable to BIM processes within the AEC industry. The existence of the NBIMS and other similar standards worldwide is evidence of the need to standardize what the AEC industry has been doing for centuries. Despite this, there is still a lack of training and education on BIM standardization. Therefore, this chapter aims to facilitate the understanding of infor-mation management in BIM processes and how it has been tackled by BIM standards from international perspectives. The information embedded in BIM processes will be discussed first to emphasize the importance of managing such information to tackle different issues raised in the construction industry. This will be followed by a discussion of BIM standardiza-tion and various BIM standards that help the management of information in BIM processes.

BIM background

There is considerable divergence among those who attempt to define the meaning of BIM. Some ambiguity is in the phrase itself. For example, is the term *modelling* intended as a noun or a verb? Does the model refer to an instantiated model or the underlying schema? BIM is usually written as *building information modelling*, with two distinct but complementary meanings: a particular engineering software, or a managing process. The latter can be characterized as the adoption of an information-centric view of the whole life cycle of a building (Watson, 2010). It is, therefore, challenging to find a single satisfactory definition of what BIM is. It is proposed that it should be considered and analysed as multidimensional information, an evolving and complex phenomenon in which the management of such information is most important. Table 10.1 shows BIM definitions arranged by year, with an emphasis on information management.

Despite the many existing definitions of BIM in the literature, construction project infor-mation management throughout the project life cycle is at the core of BIM processes. BIM dimensions (nD) reflect the extent of information embedded in BIM processes, in which they are used to manage and deliver different aspects of the construction process. For example, "the extended use of 3D intelligent design (models) has led to references to terms such as 4D (adding time to model) and 5D (adding quantities and cost of material) and on and on" (AGC, 2006, p. 3). BIM dimensions (2D, 3D, 4D, and 5D) are the only universally accepted BIM dimensions (Ahmed, 2014). However, there are more extended dimensions which are named and understood differently by different individuals and organizations. Table 10.2 shows different uses of BIM in a building construction project under each dimension.

This section discussed the information embedded in BIM processes. The following sec-tions will discuss BIM standardizations through a review of international BIM standards to manage information in BIM processes.

BIM standardization

The core of this chapter serves to facilitate the understanding of academics and students of how managing information in BIM processes has been tackled by BIM standards from inter-national perspectives. One of the main drivers that have influenced BIM implementation is the political pressure applied to the construction industry, as has occurred in many countries;

Table 10.1 Information management in various BIM definitions

Author	Year	Definition
Jung and Gibson	1999	"Integration of corporate strategy, management, computer systems, and information technology throughout the project's entire lifecycle and across different business functions".
Penttila	2006	"A methodology to manage the essential building design and project data in digital format throughout the building life cycle".
Autodesk	2008	"An innovative approach to building design, construction, and management that is characterized by the continuous and immediate availability of project design scope, schedule, and cost information that is high-quality, consistent and reliable".
London, Singh, Taylor, Gu and Brankovic	2008	"An information technology-enabled approach to managing design data in the AEC/FM (Architecture, Engineering and Construction/ Facilities Management) industry".
Kymmell	2008	"A tool helping project teams to achieve the project goals through a more transparent management process based on a three-dimensional (3D) model".
Succar	2009	"A set of interacting policies, processes and technologies generating a methodology to manage the essential building design and project data in digital format throughout the building's lifecycle".
Hardin	2009	"A revolutionary technology and process that has transformed the way buildings are designed, analysed, constructed and managed".
Weygant	2011	"A technology that allows relevant graphical and topical information related to the built environment to be stored in a relational database for access and management".
NHBC	2013	"Building Information Modelling (or 'management', more appropriately) is about identifying the important information or data that is used throughout the design, construction and operation of buildings, or any other built asset, and managing it to make it useful to all those involved".
Kim	2014	"An information management system to integrate and manage various construction information throughout the entire construction project life cycle based on a 3D parametric design to facilitate effective communication among project stakeholders to achieve a project goal(s) collaboratively".

BIM has been pushed and mandated by certain public bodies such as the UK government (Won et al., 2013).

Many researchers have investigated BIM adoption and implementation worldwide, and the main focus of these researchers was on standards, guidelines, reports, visions, and roadmaps of BIM implementation or the roles and responsibilities of stakeholders when implementing BIM. Succar (2009) listed all the reports, visions, and guides that related to BIM and which are publicly available in the United States, Denmark, Australia, Finland, Norway, the Netherlands, and a consortium of organizations in Europe. Other researchers discussed the roles of both the public and private sectors in Norway, Singapore, Finland, and Denmark in promoting and providing support for BIM implementation. Jauhiainen (2011) presented examples of BIM adoption in the public sector in three countries: the General Services

Table 10.2 Description of BIM dimensions

BIM (nD)	Capability	Description	References
2D	Drafting		(Autodesk, 2003, p. 1; Hardin, 2009, p. 253)
3D	3D Model	Project visualization, clash detection, model walkthroughs, and prefabrication	(Autodesk, 2003, p.1; Hardin, 2009, p. 253; Eastman et al., 2011)
4D	3D + Time	Schedule, visualization, construction planning, and management	(Chartered Institute of Building 2010, p. 30; Eastman et al., 2011; Hardin, 2009, p. 253)
5D	4D (3D + Time) + Cost	Quantity take-offs and real time cost estimating	(Hardin, 2009, p. 253; Eastman et al., 2011)
6D	5D (3D + Time + Cost) + Facility Management	Data capturing and monitoring (the actual data on energy efficiency and building life-cycle costs) and life-cycle management	(Hardin, 2009, p. 253; Eastman et al., 2011)
7D	6D (3D + Time + Cost + Facility Management) + Sustainability	Embodied carbon, manufacturers, and recycled content	(Hardin, 2009, p. 253)

Administration (GSA) in the United States, Senate Properties in Finland, and the Statsbygg in Norway. Wong et al. (2011) compared the governmental guidelines, standards, policies, and implementation status in the United States and Hong Kong. Cheng (2015) compared the different kinds of roles and efforts made by the public sector for BIM adoption in four main regions: Europe, the United States, Asia, and Australia.

By highlighting successful BIM implementation strategies and identifying the gaps, it was surmised that the public client has six roles to play in BIM adoption: driver and initiator, educator, regulator, researcher, demonstrator, and funding agency. Cheng and Lu (2015) and Wong et al. (2009) concluded that the public sector has a primary role to play in BIM adoption. Many countries around the globe have realized the vital role of public authorities in promoting BIM, such as in the United Kingdom and the United States (Won, 2013). Therefore, many governments including those of the United States (Wong et al., 2009), Australia (BuildingSMART, 2012), and the United Kingdom (HM Government, 2012) have established implementation strategies and standards for the use of BIM on construction projects. For instance, in the United Kingdom, BIM adoption and implementation were among the main principle objectives of the "Government Construction Strategy (GCS) 2016" to improve the national infrastructure. Thus, the government and various construction professional organizations have released standards, protocols, and guidelines for the effective management and integration of construction information. Table 10.3 shows these standards, protocols, and guidelines (NBS, 2015, 2016).

In late 2018, and in the efforts of internationalization of BIM standards, ISO 19650 Parts 1, 2, and beyond were introduced. These ISO parts have been developed based on the principles of the British standards stated in Table 10.3. Figure 10.1 illustrates how the ISO standards have been developed based on the British standards.

Table 10.3 BIM standards, protocols, and guidelines

Organizations	BIM standards and protocols
BSI	"PAS 1192-2:2013, specification for information management for the capital/delivery phase of construction projects using building information modelling".
	"PAS 1192-3:2014, specification for information management of the operational phase of assets using building information modelling BIM".
	"BS 11924-4:2014 – Collaborative production of information. Part 4: Fulfilling employer's information exchange requirements using COBie (Construction Operations Building Information Exchange) – Code of practice".
	"PAS 1192-5:2015, specification for security-minded building information management, digital built environments and smart asset management. Provides guidance on how to secure the intellectual property, the physical asset, the processes, the technology, the people, and the information associated with the asset".
	"BS 8536:2015; Facilities Management (FM) briefing for design and construction. For the building's infrastructure, guidance upon the definition of required social, environmental, and economic outcomes as well as the process of achieving those required outcomes".
	BS 8541; Range of standards for "library objects (architectural, engineering, and construction)".
CIC	BIM Protocol, Standard Protocol for use in projects using Building Information Models.
RIBA	BIM Overlay to the RIBA Plan of Work.
	RIBA Plan of Work 2013 Construction.
RIBA and NBS	"Uniclass2015. A classification system that can be used to organise information throughout all aspects of the design and construction process".
BIM Task Group	GSL (Government Soft Landings) – Developed to champion better outcomes for the UK's built assets during the design and construction stages, powered by BIM, to ensure that value is achieved in the operational life cycle of an asset.
	Construction Operations Building Information Exchange (COBie) UK 2012.

Source: NBS, 2015, 2016; Kim, 2014.

Building information modelling can also be called *building information management* or *better information management*. Whatever BIM is defined as, at the "heart" of BIM is information. Based on the above-mentioned standards, three key documents are needed to manage information in BIM and thus achieve a successful BIM project; these documents are the BIM protocol, the employer's information requirement (EIR), and the BIM execution plan (Barnes & Davies, 2015). The level of Development (LOD) and common data environment (CDE) make up essential parts of these documents. The following sub-sections will discuss these documents, aiming to facilitate the understanding of information management in BIM processes for educators.

BIM protocol

A BIM protocol aims to enhance production efficiency through adopting a consistent and coordinated approach to working within BIM (Barnes & Davies, 2015). A BIM protocol is also used to define best practices and standards that ensure the delivery of high-quality data and uniform drawings output over the entire project cycle (Ibid).

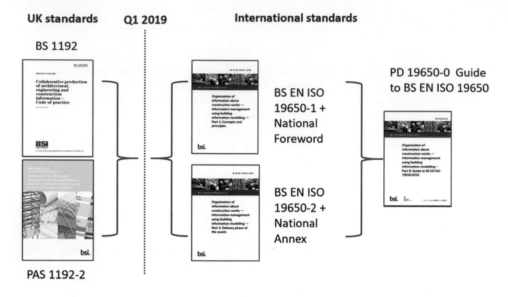

Figure 10.1 Internationalization of British Standards.

In the United Kingdom, the Construction Industry Council (CIC) BIM Protocol was issued to meet the requirements of BIM Level 2. This protocol can be used as a supplementary legal agreement that can be incorporated into a construction contract and professional service appointments by way of a simple amendment. Moreover, this protocol puts in place specific obligations, liabilities, and associated limitations on BIM model usage. In the United States, the American Institute of Architects (AIA) released its "Building Information Modelling Protocol Exhibit", which is intended to be attached to the owner–architect and owner–contractor agreements (Lowe & Muncey, 2010).

A typical BIM protocol document could include (Barnes & Davies, 2015)

- An introduction to the project;
- BIM usage extent for the project;
- How the protocol is placed in the contractual document;
- The BIM manager's details and who should appoint him/her;
- Employer information requirements (EIR);
- An organogram that shows how different stakeholders contributed to the BIM process;
- The BIM Execution Plan (BEP);
- Level of model development (LOD);
- Details of the BIM models' "data drop";
- The common data environment (CDE);
- Details of the software to be used.

Employer information requirements (EIR), the BIM execution plan (BEP), level of model development (LOD) and the common data environment (CDE) form key documents in a typical BIM protocol document. These documents are discussed in the following sub-sections.

Employer information requirements (EIR)

EIR is considered to be one of the key documents to successfully deliver BIM-based construction projects (Dwairi et al., 2016). EIR could be developed alongside the project brief, which defines the nature of the built asset that the client/developer wishes to procure. By contrast, the EIR defines the information that complies with the project/asset that the client wishes to procure, in which the design is guaranteed to be developed according to their needs (Barnes & Davies, 2015).

EIR usually forms part of the tender document on a BIM project (Barnes & Davies, 2015). It includes requirements in three main areas regarding commercial, management, and technical information. Table 10.4 shows the information embedded in an EIR document. This

Table 10.4 Aspects of the employer's information requirement (EIR) (Ashworth et al., 2017)

1. General guidance and notes (note: this section is provided as guidance and is removed on formal issue)

1. Purpose and scope
1.1 The purpose of the EIR
1.2 Use of the terms: *client, client's representative* and *contractor*

2. Client BIM and asset management strategy and objectives

3. Project details
3.1 Project information
3.2 Project contact list

4. Management Requirements	**5. Technical Requirements**	**6. Commercial Requirements**
4.1 Applicable standards and guidelines	5.1 Software	6.1 Exchange of information in line with RIBA project stages
4.2 CIC BIM protocol	5.2 IT and system performance constraints	
4.3 Project roles and responsibilities	5.3 Data exchange formats	6.2 Supplier BIM assessment form
4.4 Existing client CAFM/IWMS or enterprise asset management systems	5.4 Common co-ordinates system	6.3 BIM tender assessment
4.5 Model creation and ongoing management	5.5 Levels of definition	
4.5.1 Planning the work and data segregation	5.6 Specified model and information formats	
4.5.2 Model management plan	5.7 Site information, floor and room data information	
4.5.3 Collaboration process		
4.5.4 Model size		
4.5.5 Model viewing		
4.5.6 Volumes, zones and areas		
4.5.7 Naming conventions		
4.5.8 Model coordination, quality control and clash-detection process		
4.5.9 Use of BIM to help health and safety		
4.5.10 Delivery of asset information to the client		
4.5.11 Information publishing process		
4.5.12 Security of model information		
4.5.13 Training		
4.5.14 Model audits by the client		

EIR document has been built based on the literature review and BIM standards; a focus-group with the British Institute of Facilities Management (BIFM); case study interviews with the Glasgow Life Burrell Renaissance Project, which trialled the EIR; and peer-reviews and interviews with BIM/CAFM experts from BIM Academy1 and FM1802 (Ashworth et al., 2019).

The EIR document shown below consists of five main sections: general guidance and notes, which explain the purpose and scope of such a document; client BIM and asset management strategy and objectives, which includes details about the project information and project contact list; management requirements; technical requirements; and commercial requirements. Management, technical, and commercial requirements are stated in detail in Table 10.4.

As has been stated earlier, EIR usually forms part of the tender document on a BIM project. In order to meet the EIR, prospective suppliers prepare a BEP in which the required capacities, proposed approaches, and competencies are set out. The following section will define the BIM execution plan (BEP), its purposes, steps, and parts.

BIM execution plan (BEP)

BEP can sometimes be abbreviated as BxP. The purpose of the BEP is to manage the delivery of the project and to ensure that responsibilities and opportunities are clearly understood by all the stakeholders in a BIM-based project. The four main steps within a typical BIM execution plan procedure are as follows (Computer Integrated Construction Research Programme, 2010):

- Identifying BIM goals and uses during the project life cycle;
- Designing the BIM project execution process by creating process maps;
- Developing an information exchange by defining BIM deliverable and responsible parties;
- Defining the project infrastructure to support the developed BIM process.

A BEP comprises two parts: a pre-contract BEP and a post-contract BEP. Prospective suppliers prepare a pre-contract BEP in which the required capacities, proposed approaches, and competencies are set out to meet the EIR. Subsequently, the supplier with the awarded contract prepares the post-contract BEP to confirm the supply chain's capabilities and provides a master information delivery plan (MIDP) alongside individual task information delivery plans (TIDPs). Individual task information delivery plans include responsibilities for specific information tasks. A series of individual TIDPs comprise the MIDP, which is a primary plan that explains when the information for the project is to be prepared, the responsible parties, and the procedures and protocols to be used (Barnes & Davies, 2015). Figure 10.2 represents the information delivery cycle as mapped by the British standard (PAS1192-2:2013). This information delivery cycle illustrates the relationship between the EIR, pre-contract BEP, post-contract BEP and master information delivery plan (MIDP) in which it shows the generic process of identifying a project need, procuring and awarding a contract, mobilizing a supplier, and generating production information and asset information relevant to the need.

Figure 10.2 illustrates the information delivery cycle. However, the BIM level of information (LOI) over the information process should be discussed in more detail. The following section will discuss the BIM level of information (LOI).

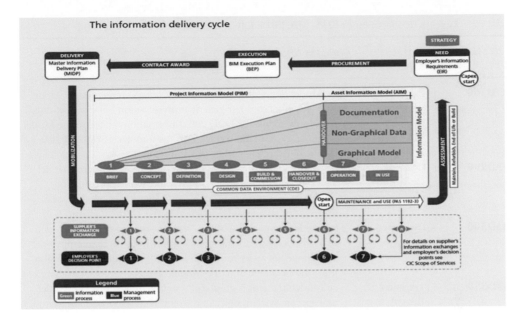

Figure 10.2 The information delivery cycle (BSI, 2013).

Level of development (LOD)

The LOD consists of two principal pieces of information: *level of details* and *level of information* (LOI). *Level of details* refers to the graphical content of a BIM model, whereas LOI refers to the non-graphical content of a BIM model. These two concepts are usually aligned as both are developed alongside each other (Barnes & Davies, 2015). LODs are identified as important and critical issues since they represent the model information at specific stages, and they are associated with the BIM implementation's practical side (Wu & Issa, 2014). Table 10.5 represents the suggested level of development by the CIC (2013) and AIA (2013).

The information delivery cycle and BIM level of information (LOD) over the information process were discussed earlier in this chapter. The following section will discuss the BIM level of information (LOD). The use of a common data environment (CDE) to exchange project information in BIM-based construction projects will be discussed in the following section.

Common data environment (CDE)

Many BIM protocols, such as the one used in the United Kingdom (CIC BIM Protocol), propose the creation of a common data environment (CDE) to exchange project information in BIM-based construction projects (McPartland, 2016). The CDE is a single source of information for the project and acts as the central repository of the project information. It is used to collect, manage, and disseminate documentation for project stakeholders; it includes graphical and non-graphical information (that is, information created in a BIM environment and in a conventional data format) (Barnes & Davies, 2015). Figure 10.2 demonstrates the position of the CDE in the overall project delivery process.

As can be seen in Figure 10.3, the CDE consists of four main areas of information: a work in progress area, in which unapproved information is held for each organization; a shared

Table 10.5 Level of development in BIM (AIA, 2013; CIC, 2013; BIM Forum, 2013, P.10)

LOD (AIA, 2013)	LOD (CIC, 2013)	Description (BIM Forum, 2013)
LOD100	1 (Preparation and brief)	"The Model Element may be graphically represented in the Model with a symbol or other generic representation but does not satisfy the requirements for LOD 200. Information related to the Model Element (i.e. cost per square foot, the tonnage of HVAC, etc.) can be derived from other Model Elements".
LOD200	2 (Concept design)	"The Model Element is graphically represented within the Model as a generic system, object, or assembly with approximate quantities, size, shape, location, and orientation. Non-graphic information may also be attached to the Model Element".
LOD300	3 (Developed design)	"The Model Element is graphically represented within the Model as a specific system, object or assembly in terms of quantity, size, shape, location, and orientation. Non-graphic information may also be attached to the Model Element".
LOD350	4 (Technical design)	"The Model Element is graphically represented within the Model as a specific system, object, or assembly in terms of quantity, size, shape, orientation, and interfaces with other building systems. Non-graphic information may also be attached to the Model Element".
LOD400	4 (Construction)	"The Model Element is graphically represented within the Model as a specific system, object or assembly in terms of size, shape, location, quantity, and orientation with detailing fabrication, assembly, and installation information. Non-graphic information may also be attached to the Model Element".
LOD500	5 (Handover and close-out)	"The Model Element is a field verified representation in terms of size, shape, location, quantity, and orientation. Non-graphic information may also be attached to the Model Elements".

area in which information is held that has been checked, reviewed, and approved for sharing with other organizations; a published area with information that the client or their representative has "signed off"; and an archive area where progress at each milestone, changed orders, and transactions are recorded.

The BIM information exchange within the CDE should be managed by an information manager (BIM manager). BIM protocols normally require the appointment of a BIM information manager by the client (CIC, 2013). The main role of a BIM information manager is to set and manage the CDE by policing it to make sure that the data are secure, and that it follows the agreed protocol. The following is a summary of the BIM information manager's other principal responsibilities (CIC, 2013):

* Managing the processes and procedures for information exchange on projects;
* Initiating and implementing the project information plan (PIP) and asset information plan (AIP);
* Assisting the preparation of project outputs, such as data drops;
* Implementating the BIM protocol, including the updating of the MPDT.

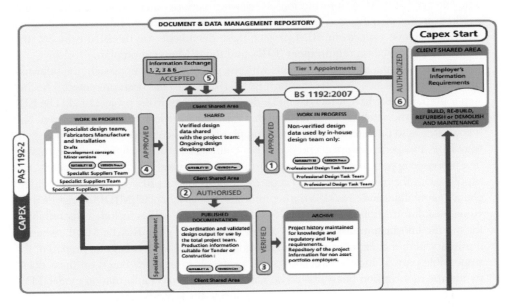

Figure 10.3 Common data exchange (CDE) (BSI, 2013).

The BIM information manager's role could be performed by different entities over the project life cycle. For example, the lead consultant or lead designer may be the information manager during the early stages, with the contractor acting as the information manager in the construction phase (Barnes & Davies, 2015).

Finally, the "heart" of BIM is information. Section 3 has focused on understanding the process side of the BIM philosophy to manage the modelling and deliver information in 3D BIM, and thus achieve a successful BIM project. Therefore, the information delivery cycle is presented and discussed. Three key documents have been found critical to manage the information in a BIM project; these documents are BIM protocol, the employer's information requirement (EIR), and the BIM execution plan. The required information in each document has been clarified through examples and mapped onto the overall information delivery cycle.

Conclusion

BIM has been adopted and implemented differently worldwide. This chapter has reviewed the various definitions of BIM, BIM dimensions, BIM standardization, and the information management process to deliver a 3D object-based BIM model. The findings of this chapter can be summarized in the following points:

- Despite the various existing BIM definitions, the important common characteristic of BIM is information management in the BIM processes.
- Governments and policy makers have an essential role in adopting and implementing BIM processes.
- Governments and policy makers are trying to standardize the BIM process worldwide through establishing international BIM standards.

- The "heart" of BIM is information. Reviewing various international standards has resulted in the identification of three key documents to manage information in BIM and thus achieve a successful BIM project; these documents are the BIM protocol, the employer's information requirement (EIR), and the BIM execution plan.
- In order to align these three documents with the project delivery approach, the information delivery cycle based on the standard PAS 1192-2 2013 has been presented. The first step is preparing the EIR, which could be developed alongside the project brief. The EIR should define the nature of the built asset that the client/developer wishes to procure. The second step is to prepare the BIM execution plans – a pre-contract BEP and a post-contract BEP. Prospective suppliers prepare a pre-contract BEP in which the required capacities, proposed approaches, and competences are set out to meet the EIR. Subsequently, the supplier with the awarded contract prepares the post-contract BEP to confirm the supply chain's capabilities and provide a master information delivery plan (MIDP) alongside individual task information delivery plans (TIPDs). Individual TIPDs include responsibilities for specific information tasks. A series of individual TIPDs comprise the MIDP, which is a primary plan that explains when the information for the project is to be prepared, the responsible parties, and the procedures and protocols to be used. Based on these documents, the mobilization phase should be started and the project information model (PIM) will be built to deliver the asset information model to the facility managers.

Questions

1. What are the main types of information embedded in BIM processes?
2. Define BIM standardization.
3. What is a BIM protocol? Give an example.
4. What are the three pillars of the employer information requirement in a BIM project? Give an example for each.
5. How does the BIM execution plan (BEP) fit into the information delivery cycle?
6. What does *level of development* (LOD) mean?
7. Define *common data environment* (CDE).

References

Ahmed, S. M. (2014, December). Barriers to BIM/4D implementation in Qatar. Paper presented at the International Conference on Smart, Sustainable and Healthy Cities. Abu Dhabi, United Arab Emirates.

AIA. (2013). *AIA G202-2013 Project Building Information Modeling Protocol*. Retrieved from http://www.aia.org/aiaucmp/groups/aia/documents/pdf/aiab099086.pdf.

Ashworth, S., Tucker, M., & Druhmann, C.K. (2019). Critical success factors for facility management employer's information requirements (EIR) for BIM. *Facilities*, 37(1/2), 103–118. https://doi.org/10.1108/F-02-2018-0027

The Associated General Contractors of America (AGC). (2006). *The Contractor Guide to BIM. The Associated General Contractors of America (AGC)*. Retrieved from https://www.engr.psu.edu/ae/thesis/portfolios/2008/tjs288/Research/AGC_GuideToBIM.pdf

Autodesk. (2003). *Building Information Modelling for Sustainable Design: Autodesk Building Solutions*. (White Paper). Retrieved from http://images.autodesk.com/latin_am_main/files/bim_for_sustainable_design_oct08.pdf.

Autodesk. (2008). *Building Information Modeling for Sustainable Design*. (White paper). Retrieved from http://images.autodesk.com/adsk/files/bim_for_sustainable_design_oct08.pdf.

Barnes, P., & Davies, N. (2015). *BIM in Principle and Practice* (2nd ed.). London: ICE Publishing.

BIM Forum (2013). *Level of Development Specification*. Retrieved from https://bimforum.org/wp-content/uploads/2013/08/2013-LOD-Specification.pdf.

Brewer, G., & Gajendran, T. (2012). Attitudes, behaviours and the transmission of cultural traits: Impacts on ICT/BIM use in a project team. *Construction Innovation: Information, Process, Management, 12*(2), 198–215.

British Standard Institution (BSI). (2013). *PAS 1192-2:2013: Specification for Information Management for the Capital/Delivery Phase of Construction Projects Using Building Information Modelling (RPRT)*. British Standard Institution (BSI). Retrieved from http://shop.bsigroup.com/Navigate-by/PAS/PAS-1192-22013/.

BuildingSMART. (2012). *National Building Information Modelling Initiative: Volume 1: Strategy*. Sydney: Building Smart Australasia.

Chartered Institute of Building. (2010). *Code of Practice for Project Management for Construction and Development*. Hoboken, NJ: Wiley-Blackwell.

Cheng, J. C. P., & Lu, Q. (2015). A review of the efforts and roles of the public sector for BIM adoption worldwide. *Journal of Information Technology in Construction, 20*, 442–478.

Clough, R. H., Sears, G. A., & Sears, S. K. (2008). *Construction Project Management: A Practical Guide to Field Construction Management*. Hoboken, NJ: Wiley.

Cohen, J. (2010). *Integrated Project Delivery: Case Studies*. Sacramento, CA: AIA California Council.

Computer Integrated Construction Research Program. (2010). *BIM Project Execution Planning Guide: Version 2.0*. University Park, PA: Pennsylvania State University.

Construction Industry Council (CIC). (2013). *Building Information Model (BIM) Protocol. Standard Protocol for Use in Projects Using Building Information Models*. Construction Industry Council (CIC). Retrieved from http://www.bimtaskgroup.org/bim-protocol/.

Davies, R., & Harty, C. (2013). Measurement and exploration of individual beliefs about the consequences of building information modelling use. *Construction Management and Economics, 31*(11), 1110–1127.

Dwairi, S., Mahdjoubi, L., Odeh, M., & Kossmann, M. (2016). Development of OntEIR framework to support BIM clients in construction. *International Journal of 3-D Information Modeling, 5*(1). 45–66. Retrieved from http://www.igi-lobal.com/gateway/article/171613.

Eastman, C., Teicholz, P., Sacks, R., & Liston, K. (2008). *BIM Handbook: A Guide to Building Information Modelling for Owners, Managers, Designers, Engineers, and Contractors*. Hoboken, NJ: Wiley.

Eastman, C., Teicholz, P., Sacks, R., & Liston, K. (2011). *BIM HANDBOOK: A GUIDE to Building Information Modeling for Owners, Managers, Designers, Engineers and Contractors* (2nd ed.). New York: Wiley.

Forcada, N., Gangolells, M., Casals M., & Macarulla, M. (2017). Factors affecting rework costs in construction. *Journal of Construction Engineering and Management, 143*(8), 445–465. https://doi.org/10.3846/13923730.2014.893917

Goodrum, P., Smith, A., Slaughter, B., & Kari, F. (2008). Case study and statistical analysis of utility conflicts on construction roadway projects and best practices in their avoidance. *Journal of Urban Planning and Development, 134* (2), 63–70.

Government Construction Strategy (GCS). (2016). Government construction strategy (2016–20). Retrieved from https://assets.publishing.service.gov.uk/government/uploads/system/uploads/attachment_data/file/510354/Government_Construction_Strategy_2016-20.pdf.

Gu, N., & London, K. (2010). Understanding and facilitating BIM adoption in the AEC industry. *Automation in Construction, 19*(8), 988–999.

Hannele, K., Reijo, M., Tarja, M., Sami, P., Jenni, K., & Teija, R. (2012). Expanding uses of building information modeling in life-cycle construction projects. *Work, 41* (Supplement 1), 114–119.

Hardin, B. (2009). *BIM and Construction Management: Proven Tools, Methods, and Workflows*. Indianapolis, IN: Wiley.

HM Government. (2012). *Building Information Modeling, Industrial Strategy: Government and Industry in Partnership* (Government Report). London: HM Government.

Hooper, M., & Ekholm, A. (2010, November). A pilot study: Towards BIM integration. An analysis of design information exchange & coordination. Paper presented at the Proceedings of the CIB W78 2010: 27th International Conference. Cairo, Egypt.

Ibrahim, M., Krawczyk, R., & Schipporiet, G. (2004). A web-based approach to transferring architectural information to the construction site based on the BIM object concept. Presented at the Proceeding of Culture, Technology and Architecture, South Korea.

Jauhiainen J. (2011). BIM maturity: The second generation. In *SOLIBRI Magazine*. Helsinki, Finland: Solibri, Inc.

Jung, Y, & Gibson, G.E. (1999). Planning for computer integrated construction. *Journal of Computing in Civil Engineering, 13*(4), 217–225.

Kim, K. P. (2014). *Conceptual Building Information Modelling Framework for Whole-House Refurbishment based on LCC and LCA*. PhD Thesis. Aston University.

London, K., Singh, V., Taylor, C., Gu, N., & Brankovic, L. (2008, September). Building information modelling project decision support framework. Paper presented at the Proceedings of the Twenty-Fourth Annual Conference Association of Researchers in Construction Management (ARCOM). Cardiff, UK, 665–673.

Loosemore, M. (2013). *Innovation, Strategy and Risk in Construction: Turning Serendipity into Capability*. London: Routledge. http://dx.doi.org/10.4324/9780203809150.

Lopez, R., Love, P. E. D., Edwards, D.J., & Davis, P. R. (2010). Design error classification, causation, and prevention in construction engineering. *Journal of Performance of Constructed Facilities, 24*(4), 399–408.

Love, P. E. D., Edwards, D.J., Smith, J., & Walker, D. H. T. (2009) Divergence or congruence? A path model of rework for building and civil engineering projects. *Journal of Performance of Constructed Facilities, 23*(6), 480–488.

Lowe, H., & Muncey, M. (2010). ConsensusDOCS 301 BIM addendum. Construction lawyer. *Associated General Contractors of America*. Retrieved from https://www.agc.org/.

McPartland, R. (2016). *What is the Common Data Environment (CDE)? National Building Specification (NBS)*. Retrieved from https://www.thenbs.com/knowledge/what-is-the-common-data-environment-cde.

National Building Specification (NBS). (2015). *National BIM Report 2015*. UK: National Building Specification (NBS). Retrieved from https://www.thenbs.com/knowledge/nbs-national-bim-report-2015.

NBS. (2016). *National BIM Report 2016*. UK: National Building Specification (NBS). Retrieved from https://www.thenbs.com/knowledge/national-bim-report-2016.

National House-Building Council (NHBC). (2013). *Zero Carbon Strategies for Tomorrow's New Homes*. Milton Keynes, UK: NHBC Foundation.

Penttila, H. (2006). Describing the changes in architectural information technology to understand design complexity and free-form architectural expression. *IT in Construction, 11*, 395–408.

Smith, D. K., & Tardif, M. (2009). *Building Information Modeling: A Strategic Implementation Guide for Architects, Engineers, Constructors, and Real Estate Managers*. Hoboken, NJ: Wiley.

Succar, B. (2009). Building information modelling framework: A research and delivery foundation for industry stakeholders. *Automation in Construction, 18*(3), 357–375.

Succar, B., & Kassem, M. (2015). Macro-BIM adoption: Conceptual structures. *Automation in Construction, 57*, 64–79. Retrieved from http://bit.ly/BIMpaperA8.

Sun, M., & Meng, X. (2009) Taxonomy for change causes and effects in construction projects. *International Journal of Project Management, 27*(6), 560–572.

Watson, A. (2010, June). BIM: A driver for change. Paper presented at the Proceedings of the International Conference on Computing in Civil and Building Engineering, W. Tizani, (Editor). University of Nottingham, Retrieved from http://www.engineering.nottingham.ac.uk/icccbe/proceedings/pdf/pf69.pdf.

Weygant, R. S. (2011). *BIM Content Development: Standards, Strategies, and Best Practices*. Hoboken, NJ: Wiley.

Won, J., Lee, G., Dossick, C., & Messner, J. (2013). Where to focus for successful adoption of building information modeling with an organization. *Journal of Construction Engineering and Management. 139*(11). 04013014. https://doi.org/10.1061/(ASCE)CO.1943-7862.0000731.

Wong, A. K., Wong, F. K., & Nadeem, A. (2009, January). Comparative roles of major stakeholders for the implementation of BIM in various countries. Paper presented at the International Conference on Changing Roles: New Roles, New Challenges. Noordwijk Aan Zee, The Netherlands.

Wong, A. K., Wong, F. K., & Nadeem, A. (2011). Government roles in implementing building information modelling systems: Comparison between Hong Kong and the United States. *Construction Innovation, 11*(1), 61–76.

Wu, W., & Issa, R. R. (2014). BIM execution planning in green building projects: LEED as a use case. *Journal of Management in Engineering, 31*(1), A4014007. http://dx.doi.org/10.1061/(ASCE)ME.1943-5479.0000314#sthash.N9eiV7VR.dpuf.

11 Scholarship of BIM and construction law

Myths, realities, and future directions

Oluwole Alfred Olatunji and Abiola Akanmu

Often the surest way to convey misinformation is to tell the strict truth – Mark Twain [1835–1910] (Twain, 2013)

Beware of ignorance when in motion; look out for inexperience when in action, and beware of the majority when mentally poisoned with misinformation, for collective ignorance does not become wisdom – William J. H. Boetcker [1873–1962] (*The William J. H. Boetcker Collections* by Bill Boetcker ([n.d.])

Introduction

BIM has been defined in different ways. Although empirical studies that articulate the definitions of BIM in detail are limited, nuances in BIM's definition can derail incipient minds unless clarified. Olatunji (2012) clarifies the confusion in some of the variants by analysing them contextually, as to whether BIM is a technology, a system, a philosophy, a software, or a platform. Olatunji argues that it is not possible for a phenomenon to mean the same thing to many disciplines. For example, in a definition by Penttilä (2006), adopted by Succar (2009) and already cited in thousands of scholarly studies, BIM is defined as a set of interacting policies, processes, and technologies producing a methodology to manage essential building design and project data in digital format throughout a building's life cycle. Race (2019) argues that BIM is not defined by the keywords in Penttilä's definition precisely, at least not at the time of the definition, nor 14 years after. Race argues that "policies", "processes", "methodology", "technologies", "manage", "project lifecycle" and "digital data" only broaden the multi-disciplinary applications of BIM, and that they are not exclusive to BIM. Race concludes that different disciplines have had specific interpretations of BIM, and that no single definition is completely satisfactory to all.

The apparent confusion triggered by the multiplicity of opinions regarding BIM's definition is not entirely inappropriate. BIM's true additionality is in its conceptual applications that extend beyond the boundaries set in extant superficial definitions. The centrality of this portrays BIM as a digital system for facilitating a data-rich, object-oriented, intelligent, and parametric representation of a construction project, and from these, views and data appropriate to various users' needs are extracted and analysed to generate information and enhance decision-making on project economics and improve project delivery processes (Olatunji, 2012:131). This version of BIM's definition embraces multiple disciplines. It defines BIM as more than a design-centric tool, and instead as design and information management practice. In addition, BIM applies beyond "building"; it applies to broad areas of building and infrastructure construction of all types. Similarly, "project data" goes beyond designing; the component of BIM relating to information management is important to both design and non-design disciplines, including contract performance and relationship management disciplines (Kagioglou et al., 2001; Meng, 2012; Stewart, 2007).

In the past two decades, BIM has remained a popular digital modelling platform. Many studies have reported BIM as the commercial reality of the future of construction – according to Luciani (2008), this future began over a decade ago. For example, Olatunji (2019) describes BIM as the commercial reality of today's construction education, in that construction graduates who are not BIM-ready are not likely to be job-ready and may have limited opportunities in a future that is driven by BIM. Similarly, some studies have portrayed BIM deployment as critical to construction's future and as the main vernacular for the survival of construction businesses in the modern world. Examples of these are the works of Aouad et al. (2006) and Volk et al. (2014), who emphasize the additionality of BIM to the future of the construction industry. They argue that BIM improves the outcomes of construction projects through collaboration, data-rich communication, and integration. A perception by Hope (2012) is that construction businesses that do not adopt BIM will die. Olatunji et al. (2010) argue against a perception that suggests BIM will lead to the extinction of traditional professional practices, based on speculations that some revolution will take place as BIM's automated processes become popular.

It is important to note that many conclusions regarding the potential of BIM are speculative. Considerable evidence has continued to emerge, showing that such speculations are both misinforming and distractive to construction scholarship. It is crucial that the future of construction education and scholarship is not fuelled further by such inaccurate speculations, in particular, in relation to the performance and relationship management of construction projects. For example, BIM has been espoused to reduce construction and associated social costs, waste, estimating errors, design errors, disputes, and project duration (Ku and Taiebat 2011; Wong and Fan 2013; Aibinu and Venkatesh 2014; Azhar 2011; Xu and Qian 2014; Bensalah et al. 2017; Wu and Issa 2012). Such claims are frequent enough to persuade clients into setting them as legal objectives of BIM to which project stakeholders must become answerable. Such claims seem harmless; however, the evidence underlying them is inconclusive and may mislead. Although there are warnings that such speculations should be interpreted with caution, speculative conclusions are still rife. They have continued to grow, but studies that critically evaluate them are not common. The purpose of this chapter is to uncover some of the myths that may jeopardize the authenticity of construction law education and scholarship in the context of BIM, and to provide useful suggestions regarding future directions.

Potentialities of BIM in contract administration and construction law

The speculations, myths, and objective realities

It is not utterly inappropriate for researchers to be positive about the potential novelty of an incipient phenomenon. Terrin et al. (2005) have captured this bias aptly by concluding that such biases often lead to overly optimistic research conclusions. It may not be the intention of researchers to misinform; it is critically important, however, that uptakers of such conclusions are cautious as they interpret and apply certain findings and suggestions from the literature. In particular, Amor and Faraj (2001) and Amor et al. (2007) have given such warnings regarding BIM. If the axiom "*Ignorantia juris non excusat*" (Latin for "ignorance of the law excuses not") holds true, scholarship of legal education regarding BIM must draw a line between speculations, misinformation, and the realities of novel knowledge around BIM. What are these speculations and the specific industry interpretations around them? Some examples are provided below:

1. **BIM works best when primed with integration, collaboration, interoperability, value sharing, robust data, and seamless communication.** This is supported by the considerable suggestions in empirical studies by Aranda-Mena et al. (2008; 2009); Azhar et al. (2012); Klaschka (2019); Manning and Messner (2008); Moon et al. (2011); Olofsson et al. (2008); Wang et al. (2014). The drawback in these suggestions is that none of these has ever been considered a hallmark of the construction industry – see Ashcraft (2008) and Olatunji and Akanmu (2014). An established framework on how the acclaimed BIM attributes apply, or are measured, is not evident. A perspective added by Olatunji (2011b; 2014; 2016) and Olatunji and Akanmu (2015) is that the uncertainties around how these expected capabilities apply are enough to derail whatever good intention any project stakeholder may have towards BIM. For example, it is still not clear whether projects that are not driven by the attributes credited to BIM cannot fail, or whether the attributes are impossible without BIM. For greater clarity, the construction literature is limited regarding the broad body of theories that shape the attributes that are speculated to promote project outcomes when BIM is deployed. For example, Kvan (2000) argues there is no clear definition of collaboration in construction literature. In addition, Ashcraft (2008) thinks trust and collaboration within or across construction disciplines is not a cultural asset. Olatunji (2011b) argues that it is possible to overcome these challenges. However, the road to desired change is not possible unless construction contract instruments are built around success enablers in BIM, in the context of owned, joint, shared, several, and accrued liabilities (see Wright [1987; 1992] and Olatunji and Akanmu [2015]). The single most significant cross-pollination between legal education and BIM knowledge is in the ability to understand these capabilities and in developing appropriate instruments for their utilization. When limited scholarship is dedicated to these, it is safe to conclude that the impact of the acclaimed success enablers in BIM is not yet scalable and is thus inconclusive.

2. **BIM offers revolutionary solutions to problems confronting the construction industry.** What are these problems and how do they relate to BIM? Flyvbjerg et al. (2018) identified construction's stubborn problems to include frequent delays and budget overruns. They conclude that most large construction projects are completed at much higher costs than their initial estimates, and as less-deserving outcomes in terms of quality, utility, and ecological value. In addition, Love et al. (2016) identified design errors as another critical challenge in major construction projects. According to Egan (1998), construction costs are high, and this requires collaborative innovations through technologies. Signor et al. (2017), Myrna and Charles (2012), and Bowen et al. (2012) have provided reasonable explanations regarding how corruption adds to the cost factors of construction. Evidence from the Arcadis (2019) report also suggests protracted disputation is another significant factor. Whilst Eastman et al. (2011), Barnes (2019), and Woodley (2019) claim that BIM reduces the propensity for contractual disputes, Hsu et al. (2015), Olatunji and Akanmu (2015), and Alwash et al. (2017) affirm that BIM's propensity for disputes is not different to traditional project environments unless BIM becomes more certain.

   Many studies have looked into the potentiality of BIM to reduce construction costs and duration and improve project safety and stakeholder satisfaction – for example, see

Bryde et al. (2013) and Cheng et al. (2015). Akanmu et al. (2016) propose an autonomous system that integrates BIM and radio frequency identification (RFID) technology to track on-site activities in real-time and to optimize remote monitoring such that project controllers are able to foresee delays and prevent them. A convenient conclusion from extant studies is that technologies can facilitate a harvest of robust information and can improve decision-making. However, they do not negate the cost burdens of amorality and complexity, as well as the cost of labour, material, safety, and project quality. Whereas BIM could help towards achieving these, its potentiality depends on target legal objectives and how such objectives are achieved. In addition, whereas certain potentialities are credited to BIM, they are meaningless legally if project contracts are not constructed around BIM deliverables as intended. This would include some clear articulation of the contract intentions of virtual models, active agents, digital scripts, hyper-models, and simulation iteration. The challenge, however, is that contract administrators of construction projects have struggled with their transitioning to BIM, from non-BIM deliverables such as 2D drawing and analogous planning tools to paper-based authoring and authorizations. It is important for legal instruments to set appropriate contexts for target performance objectives such that contractual constructs do not appear blunt and purposeless – for example, BIM does not reduce the cost of materials and labour if the client's requirements remain ambiguous and erratic. Thus, it is difficult to conclude from an informed legal perspective that BIM proffers conclusive solutions to the traditional challenges of construction project delivery; one way to facilitate the impact of BIM is to ensure BIM deliverables are trained to legal instruments and vice versa.

3. **BIM projects deliver better outcomes than where BIM was not used.**
   Examples presented in the works of Aranda-Mena et al. (2009), Azhar (2011), and Love et al. (2014) explain the benefits of developing projects with BIM, and how adopters can realize such benefits. However, according to Holzer (2007), the benefits associated with BIM do not exist in isolation; instead they are primed on speculative conditions. The conclusive empirical evidence regarding how BIM works in its best conditions, or a part thereof, is still incipient. A single most significant drawback in this is that instruments of the law are not designed to service deliverables that are academic in nature. Contracts must be definite, firm, provable, and devoid of deniability and must enforce the fault lines of risk delineations. For greater clarity – if BIM is meant to deliver outcomes that are definitely superior to non-BIM projects, how much should the difference between BIM project and non-BIM projects be before the superiority between them becomes enforceable legally? For greater clarity, the acceptance value of BIM will improve if stakeholders are able to differentiate the benefits of their investment decisions in BIM and are able to enforce them contractually. Where the promises around BIM deliverables are not clearly enforceable in the eyes of the law, it will be impossible to prove the benefits of BIM deployment using conclusive empirical evidence. In addition, it is logical to also ask: would BIM diminish the probable expectation that every non-BIM project would not perform to the level of BIM? Where a non-BIM project delivers same outcomes as a BIM project, should there be compensation for an exceptional outcome? If so, by how much? What are the characteristic attributes of BIM that make it superior to the conventional process and are impossible for traditional processes to achieve? Who pays for this superiority, and how? If BIM delivers outcomes that are less desirable than the outcomes of non-BIM projects, does that make BIM a failure? In sum, does BIM guarantee exceptional project outcomes? Where this is not possible absolutely, are there

contractual templates that enforce compensations and alignment? If not, should one be created? How, why, and how not?

4. **Policy chains geared towards forced BIM adoption propel systemic benefits to project economics.** According to Succar and Kassem (2015), BIM can be mandated by authorities across their jurisdictions. Thayaparan (2012), Dainty et al. (2017), Ho and Rajabifard (2016), Travaglini et al. (2014), Papadonikolaki (2017), Edirisinghe and London (2015), and Wortmann et al. (2016) have discussed the countries where the adoption and implementation of BIM in some categories of public projects have been mandated. It is like a race to determine which country mandates BIM deployment earliest; where a country has not made a categorical statement in favour of BIM advancement, they seem to risk their membership of BIM-policy elitists. However, evidence narrated by Edirisinghe and London (2015) is such that market forces are rather more potent in facilitating BIM's penetration than conferring an advantage to BIM through a subtle force. Though BIM is mandated in North America, Edirisinghe and London found a large portion of the industry is yet to comply.

There are few dimensions to market reaction to BIM. According to Olatunji (2014), it is important for BIM to prove its worth to the market. If a mandate is at all necessary, such action must support other autonomous tools that advance innovation in the construction industry – examples of these include dynamic modelling and remote sensing tools and exoskeletons, amongst others (see Akanmu et al., 2016; 2020). In addition, BIM adoption and implementation are measurable – see a set of metrics developed by Succar et al. (2012). Arguably, adoption and implementation planning are as important as compliance enforcement. Such planning will delineate transitions from one level of adoption to another, as well as respective expectations, deliverables, and resource requirements. Not least important however, contractual enforceability is not yet clear. Beyond the mandating, the industry needs incentives and structural support for its transition. According to Olatunji (2011a; 2019), this requires training and an appropriate sensitization of market drivers. Provisions for bifurcations such as caveats for dummy expectations should be integrated into BIM contracts such that where project outcomes are not as precise as promised they are tolerated in the form of zero vision and should add to the embodiment of the industry's transition from one level of BIM capability to another.

5. **BIM is key to the commercial reality of modern construction education and professional practice.** A study by Olatunji (2019) explains this axiom. Education providers seem to think they are inadequate without BIM content and ethos. The centrality of Olatunji's study is that BIM is not meant to be a replacement nor a substitute for existing education. However, contents that are hitherto essential must now give way to BIM. This is excusable in the context of innovation. However, a typical drawback in this is that BIM does not have to assume superiority over traditional contents; that is, students who are exceptionally capable in BIM cannot be less competent in non-BIM contents. Achieving a balance of the two paradigms is not easy to achieve. Where the industry struggles to improve BIM adoption, graduates are at the risk of seeing the industry as not being capable or ready to service their technical capability. In addition, BIM theorists seem to assume that the future of the construction industry is only in the hands of BIM and BIM only! The reference to the "future" in this is in the context of the "present", and this began about one and half decades ago! (see Ballesty et al. [2007]; Holzer [2007]) Again, at the risk of self-repetition, being positive about technological innovation is not entirely inappropriate. However, the misinformation in this is whether

the industry is ready for the future promised through BIM, though a few decades late. Commensurate constructs of the law will help demystify this. For example, contract provisos must identify specific landscapes of BIM deliverables across the many sub-disciplines in the project development process. In addition, they must articulate responsibility and limitations of sub-disciplines under specific conditions of co-creation, as well as appropriate understanding and compensation for professional services. For greater clarity, members of the project team seldom use BIM the same way and to the same degree of complexity across different project development stages – see Olatunji and Akanmu (2014). How does this affect how project teams are remunerated? In addition, should a BIM-able graduate earn the same remuneration as a BIM-unable graduate? Should BIM professional services attract the same scale of pay as traditional services? Answers to these questions are not conclusively evident in the modern construction literature, and this draws away from the established basic tenets of a contract, a consideration that is appropriate towards an exchange of services, being a fundamental condition that premises the validity of a contract.

6. **The effectuality of virtual models and consequential disclaimers** is another important example. BIM projects are still driven largely by the traditionality of construction contracts, though the premises upon which BIM deliverables are based are all enshrined in virtual models. For greater clarity, construction project contracts are still driven by drawings, rather than models; by analogous plans, rather than simulation models; by bills of quantities, rather than industry foundation classes; and by conventional cost plans and conventional estimates, rather than by hyper-models, model scripts, or schemas. BIM deliverables are desirable; however, they are unlikely to have conclusive influence unless they possess definitive contractual values. For this to happen, deliverables of digital models must be measurable with certainty. For greater clarity, an imaginative virtual process model must be enforceable on its promises in ways that are not indemnifiable by disclaimers. Modellers should be responsible for their promises, including the credits and the liabilities thereof. In an analysis by Olatunji and Akanmu (2015), disclaimers cannot excuse modellers of the consequences of the outcomes of their artefacts. When they mislead, misinform, or misguide, they should be enforceable in contract liabilities. A safe conclusion from these is that until virtual models are enforceable in construction contracts, their legal and investment worth needs to be proven. For this not to become a lacuna that will trigger disputations, it is important for construction law scholars to examine the legal efficacy of virtuality and virtualization with a view to establishing clear boundaries around their objectives and the legal ramifications of their outcomes.

7. **BIM metadata is often (mis)taken as the absolute representation of actual project artefact.** This is consistent in the definitions of BIM by Aranda-Mena et al. (2008); Love et al. (2014); Olatunji (2012); Penttilä (2006); and Thayaparan (2012). However, contrary arguments have emerged; model artefacts are shaped by the software developer's imaginative data. According to Chien and Yeh (2012b), Olatunji (2013), and Olatunji and Akanmu (2014; 2015), designers are not able to innovate beyond the remit permissible by their authoring tools. In addition, model data are not driven by actual project data, rather by data that are pre-loaded in authoring tools which may not reflect the actuality of any particular project. For example, most projects experience a significant amount of variability due to variations in siteworks, groundworks, and sub-structural work, whereas most modelling tools have limited data on soil types and geotechnical variances. In addition, according to Amor et al. (2007), BIM adoption does not

mean an absolute elimination of design errors. Models do have errors too; when they happen, they can be more misleading than the problems accurate models are meant to prevent (Love et al., 2011a, b; Love and Smith, 2016).

Legal implications of professional errors are well documented. They are not excusable in a modelling environment regardless of the good intention of using BIM. What is important is the ability of the law community to assist the construction practice community in understanding its limitations in a virtual environment, and to work out ways to protect all stakeholders. An effective BIM contract will require this to succeed.

8. **Synchronization of traditional contract instruments in BIM comes with limitations**. BIM is collaborative, whereas collaboration is not the hallmark of traditional construction contracts (Kvan, 2000; Ashcraft, 2008). Non-collaborative contracts cannot drive the expectations of BIM. Whilst attempts have been made by researchers to explore what collaboration means to BIM and construction projects, outcomes have yet to translate into an established tool, robust enough to facilitate a crisis-free BIM contract – see Alwash et al. (2017), Chong et al. (2017), and Olatunji (2016).. But what should the construct of a BIM contract look like? This is the billion-dollar question at the heart of BIM-construction law scholarship today. A section that outlines the practical implications of the current knowledge gaps in BIM literature has been created in this chapter to provide some guidance – the reader is referred to the section titled "Implications of knowledge gaps in construction legal studies and BIM".

9. **Innovation; authoring software and BIM, and bounded innovation on a software platform**. The beauty of object-oriented modelling is the ability to communicate complexity. Authors agree this requires a clear understanding of innovation (Gero and Kannengiesser, 2004). However, innovation is a question of freewill – the ability to self-express seamlessly and to communicate rigour without suppression. Some researchers have questioned whether geometric modelling or parametric modelling facilitates design innovation, as though one is better than the other – see Chien and Yeh (2012a); Yu et al. (2014). Either way, a notable constraint in a virtual modelling environment is in the authoring platform; model authors are not able to innovate beyond the platform they operate in. This is not the only limitation; the ownership and the warehousing of their product is embedded in the copyright of their authoring platform. The right they have to create, transmit, market, use, and apply their work variously on the software platform is neither absolute nor exclusive to them (such rights can be withdrawn, disabled, restricted, ceased, or taken away from them in different ways and forms without an apology or prior notice or revocation). Model authors are not able to transmit their shared ownership of or transmit their access authority to software platforms to their clients. For greater clarity, modellers require licencing authorization to access and create models. Their access can neither be transmitted to their clients nor remain permanent for them to retain their creation or represent their work as long as they desire. A bug on the software could disrupt the integrity of a model – they are neither able to prevent nor control this, unless through substitution. The security of the model could be compromised because of sinister actions or a compromise from sources other than the modeller. Modellers can only guarantee their own access; they cannot shape or facilitate how their clients use their work and where they deploy it. Not least important, do they truly own their work, or does whatever they create on a software belong only to the software? If model ownership and intellectual properties are valid questions in the construction of an adequate BIM contract, construction lawyers must find a rounded answer to this question.

10. **BIM, authoring authority, and professional boundary spanning**. Researchers have discussed the importance of collaboration variously (Bainbridge et al., 2010; Thomson and Perry, 2006; Wood and Gray, 1991). For BIM and traditional construction, it means disciplinary boundaries are weakened and people are able to work and co-produce across and beyond their disciplines (Olatunji et al., 2010). This is one phenomenon that has continued to grow in popularity with BIM; people with functional or theoretical knowledge of BIM seem to assume the knowledge that belongs to other spaces. They have done this in the name of boundary spanning rather than exercising reasonable professional judgement. Apparently, because construction audience are not able to tell the difference between *who says what* and *who has the authority to say what*, BIM has been espoused wrongly – many aspects of these have been dealt with in this section. Such misinformation should not prevent construction lawyers from distinguishing speculation from reality. To simplify further, who can be a modeller or a model manager? What professional qualifications and liabilities define their roles? Can a modeller or a model manager who is not licenced to authorize a design issue a model with absolute authority? If every author cannot be an authoring authority, what manner or level of authority should reside in boundary spanning? The answers to these are important for the legal validity of model authoring. They also inform professionalism, that is, whether owners of disciplinary knowledge have reasons not to see boundary spanning as invasive and antithetical. They may choose to resist BIM for this reason. Not least important, it is important to clarify whether advice issued by boundary spanners could be taken as valid and conclusive.

The additionality of these speculations is that they have far-reaching implications for legal constructs. This is because they shape professional and scholarly misinformation on the subject. Until there is clarity around them, there are significant limitations regarding the objective reality around the intents of legal education in relation to BIM. In summary, it is appropriate and convenient to conclude that BIM is a positive addition to the construction industry. The inclination to adopt and implement BIM protocols in projects must be as though BIM capacities are to specific levels, and that the capabilities often credited to BIM in literature as still incipient. Whilst it is unprofitable to rule out the possibility of achieving the positive attributes expected of BIM, the law must take a different perspective. The law is meant to protect the good but with the strength to anticipate the risks of negativity and to possess the capacity to attenuate their effects. BIM may not have had many disputations historically; in part, the non-contractual nature of BIM deliverables explains this. In an era when BIM rules construction professional practice, BIM disputes do not have to be protracted and heavily damaging if BIM dispute scenarios are understood and are anticipated as appropriate. Where BIM deliverables are espoused in relation to the management of costs, durations, and contract relationships, it is important to consider digital information management in contexts that are specific to discipline applications of BIM, including the broad areas of quantity surveying, management, and administration of construction projects, costs, contracts, and law.

Apparently, following rife speculation about BIM, certain myths have permeated the construction literature, which should be debunked so they do not draw integrity away from BIM education. They need to be understood appropriately such that they are not confused with the alternative realities about them. Examples of such myths are given below:

1. *BIM facilitates accurate estimates* (Choi et al., 2015; Kehily and Underwood, 2017). The alternative reality is that BIM metadata is not structured to meet the requirement of any

specific discipline, nor any particular estimating standard (Amor et al., 2007). Whereas model data are structured as product models, estimates are based on process methodologies, requiring allowances for considerations beyond the outturn product.

2. *Traditional contracts are confrontational. BIM facilitates dispute avoidance* (Aidibi, 2016). This popular perception is difficult to prove as there are no ways of measuring BIM's ability to prevent disputes when BIM deliverables are hardly enforceable in construction contracts.

3. *BIM's robust data repository elicits openness, holding just about all the information required for a project to succeed* (Ismail et al., 2016). However, evidence by Chu et al. (2018) suggests this is of limited help as model data can be excessive and often do not offer clear directions regarding their best impacts.

4. *BIM requires collaboration and enforced policies to succeed* (Aranda-Mena et al., 2008). However, there is limited evidence to conclude that traditional methods will not achieve exceptional outcomes if supported with the instrumentality of collaboration and forced macro-policies.

5. *BIM is akin to integrated project delivery; it facilitates multidisciplinary integration; it is a repository for life-cycle data* – see Glick and Guggemos (2009). According to Amor and Faraj (2001), this is often misconceived. Model development protocols are still fragmented; interoperability and value sharing issues are rife.

Implications of knowledge gaps in construction legal studies and BIM research

The single most important challenge of BIM in construction law is in the need to identify the challenges of BIM's contract language design. Apparently, extant contract standards still grapple with this. An appropriate legal instrument in BIM needs to address the missing link between BIM's capacity for co-production as against traditional contracts' fragmentation bias. The right instrument for a BIM contract will develop an articulation of contract condition and support infrastructure, and review them as to whether they support BIM's actual attributes or not, now or whether there is a future for such. In particular, such an instrument should have a place for the following:

- **Shared ownership (collaboration):** digital models are co-produced virtual artefacts intended to replicate real-life potency, the emergence of which involves the participation of multiple disciplines and different contract parties. The outcome of this is a unitary artefact – a repository of design or modelling data, as well as robust management and lifecycle data. Researchers have often sought to know who the true owner of a shared digital artefact is – for example, see Bloomberg et al. (2012), Olatunji (2011b) and Wong et al. (2014). Some have argued that the ultimate owner of a digital artefact is the client because they pay for the model-authors' services. However, clients, the supposed owners of their digital artefact, have no authority to use the model they own for reasons beyond the intentions permitted by their designers, unless they are so authorized. Some studies have also argued that model authors do not own their contributions. This is because all the data they use and the platform on which they express their work are owned by a different party, to whom they are licensees. A convenient neutral point is to assume parties who co-contribute to a unitary artefact should own the part they contribute, and such ownership should be absolute except surrendered wilfully. The practicality of this is often questionable; rights buyout should specify the elements of the rights under exchange, and rights and events or deliverables preceding and beyond a buyout should be specified, to which each party must understand and agree.

- **Intellectual properties (integration, interoperability, and platform issues):** there are different aspects to the property and the propertization of the intellectual assets underlying a virtual artefact – propertization being the process of establishing a property (see Radin, 2006 and Olatunji, 2013). Model authors co-create a joint digital artefact, requiring seamless alignment and integration of platforms and processes. Whilst they own their contributions, the attribution of the outcome of their co-creation would have both soft and hard boundaries. To clarify – suppose a project team involving the contractor, clients' cost consultants, designers, construction law team and other stakeholders choose to integrate their contributions into a digital model. Each member of the team will provide data – for example, the contractor, some clarity on construction methodology and risk considerations; the cost consultant, some objective consideration around cost and value dynamics; the designers, model data of the considered designed outcome; the construction lawyer, the framing of contracts and the execution of same. When these disciplinary inputs converge, there are cross-boundary effects that members of the project team could claim as their rights. For example, contractors and cost consultants would converge their inputs to develop the 5D model; both parties and the client own the right to the outcome. The gap is: at what point does such an item becomes property of a party, and what are the processes of establishing such rights when deliverables are integrated? Thus, legal constructs around these should be established, and such constructs must identify liabilities and compensations as appropriate. Similarly, where properties and propertizations of intellectual rights are shared, determinate, or transferable, it is important that contract constructs can specify the proportionality of liabilities and compensations where necessary.

- **Data security (transferability):** model data are often not owned by modellers. It is impossible for modellers to verify their sources and the integrity thereof, all the time. In addition, such data are for an assumed reality, to which an actual reality may have no indemnity. The sense of liability in this is whether uncertainty around the integrity of model data is excusable – for example, with disclaimer clauses. Furthermore, an integrated platform means multiple access points, requiring multiple layers of security consciousness. This does not only include transmission of vulnerabilities but also whether open-source modelling data or unfettered access to data repositories by a wide range of parties does not increase models' exposure to vulnerabilities. For example, suppose a safe house or military facility is modelled, and all contributors have access to the model. They are able to keep their version of the model; if the model becomes compromised as a result of distributed access, will it be possible to trace the source of the compromise and who is liable? If the compromise is not about data security but about the integrity of the data that shaped professional judgements made by others, what remedies are available to the end-users or other team members who suffer a loss as a result of the compromise?

- **Buyouts and valuation of services (what is the right value/consideration for BIM services?):** BIM is espoused to revolutionize traditional professional processes within the construction industry (Boon, 2009; Luciani, 2008; Succar, 2009). According to Olatunji et al. (2010), this means objects are used in place of line, and estimators will be able to export modelling data for costing – in the form of information management rather than information creation. The resource and skill requirements of BIM are different to traditional practices (Olatunji, 2012; Sher et al., 2009). Whilst traditional professional services have been evaluated and established in line with specific scales of fees, no such tool is yet popular with BIM. An example of a knowledge gap is in how professional services should be remunerated in BIM. What exactly does the client pay for in a buyout, and what constitutes a fair buy? What limits and liabilities are attributable to individual efforts and co-creation of digital artefacts?

- **Contractual and non-contractual deliverables (is a BIM contract for all deliverables? What are the exclusions?):** certain elements of simulation models, augmented deliverables, and metadata are not contractual definitively. Whilst some elements of design models may be for demonstration only, they relate to actual deliverables that are contractual. If the boundaries between contractual and non-contractual elements are unclear, this could provide a fertile ground for disputes. For example, whilst project duration is contractual, traditional contracts may be silent on simulation of alternative work methodologies and augmented resourcing. Where a BIM contract does not recognize this, conflicts could arise.

Conclusions

BIM is a positive addition to the construction industry. However, legal constructs that support its deliverables are still incipient. Much of the knowledge gaps in BIM's legal deliverables are in theoretical biases around what BIM truly is and speculations about BIM deliverables. It is important that BIM up-takers understand what these are and how they apply to construction. As argued in the chapter, the legal constructs of BIM must be clear about shared liabilities in collaborative environments. This involves ingraining clarity as to the objectives of virtual modelling and setting up appropriate instrumentalities that ensure that expectations and outcomes from BIM are measurable and justiciable. This goal is achievable; the main constraint is that construction law scholars must first understand the implications of theoretical misinformation about BIM. This must happen in consideration of other autonomous technologies that work like, and work with, BIM.

The following questions will assist readers to apply this chapter for the purposes of teaching and learning:

1. What are the challenges of current BIM scholarship to contract administration?
2. What are the benefits of BIM to project development?
3. Explain the implications of shared liabilities in collaborative project platforms.

References

Aibinu, A., and Venkatesh, S. (2014). Status of BIM adoption and the BIM experience of cost consultants in Australia. *Journal of Professional Issues in Engineering Education and Practice* 140, 04013021.

Aidibi, H. I. (2016). Studying the effect of BIM on construction conflicts and disputes using agent-based modeling, America University of Beirut, Beirut, Lebanon.

Akanmu, A., Olatunji, O. A., Love, P. E. D., Nguyen, D., and Matthews, J. (2016). Auto-generated site layout: An integrated approach to real-time sensing of temporary facilities in infrastructure projects. *Structure and Infrastructure Engineering* 12, 1243–1255.

Akanmu, A., Olayiwola, J., and Olatunji, O. A. (2020). Musculoskeletal disorders within the carpentry trade: Analysis of timber flooring subtasks. *Engineering, Construction and Architectural Management*. Accepted, https://doi.org/10.1108/ECAM-08-2019-0402.

Alwash, A., Love, P. E., and Olatunji, O. (2017). Impact and remedy of legal uncertainties in building information modeling. *Journal of Legal Affairs and Dispute Resolution in Engineering and Construction* 9, 04517005.

Amor, R., and Faraj, I. (2001). Misconceptions about integrated project databases. *Journal of Information Technology in Construction* 6, 57–68.

Amor, R., Jiang, Y., and Chen, X. (2007). BIM in 2007: Are we there yet? *In* CIB International Conference on applications of IT in Construction (CIB-W78), available at http://www.cs.auckland .ac.nz/~trebor/papers/AMOR07B.pdf, Auckland, New Zealand.

Aouad, G., Lee, A., and Wu, S. (2006). *Constructing the Future: nD Modelling*, Taylor & Francis.

Aranda-Mena, G., Crawford, J., Chevez, A., and Froese, T. (2009). Building information modelling demystified: Does it make business sense to adopt BIM? *International Journal of Managing Projects in Business* 2, 419–434.

Aranda-Mena, G., Succar, B., Chevez, A., Crawford, J., and Wakefield, R. (2008). BIM National guidelines and case studies. *In* Cooperative Research Centres (CRC) for Construction Innovation (2007–02-EP), Melbourne, Australia.

Arcadis (2019). *Global Construction Disputes 2019: Laying the Foundation for Success*. Arcadis NV, Zuidas, Amsterdam, the Netherlands.

Ashcraft, H. W. (2008). Building information modeling: A framework for collaboration. *Construction Lawyer* 28, 1–14.

Azhar, S. (2011). Building information modeling (BIM): Trends, benefits, risks, and challenges for the AEC industry. *Leadership and Management in Engineering* 11, 241–252.

Azhar, S., Khalfan, M., and Maqsood, T. (2012). Building information modelling (BIM): Now and beyond. *Construction Economics and Building* 12, 15–28.

Bainbridge, L., Nasmith, L., Orchard, C., and Wood, V. (2010). Competencies for interprofessional collaboration. *Journal of Physical Therapy Education* 24, 6–11.

Ballesty, S., Mitchell, J., Drogemuller, R., Schevers, H., Linning, C., Singh, G., and Marchant, D. (2007). *Adopting BIM for Facilities Management: Solutions for Managing the Sydney Opera House*. Cooperative Research Centre (CRC) for Construction Innovation, Brisbane, Australia.

Barnes, P. (2019). *BIM in Principle and in Practice*, Institution of Civil Engineers Publishing.

Bensalah, M., Elouadi, A., and Mharzi, H. (2017). Optimization of cost of a tram through the integration of BIM: A theoretical analysis. *International Journal of Mechanical and Production Engineering* 5, 138–142.

Bill Boetcker (n.d.). Working notes of Bill Boetcker, containing outlines, chronology, quotes, and events of the life of William J. H. Boetcker. *In*: *The William J. H. Boetcker Manuscript Collection*, pp. 24 boxes, available at https://princetonseminaryarchives.libraryhost.com/repositories/2/archival_object s/133581 Accessed May 28, 2020. Special Collections of the Princeton Theological Seminary Library.

Bloomberg, M. R., Burney, D., and Resnick, D. (2012). *BIM Guidelines*, New York City Department of Design and Construction, 1–57.

Boon, J. (2009). Preparing for the BIM revolution. *In*: 13th Pacific Association of Quantity Surveyors Congress, Kuala Lumpur, Malaysia, pp. 33–40. Pacific Association of Quantity Surveyors.

Bowen, P. A., Edwards, P. J., and Cattell, K. (2012). Corruption in the South African construction industry: A thematic analysis of verbatim comments from survey participants. *Construction Management and Economics* 30, 885–901.

Bryde, D., Broquetas, M., and Volm, J. M. (2013). The project benefits of building information modelling (BIM). *International Journal of Project Management* 31, 971–980.

Cheng, J. C., Won, J., and Das, M. (2015). Construction and demolition waste management using BIM technology. *In*: 23rd Annual Conference of the International Group for Lean Construction, Perth, Australia, pp. 381–390.

Chien, S.-F., and Yeh, Y.-T. (2012a). On creativity and parametric design: A preliminary study of designers' behaviour when employing parametric design tools. *In*: Proceedings of the 30th International Conference on Education and Research in Computer Aided Architectural Design in Europe, Prague (eCAADe), Czech Republic.

Chien, S.-F., and Yeh, Y.-T. (2012b). On creativity and parametric design: A preliminary study of designers' behaviour when employing parametric design tools. *eCAADe 30: Digital Aids to Design Creativity* 1, 245–253.

Choi, J., Kim, H., and Kim, I. (2015). Open BIM-based quantity take-off system for schematic estimation of building frame in early design stage. *Journal of Computational Design and Engineering* 2, 16–25.

Chong, H.-Y., Fan, S.-L., Sutrisna, M., Hsieh, S.-H., and Tsai, C.-M. (2017). Preliminary contractual framework for BIM-enabled projects. *Journal of Construction Engineering and Management* 143, 04017025.

Chu, M., Matthews, J., and Love, P. E. D. (2018). Integrating mobile building information modelling and augmented reality systems: An experimental study. *Automation in Construction* 85, 305–316.

Dainty, A., Leiringer, R., Fernie, S., and Harty, C. (2017). BIM and the small construction firm: A critical perspective. *Building Research & Information* 45, 696–709.

Eastman, C., Liston, K., Sacks, R., and Teicholz, P. (2011). *A BIM Handbook: A Guide to Building Information Modeling for Owners, Managers, Designers, Engineers and Contractors*, 2nd Ed. Wiley, New York.

Edirisinghe, R., and London, K. (2015). Comparative analysis of international and national level BIM standardization efforts and BIM adoption. *In:* Proceedings of the 32nd CIB W78 Conference, Eindhoven, The Netherlands, pp. 149–158.

Egan, J. (1998). Rethinking construction. *In: Department of the Environment Transport and the Regions*, HMSO, London.

Flyvbjerg, B., Ansar, A., Budzier, A., Buhl, S., Cantarelli, C., Garbuio, M., Glenting, C., Holm, M. S., Lovallo, D., Lunn, D., Molin, E., Rønnest, A., Stewart, A., and van Wee, B. (2018). Five things you should know about cost overrun. *Transportation Research, Part A: Policy and Practice* 118, 174–190.

Gero, J. S., and Kannengiesser, U. (2004). The situated function–behaviour–structure framework. *Design Studies* 25, 373–391.

Glick, S., and Guggemos, A. (2009). IPD and BIM: Benefits and opportunities for regulatory agencies. *In:* Proceedings of the 45th ASC National Conference, Gainesville, Florida, April, Vol. 2. http://asc pro0.ascweb.org/archives/cd/2009/paper/CPGT172002009.pdf.

Ho, S., and Rajabifard, A. (2016). Towards 3D-enabled urban land administration: Strategic lessons from the BIM initiative in Singapore. *Land Use Policy* 57, 1–10.

Holzer, D. (2007). Are you talking to me? BIM alone is not the answer. *In:* Association of Architecture Schools Australasia Conference, University of Technology Sydney, Australia.

Hope, G. (2012). Contractors must 'adapt or die' in deploying BIM. *Construction Week Online*, available at https ://www.constructionweekonline.com/article-16933-contractors-must-adapt-or-die-in-deploying-bim

Hsu, K.-M., Hsieh, T.-Y., and Chen, J.-H. (2015). Legal risks incurred under the application of BIM in Taiwan. *Proceedings of the Institution of Civil Engineers: Forensic Engineering* 168, 127–133.

Ismail, N. A. A. B., Drogemuller, R., Beazley, S., and Owen, R. (2016). A review of BIM capabilities for quantity surveying practice. *In:* Proceedings of the 4th International Building Control Conference, Kuala Lumpur, Malaysia, Vol. 66, pp. 1–7. EDP Sciences, France.

Kagioglou, M., Cooper, R., and Aouad, G. (2001). Performance management in construction: A conceptual framework. *Construction Management and Economics* 19, 85–95.

Kehily, D., and Underwood, J. (2017). Embedding life cycle costing in 5D BIM. *Journal of Information Technology in Construction* 22, 145–167.

Klaschka, R. (2019). *BIM in Small Practices: Illustrated Case Studies*, Routledge.

Ku, K., and Taiebat, M. (2011). BIM experiences and expectations: The constructors' perspective. *International Journal of Construction Education and Research* 7, 175–197

Kvan, T. (2000). Collaborative design: What is it? *Automation in Construction* 9, 409–415.

Love, P., Edwards, D. J., and Han, S. (2011a). Bad apple theory of human error and building information modeling: A systemic model for BIM implementation. A paper presented at the 28th ISARC, Seoul, Korea.

Love, P. D., Edwards, D., Han, S., and Goh, Y. (2011b). Design error reduction: Toward the effective utilization of building information modeling. *Research in Engineering Design* 22, 173–187.

Love, P. E., Matthews, J., Simpson, I., Hill, A., and Olatunji, O. A. (2014). A benefits realization management building information modeling framework for asset owners. *Automation in Construction* 37, 1–10.

Love, P. E., and Smith, J. (2016). Error management: Implications for construction. *Construction Innovation* 16, 418–424.

Love, P. E. D., Zhou, J., Matthews, J., and Edwards, D. (2016). Moving beyond CAD to an object-oriented approach for electrical control and instrumentation systems. *Advances in Engineering Software* 99, 9–17.

Luciani, P. (2008). Is a revolution about to take place in facility management procurement? *In European FM Insight*, pp. 1–3. EuroFM.

Manning, R., and Messner, J. (2008). Case studies in BIM implementation for programming of healthcare facilities. *Electronic Journal of Information Technology in Construction* 13, 446–457.

Meng, X. (2012). The effect of relationship management on project performance in construction. *International Journal of Project Management* 30, 188–198.

Moon, H. J., Choi, M. S., Kim, S. K., and Ryu, S. H. (2011). Case studies for the evaluation of interoperability between a BIM based architectural model and building performance analysis programs. *In:* Proceedings of 12th Conference of International Building Performance Simulation Association, Sydney, 14–16 November, pp. 1521–1526.

Myrna, A., and Charles, F. I. (2012). *Review of the World Bank's Procurement Policies and Procedures: Analysis of World Bank Completed Cases of Fraud and Corruption from the Perspective of Procurement*, The World Bank Group, Washington, DC.

Olatunji, O. (2016). Constructing dispute scenarios in building information modeling. *Journal of Legal Affairs and Dispute Resolution in Engineering and Construction* 8, C4515001.

Olatunji, O. A. (2011a). Modelling the costs of corporate implementation of building information modelling. *Journal of Financial Management of Property and Construction* 16, 211–231.

Olatunji, O. A. (2011b). A preliminary review on the legal implications of BIM and model ownership. *Journal of Information Technology in Construction* 16, 687–696.

Olatunji, O. A. (2012). The impact of building information modelling on estimating practice: Analysis of perspectives from four organizational business models. Ph.D. Dissertation, University of Newcastle, Newcastle, Australia.

Olatunji, O. A. (2013). Building information modelling and intellectual propertization: A revolutionary nirvana or a disillusionment? *In: eBook of the Society of Construction Law*, pp. 77–85. Society of Construction Law, Australia.

Olatunji, O. A. (2014). Views on building information modelling, procurement and contract management. *Proceedings of the Institution of Civil Engineers: Management, Procurement and Law*, 167(3): 117–126.

Olatunji, O. A. (2019). Promoting student commitment to BIM in construction education. *Engineering, Construction and Architectural Management* 26, 1240–1260.

Olatunji, O. A., and Akanmu, A. (2014). Latent variables in multidisciplinary team collaboration. *In:* International Conference on Construction and Real Estate. Harbin Institute of Technology, Kunming, China.

Olatunji, O. A., and Akanmu, A. (2015). BIM-FM and consequential loss: How consequential can design model be? *Built Environment Project and Asset Management* 5, 304–317.

Olatunji, O. A., Sher, W. D., and Gu, N. (2010). Building information modeling and quantity surveying practice: Whatever you thought, think again. *Emirate Journal of Engineering Research* 15, 67–70.

Olofsson, T., Lee, G., and Eastman, C. (2008). Editorial: Case studies of BIM in use. *IT in construction – Special Issue Case studies of BIM use* 13, 244–245.

Papadonikolaki, E. (2017). Grasping brutal and incremental BIM innovation through institutional logics. *In:* Proceedings of the 33rd Annual ARCOM Conference, 4-6 September 2017, Cambridge, UK, Vol. 33, pp. 54–63. Association of Researchers in Construction Management (ARCOM).

Penttilä, H. (2006). Describing the changes in architectural information technology to understand design complexity and free-form architectural expression. *Journal of Information Technology in Construction* 11, 395–408.

Race, S. (2019). *BIM Demystified*, Routledge, London.

Radin, M. J. (2006). A comment on information propertization and its legal milieu. *Cleveland State Law Review* 54, 23–40.

Sher, W., Sheratt, S., Williams, A., and Gameson, R. (2009). Heading into new virtual environments: What skills do design team members need? *Journal of Information Technology in Construction* 14, 17–29.

Signor, R., Love, P. E., Olatunji, O., Vallim, J. J., and Raupp, A. B. (2017). Collusive bidding in Brazilian infrastructure projects. *Proceedings of the Institution of Civil Engineers: Forensic Engineering* 170, 113–123.

Stewart, R. A. (2007). IT enhanced project information management in construction: Pathways to improved performance and strategic competitiveness. *Automation in Construction* 16, 511–517.

Succar, B. (2009). Building information modelling framework: A research and delivery foundation for industry stakeholders. *Automation in Construction* 18, 357–375.

Succar, B., and Kassem, M. (2015). Macro-BIM adoption: Conceptual structures. *Automation in Construction* 57, 64–79.

Succar, B., Sher, W., and Williams, A. (2012). Measuring BIM performance: Five metrics. *Architectural Engineering and Design Management* 8, 120–142.

Terrin, N., Schmid, C. H., and Lau, J. (2005). In an empirical evaluation of the funnel plot, researchers could not visually identify publication bias. *Journal of Clinical Epidemiology* 58, 894–901.

Thayaparan, G. (2012). *Building Information Modelling (BIM): Australian Perspectives and Adoption Trends,* Centre for Interdisciplinary Built Environment Research, University of Newcastle, Australia

Thomson, A. M., and Perry, J. L. (2006). Collaboration processes: Inside the black box. *Public Administration Review* 66, 20–36.

Travaglini, A., Radujković, M., and Mancini, M. (2014). Building information modelling (BIM) and project management: A stakeholder's perspective. *Organization, Technology & Management in Construction: An International Journal* 6, 1001–1008.

Twain, M. (2013). *Delphi Complete Works of Mark Twain (Illustrated),* Delphi Publishing Limited, Hastings, East Sussex, UK.

Volk, R., Stengel, J., and Schultmann, F. (2014). Building information modeling (BIM) for existing buildings: Literature review and future needs. *Automation in Construction* 38, 109–127.

Wang, X., Yung, P., Luo, H., and Truijens, M. (2014). An innovative method for project control in LNG project through 5D CAD: A case study. *Automation in Construction* 45, 126–135.

Wong, K. d., and Fan, Q. (2013). Building information modelling (BIM) for sustainable building design. *Facilities* 31, 138–157

Wong, J., Wang, X., Li, H., and Chan, G. (2014). A review of cloud-based BIM technology in the construction sector. *Journal of Information Technology in Construction* 19, 281–291.

Wood, D. J., and Gray, B. (1991). Toward a comprehensive theory of collaboration. *The Journal of Applied Behavioral Science* 27, 139–162.

Woodley, C. (2019). Will digitalisation end construction disputes? *Construction Research and Innovation* 10, 15–17.

Wortmann, A., Root, D., and Venkatachalam, S. (2016). Building information modelling (BIM) standards and specifications around the world and its applicability to the South African AEC sector: A critical review. *In:* Proceedings of the 1st International BIM Academic Forum (BAF) Conference, 13-15 September 2015. Glasgow Caledonian University, Scotland.

Wright, R. W. (1987). Allocating liability among multiple responsible causes: A principled defense of joint and several liability for actual harm and risk exposure. *University of California Davis Law Review* 21, 1141–1212.

Wright, R. W. (1992). The logic and fairness of joint and several liability. *Memphis State University Law Review* 23, 45–84.

Wu, W., and Issa, R. R. (2012). BIM-enabled building commissioning and handover. International Conference on Computing in Civil Engineering June 17-20, 2012, Clearwater Beach, Florida, United States, American Society of Civil Engineers , 237–244.

Xu, Y. G., and Qian, C. (2014). Lean cost analysis based on BIM modeling for construction project. *Applied Mechanics and Materials* 457–458, 1444–1447

Yu, R., Olatunji, O. A., and Akanmu, A. (2014). An ontology for analysing cognition in geometric and parametric design platforms: A review. *In:* International Conference on Construction and Real Estate Management, Kunming, China.

12 Interoperability and emerging smart technologies

Gökhan Demirdöğen, Zeynep Işık, Yusuf Arayici, and Hande Aladağ

Introduction

In the architecture, engineering, and construction (AEC) industry, construction projects are developed and conducted under highly complex and ambiguous conditions. Every activity, whether design and construction activities or operation and maintenance activities, needs different expertise. In other words, small construction projects are even beyond the capability of a single company. In this context, many stakeholders participate in the construction project processes, which are initiation, planning, execution of construction activities, monitoring and controlling, closure, operation and maintenance and demolition. Data produced during the construction project processes cannot be managed with any unique software tool (Ozturk, 2020). Thus, every stakeholder uses different technologies to maximize the profitability and efficiency of their activities.

The use of different information and communication technologies (ICT) by different stakeholders induces collaboration issues when managing business and processes. Interoperability, which arises from incompatibility with the reference models adopted by the various software applications, is one of the collaboration issues. Basically, interoperability does not only consider information systems or technology, but also covers business processes, employees and culture, and the management of external relationships (Grilo & Jardim-Goncalves, 2010). The ability of enhancement in collaboration is the most important feature of BIM (Oraee et al., 2019).

Building information modelling (BIM) has been developed as a result of the imitation of more stable and advanced sectors solutions such as the aeronautics industry to solve fragmentation and heterogeneity issues in the AEC industry. BIM incorporates a technology facet through which stakeholders interact with each other intensively to provide an exchange of information by using standards such as IFC, COBie, XML, IDM (Grilo & Jardim-Goncalves, 2010; Ozturk, 2020; Steel et al., 2012). During this process, a great deal of information, such as architectural, structural, heating ventilation and air-conditioning (HVAC), mechanical systems, and electrical systems information, is combined in a common BIM model. Therefore, this does not only necessitate information readability on the part of every stakeholder but also necessitates performing further analysis such as artificial lighting analysis, daylight analysis, and energy analysis (Steel et al., 2012; Taha et al., 2020).

Interoperability allows the sharing of information from the BIM model and helps to manipulate them within different software environments from different perspectives in a structured and common way. Information transfer from one software programme to another must take place without information loss and it must be readable from another software via different protocols and standards such as IFC and COBie (Ozturk, 2020). Goulding et al. (2014) defined interoperability as "incompatibility between inter-products and software

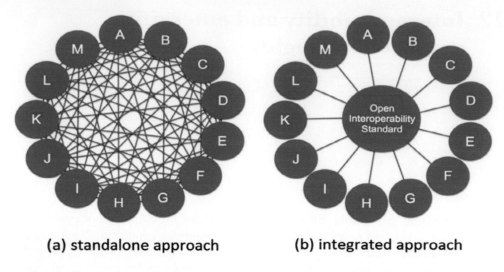

(a) standalone approach   (b) integrated approach

Figure 12.1 The ideal interoperability environment for the AEC industry (Arayici et al., 2018).

applications". Therefore, achieving full interoperability between BIM tools is a challenging task in the AEC industry.

To cope with interoperability issues, some efforts, such as the development of the Industry Foundation Class, have been made. However, the National Institute of Standards (NIST) has reported that issues originating from interoperability, such as information losses, cost approximately $16 billion per year (Gallaher et al., 2004). The solution is smooth interoperability in which clients, architects, engineers, financiers, builders, subcontractors, local authorities, and consultants can collaborate without information loss (Figure 12.1).

Although there are already many interoperability research efforts in BIM, interoperability remains problematic and focuses on 3D coordination. Furthermore, increasing BIM usage in the AEC industry brings more interoperability needs. Therefore, an integrated approach, which "uses a translator tool to convert the proprietary format into open data readable by any software", needs to be followed (Arayici et al., 2018; Oraee et al., 2019). Even though there are some standards such as IFC and IDM, there are some issues in the practical implementation of these standards, such as performance-based design (Arayici et al., 2018). Moreover, interoperability between BIM and asset management software is not at the intended level (Farghaly et al., 2018). Therefore, an introduction to the basic concepts, challenges, and research efforts on interoperability for graduates and educators will help to demonstrate the importance of interoperability for collaborative project management and building life cycle management with BIM. Additionally, readers will find in this chapter a discussion of the development of interoperability efforts and the common data exchange formats which are used in the AEC industry. This chapter helps graduates be better prepared to transfer and use recently developed (Lidar, Blockchain etc.) concepts, which can be used to solve interoperability issues, into the AEC industry by laying a bridge between academia and industry.

Current state of interoperability in BIM from the past to today

Interoperability issues in integrated project delivery processes emerged in the 1970s in aerospace manufacturing technologies due to the fact that every stakeholder was working on different CAD systems (Liao et al., 2017; Szeleczki, 2019). The same issue is starting to be increasingly evident in AEC in the usage of CAD systems such as Intergraph for data transfer between design and bills of material. DXF and IGES were commonly used to solve interoperability problems for CAD systems in the 1990s. However, demands for complex models and more expectations from the AEC industry have created a need for innovation in terms of data interoperability standards and schemas.

STEP is the first effort to solve interoperability issues as an international standard (Sacks et al., 2018). STEP is known as ISO 10303, which is used to define product data and geometric data (El Asmi et al., 2015). Since then, different schema languages have been developed. Among these are EXPRESS and XML. The EXPRESS language was developed according to Part 11 of STEP (ISO 10303). EXPRESS helps to represent product data through schemas and constraints. It is not a programming language (Sustainability of Digital Formats: Planning for Library of Congress Collections, 2016). CIS/2 and an earlier version of IFC were developed using the EXPRESS data language (Kamel & Memari, 2019). Extensible Markup Language (XML), which "is a set of rules for designing text format specification", is used in BIM to define information. With the invention of XML standards, successful schemas were created for the AEC industry such as CityGML, landXML, gbXML, ifcXML, and aecXML (El Asmi et al., 2015).

Interoperability should also provide automation during the process. If engineers or architects face challenges during the process, such as manual data entry, it is disappointing from their point of views. Interoperability issues are dealt with by organizations and researchers, which will be shown with examples in the following sections.

Data exchange formats

Arayici (2015) stated that there are three main ways to exchange data between software tools: direct links, proprietary exchange formats, and public product data model exchange formats. Eastman et al. (Sacks et al., 2018) added one more way to exchange data: model–server-based data exchange.

- **Direct links:** The receiver programme re-writes information, which is extracted from a sender programme, through the API (application programming interface).
- **Proprietary exchange format:** This type of data exchange depends on file-based data exchange. The underlying reason behind a proprietary exchange format is that every commercial organization can create its own product. To support them, the specific file format is produced by organizations. Some important examples of proprietary exchange formats are RVT (Revit), DWG (AutoCAD), DGN (Bentley Systems), etc. (Sacks et al., 2018).
- **Public product data model exchange formats:** This type of format depends on open and publicly managed language and schemas. Different product data models refer to different types of data structure in terms of geometry, relations, process, material, and performance. It means that if there is a data definition difference in the same object between two product data models, interoperability issues will be observed during the process (Sacks et al., 2018).
- **Model-server based data exchange:** Data exchange between stakeholders is performed on a database management system (DBMS). This type of data exchange can also be called a *common data environment* or *BIM server* (Sacks et al., 2018).

Within this section, product data model exchange standards or schemas and modelling languages will be intensively introduced.

Developments in data exchange standards

The history of standards development for product data exchange started in the 1950s. Ad-hoc solutions (first-generation exchange formats) existed between the 1950s and 1980s. These solutions are called *closed* and *proprietary* solutions. In the 1970s, heavy industry CAD users and CAD vendors made an agreement to come together to develop an open exchange mechanism (second-generation exchange formats).

The first version of the Initial Graphics Exchange Specification (IGES) was later established. In these third-generation exchange formats, the developed formats were planned to serve multiple industrial and manufacturing industries. The development of the third generation of standard efforts began in the 1980s and continued until the mid-1990s due to the development of the EXPRESS language, which was based on the STEP standard, or Standard for the Exchange of Product Model Data. However, this developed standard was valid for all industries. Accordingly, there was a need for a specific standard to support the AEC industry. To this end, the IFC standard, which is used today, was developed (Laakso & Kiviniemi, 2012).

Modeling languages

The existing data models or data schemas are different from each other and they are used to share and exchange manufacturing data in the model. Data models also define data objects and their relationships. The development of data models depends on the modelling language used. Different modelling languages are available, such as EXPRESS, XML (eXtensible Markup Language), etc.

The EXPRESS language, which is a data modelling language, was identified in the ISO 10303-Part 11. EXPRESS language contains a set of conditions which are used to establish a domain. According to the constraints, instances are evaluated to determine whether they are in the domain or not. However, EXPRESS contains more sets of constraints than XML for validation. "The language elements are formed into a stream of text, typically broken into physical lines. A physical line is any number (including zero) of characters ended by a newline" (Sustainability of Digital Formats: Planning for Library of Congress Collections, 2017). XML is another language which is used to create data schemas. XML is a very flexible text format. XML language was developed based on ISO 8879. XML helps to store information and exchange a wide variety of data via the Web or elsewhere (W3C, 2016). XML was developed by the World Wide Web Consortium. However, XML files contain excessive verbosity, which causes issues in the exchange of large data sets (Murphy et al., 2017). Data schemas developed with EXPRESS and XML languages are given in the next section.

Current standardization efforts

The most well-known standards or schema types (IFC, gbXML, COBie, Omniclass, and XML-based schemas) have been summarized under this section. However, the schema types or standardization efforts are not limited to IFC, gbXML, COBie, Omniclass, and XML-based schemas. For instance, CIS/2 is used as a steel integration standard.

Industrial Foundation Classes (IFC): Industrial Foundation Classes (IFC) provide an organized data model in terms of geometric and non-geometric information on building elements and their relationships (Zheng, 2014). IFC files do not just enable the attachment of information about building elements such as walls, columns, and beams, but it also enables the attachment of specific attributes such as material type and vendor information (Kim et al. 2016). IFC is a schema type and open standard (ISO 16739-1, 2018) widely accepted by AEC industry practitioners. IFC was initially developed in 1996 by the International Association for Interoperability (IAI), which is rebranded today as BuildingSMART alliance. BuildingSMART alliance produces data format technology and provides standardization of processes, workflows, and procedures (Ozturk, 2020).

IFC has been continuously developed. Today, the latest version is IFC4.3, which is a potential standard and currently under review. As a result of the review process, if there is no issue with the applicability of IFC 4.3, it will become an official standard (BuildingSMART, 2020). In the IFC schema, there are four layers: resources, core, interoperability, and domain. These layers are used to describe information and relationships about building objects in the BIM model. In other words, metadata rather than 3D geometry information or surface information can be embedded with the help of the IFC schema (Arayici et al., 2018; Sadeghi et al., 2019a; Steel et al., 2012; Zheng, 2014). An earlier version of IFC was developed by using EXPRESS language. However, later versions of IFC are based on both XML and EXPRESS data specification language (ISO, 2018).

After progressing data exchange between two BIM applications, there is a need to support workflows and processes because every stakeholder needs different data aspects from the BIM model (Sacks et al., 2018). Depending on the IFC file format, the information delivery manual (IDM) and model view definition (MVD) are used to identify data exchange requirements. While IDM is used to elaborate information exchange and quality requirements and process maps (Ozturk, 2020; Zheng, 2014), MVD is used to show additional constraints of the subsets of IFC schema, such as structural design or energy analysis (Kamel & Memari, 2016; 2019; Sacks et al., 2018). The specific focus of MVD is energy simulations (Kamel & Memari, 2019). In the implementation of IFC schema, there are some problems (Arayici et al., 2018; Ozturk, 2020; Steel et al., 2012; Zheng, 2014) such as

- Lack of information creator in the schema;
- Lack of specific information included in the data exchange procedure;
- Lack of information granularity included in the data exchange procedure;
- Missing information in the exchange process;
- Restriction of the number of model objects (memory consumption issue);
- IFC opening;
- Modelling style;
- General data structure;
- Inconsistent naming conventions;
- A myriad of bespoke facilities management (FM) information requirements;
- Inadequate data categorization;
- Poor information synchronization;
- Access to accurate data information and knowledge.

Green Building Extensible Mark-up Language (gbXML): The gbXML schema type is very useful for performing data exchange between BIM and building performance simulations (BEPS) (Arayici et al., 2018). The gbXML format includes information about building zones, surfaces, fenestration, and environment data. The most prominent feature of gbXML is to give location data, which is not given in other schema types. gbXML is created based on XML language, with a top-down data structure. A top-down structure is a relatively complex schema with large file size. Coding of this structure in software is also difficult. However, semantic changes can be easily observed within the schema (Kamel & Memari, 2019).

The gbXML data schema is in its infancy due to the readability of information of complex systems (Arayici et al., 2018). gbXML has an issue in data transfer such as HVAC. It provides a database that contains building elements' geometry and its explanations. However, this schema blocks the remodelling of the building in energy simulation models (Abanda & Byers, 2016). Rectangular geometry is only allowed in the representation of building geometry in gbXML (Kamel & Memari, 2019).

Construction Operation Building Information Exchange (COBie): COBie is a formal scheme for the organization of data from design and construction to facility management purposes, and it provides a standardized level of detail for materials, maintenance information, serial numbers, location, tag, and performance data (Gholami, 2015). COBie is specifically designed to solve facility or asset management interoperability issues, and it is related to asset data rather than geometric information.

COBie helps to reduce cost items originating from data loss or data unavailability for the FM stage. COBie is based on model view definition (MVD). COBie aims to facilitate the transfer of as-built models into computer-aided facility management (CAFM) platforms (Farghaly et al., 2018). COBie presents numerous spreadsheets featuring a great deal of information. COBie spreadsheets are very convenient to implement sorting, querying, and basic formula applications. However, COBie does not include information related to geometric and architectural objects such as walls and doors (Gholami, 2015).

Nonetheless, this leads to an unintended data burden to interpret data and understand data dependency. Therefore, memory overload is commonly encountered as an issue. Access to specific data between workbooks with COBie is also very challenging. Users have also a problem with querying spreadsheet data (Yalcinkaya & Singh, 2019). COBie implementation is not user friendly in terms of identifying information requirements (Sadeghi et al., 2019b).

To date, COBie usage in the construction industry has not been fully engaged. The reason for its non-usage in the industry has been attributed to the conflict between COBie requirements and industry requirements and the rigid structure of COBie for unexpected situations. COBie is not comprehensive for use in asset management (Abdirad & Dossick, 2019). COBie implementations also represent a "complex, fragmented and labor-intensive process" and COBie can present incomplete, unnecessary, low-quality data" (Sadeghi et al., 2019b).

COBie also presents a document in which information types are identified. To use COBie in a project, the owner needs to choose the necessary information types for FM use. However, the lack of knowledge on the owner's side in terms of determining necessary information types leads to data loss or unnecessary data collection for the FM stage (Abdirad & Dossick, 2019). There are many stakeholders in a project. For the delivery of the necessary information to the owner's side, stakeholder roles and responsibilities need to be identified before the project starts (Alnaggar & Pitt, 2019b).

Omniclass: Omniclass was developed in the early 1990s by considering internationally accepted standards (International Organization for Standardization-ISO and the International Construction Information Society-ICIS). In particular, ISO 12006-2, which is used as a standard for classification framework, and ISO 12006-3, which is used for tagging and managing objects and their attributes, played an important role in the development of Omniclass (CSI, 2020).

Omniclass is used for organizing and retrieving information, which is achieved in the construction industry by segregating objects' information into discrete and coordinated tables. Additionally, filtering and sorting can be performed with Omniclass. Also, Omniclass can be used to communicate with other schemas such as COBie. Omniclass can be used throughout a facility's life cycle (Ceton, 2019).

XML-Based Schemas: XML-based schemas are an alternative to IFC models. They aid data exchange by simplifying data in AEC applications. The simplification is made by converting building data (sites, buildings, floors, spaces, and equipment and their attributes) to a spatial building model (extruded shapes and spaces) (BIMXML, 2020). For instance, CityGML, which was developed by the Open Geospatial Consortium (OGC), was created by using XML-based schemas. City GML is an open data model to store and exchange 3D city models. The CityGML schema types aim to create common definitions for entities, attributes, and relations of 3D city models (OGC, 2020). Furthermore, in the AEC industry, different XML schemas are available. OpenGIS, gbXML, ifcXML, aecXML, agcXML, BIM collaboration format, and CityGML are examples of XML schema types (Sacks et al., 2018).

Interoperability problems and advanced solutions in BIM applications

This section aims to explain interoperability issues and advanced methods to solve these interoperability issues from the existing studies, specific to building energy performance simulations, facility management, and geographical information systems.

BIM servers

A BIM server is a collaboration platform that maintains a repository of the building data, and allows native applications to import and export files from the database for viewing, checking, updating, and modifying the data. A BIM server also serves as a multidisciplinary platform for project teams (Singh et al., 2010). BIM servers are used to manage building life-cycle data on a server or database specific to a single project (Graphisoft, 2017). The advantages of the BIM server are

- Partial models/views;
- Ad hoc queries;
- Merge function;
- Concurrent usage;
- Team members' rights/security, and speed/performance/integrity;
- Version control;
- Transaction processing;
- Audit (users' roles, decisions, and issue tracking);
- Data protection (mirroring/back-up);
- Storage (Singh et al., 2010).

Different BIM servers published by research institutions and software vendors are available. Some of them are summarized below:

- IFC Model Server: IFC Model Server was established by SECOM. The developed service helps to manage and share BIM models on the internet. To use the IFC model server, first of all the user needs to upload the BIM model to the server. After that, the users can access the model on the internet by using APIs, which are provided by the IFC Model server. The stored BIM model can be also be used with augmented reality and virtual reality (AR/VR) simulations, etc. (Secom, 2020).

- Bimserver.org: Bimserver.org is an IFC STEP Express based open-source BIM server which has a multi-layer design with a generative, model-driven architecture. Bimserver .org was developed by TNO Netherlands and TU of Eindhoven. By using Bimserver .org, users can import, export, modify, track, filter, and query on an IFC model (Beetz, 2010; Zhang et al., 2014).
- Express Data Manager (EDM) model server (IFC): This is an IFC and ifcXML-based source server. It was developed by Jotne EPM Technology. The EDM model server allows the reading and writing of an IFC model (Zhang et al., 2014). An EDM model server (IFC) is also able to check-out and check-in partial data (IT, 2020).
- Bentley I-Model: Bentley-I Model is used to exchange lifecycle information of assets. Bentley I-Model is open source. It enables knowledge to be retained about their source. By using Bentley I-Model, users can query and filter a model. Additionally, component information, business property, geometry, and relationship can be observed on Bentley I-Model (Bentley, 2020).

BIM cloud

Data which is available in BIM models are still stored and transferred in the form of either files such as rvt. or in a neutral format such as IFC. Nowadays, cloud computing technologies have developed rapidly. Due to technological breakthroughs in cloud computing, BIM modelling has recently started to move to cloud servers. The transformation from a single file or neutral format to BIM cloud helps mitigate performance issues, high costs, and interoperability between stakeholders (Zhang et al., 2014). Cloud-based BIM has a positive effect on data availability and accessibility, scalable storage, the effective use of data, and environmental effects (due to reduced energy consumption) (Alreshidi et al., 2017).

Cloud-based BIM helps multinational or large companies to manage multiple and complex projects using a cloud-based technology by dividing projects on a BIM cloud server. Also, the number of users on the BIM cloud is more than on BIM servers (Graphisoft, 2017). Therefore, cloud technology is used to provide a multi-user network and central repository. Three types of cloud technology model can be used in the AEC industry: software as a service (SaaS), platform as a service (PaaS), and infrastructure as a service (IaaS).

The fundamental difference between these cloud technology models is between renting software on the cloud platform (SaaS), interacting with existing software (PaaS), and using only servers (IaaS) (Zheng, 2014). In other words, while IaaS allows usage of servers, storage, network security, and data centres, PaaS adds more services (operating system and database management, business analytics) to the IaaS service. In SaaS, applications which are used by stakeholders are added to PaaS services (Microsoft Azure, 2020). The idea of an open BIM was developed to solve the below issues, which are reported by (Ding & Xu, 2014):

- Numerous stakeholders participating in the project life cycle;
- The amount of information;
- Large investment needs for BIM;
- Life-cycle perspective;
- Security issues.

Different BIM cloud solutions published by various research software vendors are available. Some of them are summarized below:

- **BIMcloud:** BIMcloud was developed by Graphisoft. The company aimed to enable "real-time, secure teamwork between architects, regardless of the size of the design project, the location of the offices, or the speed of the Internet connection" (Graphisoft, 2020).
- **BIMPLUS:** This is a cloud-based building life-cycle management platform. The solution enables different teams on models such as architects, engineers, BIM coordinator, building contractor, MEP engineer, facility manager, etc. By using BIMPLUS, model management, document management, filtering, collaboration, clash detection can be performed (BIMPLUS, 2020).
- **Autodesk BIM 360:** BIM 360 is a unified platform on which project stakeholders can make arrangements on model or make decisions by using available data from BIM servers. By using Autodesk BIM 360, users can benefit from services related to RFIs and submittals management, document management, communication between stakeholders, safety management, quality management, constructability, design collaboration, and data and analytics in real time.

However, there are some worries about the usage of cloud BIM: security and privacy, internet connection dependency, lack of legal considerations (related to laws of countries where data centres are found), anonymous control (data management transparency), physical location of data storage concerns, and initial set-up costs (Alreshidi et al., 2017).

Energy analysis simulations

The construction industry is responsible for 23% of energy consumption and 40% of CO_2 emissions globally. Therefore, keeping energy consumption under control is important for the AEC industry (Choi et al., 2016). To enable control and optimize energy consumption in the building environment, building energy performance simulations (BEPS) are intensively used, especially at the design stage. While BEPS software helps to reduce energy consumption, it also helps to increase occupancy comfort at the same time (Garwood et al., 2018). However, issues related to the manual creation of a building energy model, data replication, data leaks, and redundant data can be faced in conventional building energy analysis.

BIM can provide 70% of the needs for BEPS when performing building energy analysis via BEPS (Choi et al., 2016). In other words, today's 3D-CAD/BIMs provide users with an opportunity to explore different energy-saving alternatives in the early design stage while avoiding the time-consuming process of re-entering all the building geometry, enclosure, and HVAC information necessary for a complete energy analysis (Stumpf et al., 2009). Moreover, BIM facilitates users' understanding in terms of the building environment and components.

Data transfer from BIM to BEPS can be realized with data schema types such as COBie, IFC, or gbXML to enable interoperability among them. Geometric and property data are transferred using these schema types. However, gbXML is the most common schema type when transferring building data to energy analysis simulations (Gerrish et al., 2017). The data requirements and the place interoperability schema types are simply summarized in Figure 12.2, adapted from (Choi et al., 2016).

The AEC industry has a wide range of BEPS software tools such as BLAST, BSim, DOE-2, DeST, Ecotect, eQuest, EnergyPlus, etc. (Crawley et al., 2008; Kim et al., 2013). These

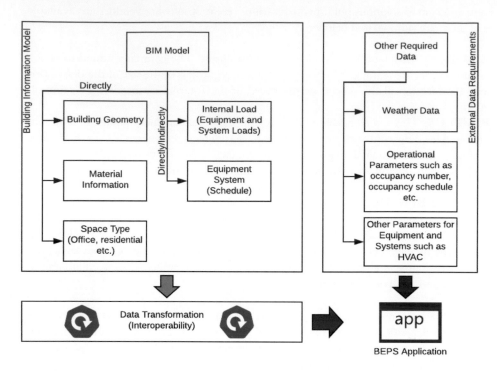

Figure 12.2 Interoperability in the integration of BIM and BEPS (adapted from [Choi et al., 2016]).

BEPS tools have pros and cons. Therefore, the user needs to be careful when choosing BEPS software. In other words, every data transfer process could be different for different BEPS software tools. For example, Choi et al. (Choi et al., 2016) stated that the geometry simplification tool (GST) which is used for data transformation from BIM to input data format (IDF, used in EnergyPlus) has errors when exporting an IFC file from BIM software, except from ArchiCAD.

When the integration of BIM data into BEPS is performed, some barriers are also observed in practical implementation. Choi et al. (2016) summarized some of them: (i) some building elements in BIM have no corresponding element in the BEPS data structure, such as columns and beams. This creates an oversimplification issue when making BIM elements; (ii) the unavailability of data in BIM models requires BEPS analysis, such as heat transfer coefficient and (iii) there are data exchange issues related to interoperability.

Low interoperability also induces low implementation of building energy analysis due to the fact that it necessitates the rework and correction of transferred data after the data exchange is performed. High deviations between predicted energy consumption and the actual energy consumption can be observed since energy simulations depend greatly on assumptions such as occupancy schedule (Ham & Golparvar-Fard, 2015). To overcome these issues, external databases (Kim et al., 2016) and data extraction using Dynamo (Gerrish et al., 2017) can be used. Dynamo helps to extract geometric and material information from BIM, and the extracted information is converted into the JavaScript Object Notation (JSON) lightweight data-interchange format (Gerrish et al., 2017). Briefly, Dynamo is an open-source visual programming application that interacts with Revit to extend its parametric capabilities to the Revit projects (Rahmani Asl et al., 2015).

Facility management

Facility management (FM) corresponds to the longest duration in the building life cycle (Koch et al., 2014). Also, studies have acknowledged that OPEX (operational expenditures – 85% of building life-cycle cost) is more costly than CAPEX (capital expenditures – 15%) (Edirisinghe et al., 2017; Koch et al., 2014). Additionally, financial losses due to interoperability were found to be $15.8 billion annually in the United States. Furthermore, 67% of financial losses is related to interoperability issues at the FM stage (Terreno et al., 2019).

FM consists of the integration of people, place, process, and technology (Gao & Pishdad-Bozorgi, 2019). Under the FM process, cleaning services, support services, property services, energy, catering services, and security services are performed (Potkany et al., 2015). Therefore, FM requires data-driven decisions, which are performed by facility managers who collect, analyze, store, exchange, and manage facility data (Alnaggar & Pitt, 2019a). To provide efficient facility management, FM systems such as building automation systems (BAS), building management systems (BMS), building energy management systems (BEMS), computerized maintenance management systems (CMMS), building information models (BIM), or simulation tools are used (Kučera et al., 2013; Kučera & Pitner, 2018; Lee et al., 2018). These management systems can be integrated with BIM because BIM not only provides geometric data but also provides databases to show and store processed data such as alarms (Shalabi & Turkan, 2016).

Practitioners believe that the integration of BIM and FM systems helps to create a single management source for FM decisions since BIM can host as-built information, records of maintenance, details of warranty and service, assessment and monitoring, space making and energy monitoring, emergency procedures, retrofit planning, reduced implementation costs, the provision of feedback to eliminate design-related performance issues, and visualization data. Terreno et al. (2019) explained five different methods that can be followed for data transfer between BIM and FM software (Table 12.1).

However, Dixit et al. (2019) stated that interoperability issues between BIM and FM technologies hinder BIM usage in FM. During the data exchange between BIM and FM systems, COBie and IFC are intensively used in the industry (Alnaggar & Pitt, 2019b). However, COBie has not reached sufficient maturity. In other words, organizations need guidance

Table 12.1 Methods and approaches to link BIM with FM software

Approach	Methods/Approaches for Linking Information	
Manual; spreadsheets	Extract, transform, and load (ETL); Data warehouse (DW)	Hyperlinking
COBie spreadsheets	BIM-based neutral file format; Design pattern and application programme interface (API); ETL; DW	Hyperlinking, exchanging, and synchronizing data
IFC format	BIM-based neutral file format	Exchanging and synchronizing data
Application programme interface (API) coupling	Design and API coupling	Portal solution
Proprietary middleware	BIM-based neutral file format; web service; ETL and DW; IDM and MVD	Portal solution using middleware such as EcoDomus, FM: Interact and Onuma Systems

as to how COBie will be implemented in the data transfer between BIM and FM software. When users follow COBie as a data exchange format, unnecessary data such as architectural and structural information can be seen in spreadsheets (Terreno et al., 2019). A data burden in COBie spreadsheets induces filtering issues for the building operation and maintenance requirements. Alnaggar and Pitt (Alnaggar & Pitt, 2019a) found that COBie may not be the future of the industry due to its rigid data structure, ambiguity, and data exchange process.

There is also a lack of clarity about roles and responsibilities, a lack of a mapping process between COBie and CAFM systems, and a lack of precise COBie requirements. In addition to COBie, IFC presents only a partial interoperability solution for data exchange between BIM and FM software, since it causes data complication and data losses (Pärn et al., 2017; Terreno et al., 2019). Implementing changes on the schema with MVD is not a smooth process due to the lack of a user-friendly interface. Therefore, data transfer between BIM and FM systems is not an error-free process due to the interoperability problems (Dixit et al., 2019; Heaton et al., 2019; Sadeghi et al., 2019a; Shalabi & Turkan, 2016). The interoperability issues are not only seen in BIM implementations. They are also seen between CMMS and other FM systems (Shalabi & Turkan, 2016). Manual and ad-hoc solutions are produced and followed to solve the interoperability issues by AEC industry stakeholders (Heaton et al., 2019).

However, facility managers need to give a quick response to changes and issues in the facility environment. In the literature, interoperability solutions for effective FM were developed. To solve interoperability issues, model view definition (MVD) was used to find the necessary information in models (William East et al., 2012); Naviswork DataTools (Lucas & Thabet, 2018); IFCOpenShell, which is a Python module to manipulate IFC file (Heaton et al., 2019); an application programming interface (API) plug-in, in which an as-built BIM model is updated and maintained (Pärn & Edwards, 2017); Cloud-BIM (Ding & Xu, 2014); etc. can be used.

Geographical information systems (GIS)

A combination of BIM and geographical information systems (GIS) is commonly used due to the fact that GIS can support resource arrangements, the selection of appropriate crane location, and safety analysis. GIS helps to manage georeferenced data, used in 3D analysis, spatial analysis, queries such as distance calculation, and optimal location, for which BIM has a lack of capability (D'Amico et al., 2019; Sani & Rahman, 2018). While GIS is used to manage building geographical data, BIM focuses more on building elements and project life-cycle data (Sani & Rahman, 2018).

GIS systems has also great flexibility to interoperate CMMS, CAFM, integrated workplace management systems (IWMS), and Electronic Document Management System (EDMS). Together, the integrated GIS–BIM platform offers the total scalability from which geospatial data at the building environment to building system information can be elaborated. Therefore, desired business needs such as energy management, facility management, space management, etc. can be provided with the integration of BIM and GIS (Wu et al., 2014).

Amirebrahimi et al. (2015) defined three levels for the data integration process: the data, application, and process levels. In the first level, the data level, the requirements of the application are met by the manipulation of the data models and structure. In the second level, the application level, the adoption of new applications is performed via successful and efficient data interoperability. In the last level, the process level, a workflow is provided between BIM and GIS.

The integration of BIM into GIS or vice versa requires a comprehensive data exchange. However, while BIM is generally used with IFC schemas, GIS is used with the City Geography Markup Language (CityGML) or shapefile (D'Amico et al., 2019). In the literature, the researchers developed lots of solutions to solve the integration issue between GIS and BIM. To do this, unidirectional approaches (GIS to BIM or BIM to GIS), new tools, or CityGML and IFC were commonly observed studies in the literature. However, information loss due to schema types can be also seen in the integration of BIM and GIS.

BIM (a local placement system) and GIS (a geographic coordinate system) use different coordinate systems. The two systems have different views in terms of the definition of level of detail. CityGML has fewer classes compared to IFC classes. For instance, *stair* is not defined in CityGML (Sani & Rahman, 2018). There is a limitation in representing utility components in IFC and CityGML. Some information such as inspection history is also not included in both schemas (Wang et al., 2019). Interoperability issues can be solved with the implementation of cloud systems (Wu et al., 2014) and the extract, transform, and load (ETL) method (Kang & Hong, 2015).

Light detection and ranging (Lidar)

The construction industry is less innovative than other industries. However, this state of affairs has started to change with the increase in the application of remote sensing technologies such as Lidar technologies, photogrammetry, etc. These technologies facilitate as-built model creation. Therefore, some issues such as outdated data, manual data entry, etc. can be eliminated. Lidar is one technology which takes a digital snapshot from the building environment. With the help of Lidar, a point cloud (3D geometric information) is gathered by measuring the distance between the sensor and the measured surfaces (Puri & Turkan, 2020; Wang et al., 2015). Lidar can perform the same work with laser scanning technologies. Furthermore, they can be used as synonyms in some resources. However, there is a small difference between them. According to (Wang et al., 2015), laser scanning has a limited capability to perform scanning activity in wide-ranging construction conditions. Lidar technology can also support real-time data collection from a site.

Laser scanning is an important technology to "gather in-situ geometric data" from the building. This technology has a wide range of application areas such as damage assessment, construction monitoring, retrofitting, energy analysis, and creating blueprints for an existing building (Cabaleiro et al., 2014; Sanhudo et al., 2020). Lidar scanning provides point clouds, which are datasets representing the surface of objects using Cartesian coordinate system (X, Y, and Z) points. With the help of datasets, topography, building elements, and equipment/systems can be modelled. The usage of Lidar technology helps to produce millions of data points with high precision (Garwood et al., 2018). The superiority of this technology is its automation, shorter duration, and accuracy. Transferring the gathered data from the laser scanning technology into BIM is an important way to solve interoperability issues. Thus, scanning to BIM commercial services are developed with a combination of machine learning technologies (Sanhudo et al., 2020).

The integration of BIM and Lidar technologies is used in a variety of areas. For instance, construction monitoring is one challenging process for construction projects. Progress tracking in construction projects necessitates the comparison of the planned and actual project status. In this process, Lidar technology can be used to detect the actual progress status of an ongoing construction project. The comparison of the output from laser scanning with 4D design helps to evaluate project performance rapidly (Puri & Turkan, 2020). Additionally, real-time quality control is very important in order to detect defects earlier.

Current approaches are time consuming and less effective than automated process due to their openness to human error (Wang et al., 2015). Therefore, the integration of BIM and Lidar technology enables real-time construction quality control (Wang et al., 2015). Furthermore, building performance analysis for existing buildings has attracted more attention with the increase in the energy consumption rate of building industry as a fraction of global energy consumption. The manual creation of building model is labour-intensive. Therefore, laser scanning and Lidar technologies can be used as a solution. Lidar technologies help to create an as-is BIM. The developed model can be also saved in gbXML format, so that interoperability issues can be eliminated (Wang & Cho, 2015).

Big Data analytics

Even though BIM involves data related to the building life cycle, these data are rarely used to increase productivity. Therefore, the application of business intelligence and analytics (BI&A) such as Big Data analytics has gained importance in the AEC industry. BI&A is an umbrella term which encompasses techniques, technologies, systems, practices, methodologies, and applications to retrieve necessary business information from raw data sources (Ahmed et al., 2018). The need has arisen for data analysis techniques with the diffusing of BIM, databases, and the Internet of Things (IoT) into the AEC industry. Additionally, data sources such as IoT devices necessitate the dynamic analysis of a huge data stack. Therefore, Big Data analytics have promising features for data analysis. Farghaly et al. (2017) stated that Big Data analytics have a positive impact on interoperability issues.

Big Data applications have grown tremendously while data collection is beyond the capability of commonly used software tools to capture, manage, and process within a tolerable elapsed time (Wu et al., 2013). Also Big Data is defined with the V's concepts (Assunção et al., 2015). These concepts are;

- Variety: This refers to multiple data types.
- Velocity: This refers to data production and processing speed.
- Volume: This refers to data size.
- Veracity: This refers to the data reliability of processed data.
- Value: This refers to the worth of data derived from data analysis.

According to Kambatla et al.'s study (2014), there are four types of data structure:

- Structured data, such as financial data, electronic medical records, and government statistics;
- Semi-structured data, such as texts, tweets, and emails;
- Unstructured data, such as audio and video;
- Real-time data, such as network traces and generic monitoring logs.

Big Data analytics are used in data monitoring, collection, analysis, and prediction. As a result of analysis, the users can create reports, graphs, and charts to retrieve information easily (Al-Ali et al., 2017).

Nowadays, the size of BIM models is increasing depending on project size. When increasing the size of a project, stand-alone BIM software remains incapable of managing building information, multiple project information, unfriendly information-sharing environments, and data analysis on model information. On the contrary, cloud BIM comes into prominence, with its huge data storage and powerful computing capabilities, and the multiple working environment it offers for the stakeholder.

Cloud service providers also offer a service for Big Data application on cloud BIM. With the implementation of Cloud-BIM, queries about the facility and operational data (such as energy management, emergency management etc.) in the facility can be more effectively performed (Arslan et al., 2017; Chen et al., 2016; Kang & Choi, 2018; Stavropoulos et al., 2015). However, Farghaly et al. (2017) stated that "a suitable schema that allows fast and efficient queries, and deals with graphical information and geographical reference location is required" when data transfer between stand-alone BIM and Big Data analytics is performed.

IFC has a lack of capability in terms of dynamic data while the implementation of Big Data analytics is not limited to the FM stage. Big Data analytics can be also used at the construction stage. For instance, lots of image and videos are taken from the construction site. For example, 325,000 images by a photographer or 95,400 images by webcams can be collected from construction sites for a typical project. However, it is not possible to use them effectively without using an analysis tool. These data sources can be effectively used when comparing as-planned and as-built projects. Therefore, big visual data needs to be analyzed with Big Data analytics (Han & Golparvar-Fard, 2017).

Blockchain

A BIM model is created by entering information from different areas of expertise. However, this brings new issues in the AEC industry, because of ownership of the model. For instance, different design groups share companies' knowledge on a BIM model. However, there is a legal gap to protect companies' rights. Therefore, disputes can arise over ownership of the model. Also, the need for review records in BIM model is another issue (Hargaden et al., 2019).

Due to technological developments and the above issues, blockchain technology has been recently implemented in the construction industry. Blockchain technology can be thought of as a distributed database which is used for data replication, sharing, and synchronizing. Also, blockchain technology helps users govern data from different geographical locations. "Equality, direct dealing, openness, consensus and mutual trust" play an important role under the blockchain logic (Perera et al., 2020). More technically, blockchain consist of blocks in which every hash contains a cryptographic hash and every hash contains operation sets (Alharby & Moorsel, 2017). This architecture helps to store the transactions of stakeholders reliably and synchronously. If any stakeholder wants to make a transaction on a project network, other stakeholders who are affected by this transaction need to approve the transaction (Ahmadisheykhsarmast & Sonmez, 2018). Furthermore, records cannot be changed to record historical records (Ahmadisheykhsarmast & Sonmez, 2018).

BIM provides important opportunities for digital management of the building's product and process information. Blockchain technology allows more data storage than BIM. With the integration of BIM and blockchain technology, data exchange and communication between the stakeholders depending on time and the relationship between 3D model and paper-based data can be solved (Hargaden et al., 2019).

Depending on the use of blockchain technology in the AEC industry, smart contracts are now begin used to implement building life-cycle stages. Smart contracts help to monitor and control contract protocols with the help of blockchain codes (Ahmadisheykhsarmast & Sonmez, 2018; Alharby & Moorsel, 2017). In other words, contract clauses are managed via codes. If the contract clauses are fulfilled by any stakeholder, predefined actions are performed via codes. For instance, if a subcontractor performs his work package on the construction site, crypto currency is transferred to the subcontractor. Every contract is stored in a file which has a 20-byte storage capacity. Also, if transactions cause a change in other contracts, the change can be reflected on the affected document automatically (Alharby & Moorsel, 2017).

Conclusion

BIM or CAD systems are an ICT solution to eliminate inefficiency related to conventional work processes in the AEC industry (Wong et al., 2013). BIM helps to handle building design holistically. Therefore, BIM helps to improve collaboration and communication issues among stakeholders. It is believed that project quality, time and cost control and errors can be more strictly performed with the implementation of BIM in construction projects (Ratajczak et al., 2015).

Effective stakeholder management and smooth information sharing between stakeholders are the key factors in a project's success (Jergeas et al., 1986). Therefore, the application of BIM or CAD systems in construction projects is inevitable. The nature of construction projects requires the participation of numerous stakeholders. Also, they are not limited to design and construction but also involve the demolition stage.

In the building life cycle, stakeholders use different ICT technologies to perform their activities. However, the output of one stakeholder can be the input of another stakeholder, or outputs can be used in another software programme. This requires a common language which can be readable and writable by stakeholders. Therefore, this chapter aims to explain interoperability, which provides for data exchange between stakeholders. Additionally, standardization efforts, the challenges faced in interoperability, and some examples from the literature were summarized in the chapter. This chapter will help to improve the skills of graduates in terms of the technical side of BIM. Thus, the knowledge from this chapter will familiarize graduates with new project delivery approaches (integrated project delivery). Moreover, the adaption or transfer of recent developments and solutions developed in universities into industry can be facilitated.

The history of the interoperability effort shows that interoperability cannot be perfectly solved by AEC industry practitioners. The underlying issue can be attributed to the demand and complexity of projects, since these constraints require new solutions specific to the AEC industry. Therefore, new software solutions have been produced and will be produced with different requirements. Also, software vendors present a wide range of products. To actually penetrate the use of the same vendors' products, the vendors hampers the users with interoperability issues.

Nonetheless, the literature reviews specific to BEPS, facility management, GIS, Lidar technology, Big Data analytics, and Blockchain technology do not just show the development of interoperability solutions, but also that interoperability or data loss is not an unsolvable issue. New technologies such as BIM servers offer promising features to solve interoperability issues. The introduction of these concepts to AEC graduates will facilitate them in finding new career opportunities in the AEC industry. Also, the adoption of new technologies with the transfer of skilled graduates into industry will help to eliminate and remedy the low productivity, poor functionality, rework, and wastage that are observed in construction projects.

Questions

1. What is interoperability and what is its importance for the AEC industry?
2. What is the main difference between the EXPRESS and XML languages?

3. Please explain why the AEC industry practitioners need the development of IDM and MVD.
4. Please explain the data formats which are used in energy analysis. Why is BIM used with building energy performance simulation programmes? And what should be considered when using BIM and BEPS together?
5. What is the main difference between BIM servers and BIM cloud technology? Please explain under which conditions a BIM server should be chosen.
6. Please discuss the benefits of the collocation of BIM and Lidar technologies.
7. Please explain the relationship between data formats and Big Data analytics. What is the benefit of analyzing existing data that is stored in BIM?

References

Abanda, F. H., & Byers, L. (2016). An investigation of the impact of building orientation on energy consumption in a domestic building using emerging BIM (Building Information Modelling). *Energy*, *97*, 517–527. https://doi.org/10.1016/j.energy.2015.12.135

Abdirad, H., & Dossick, C. S. (2019). Normative and descriptive models for COBie implementation: Discrepancies and limitations. *Engineering, Construction and Architectural Management*. https://doi.org/10.1108/ECAM-10-2018-0443

Ahmadisheykhsarmast, S., & Sonmez, R. (2018). Smart contracts in construction industry. In 5th International Project and Construction Management Conference (IPCMC2018), Cyprus International University, Faculty of Engineering, Civil Engineering Department, North Cyprus, December, 767–774.

Ahmed, V., Aziz, Z., Tezel, A., & Riaz, Z. (2018). Challenges and drivers for data mining in the AEC sector. *Engineering, Construction and Architectural Management*, *25*(11), 1436–1453. https://doi.org/10.1108/ECAM-01-2018-0035

Al-Ali, A. R., Zualkernan, I. A., Rashid, M., Gupta, R., & Alikarar, M. (2017). A smart home energy management system using IoT and Big Data analytics approach. *IEEE Transactions on Consumer Electronics*, *63*(4), 426–434. https://doi.org/10.1109/TCE.2017.015014

Alharby, M., & van Moorsel, A.. (2017). Blockchain based smart contracts : A systematic mapping study. *Computer Science and Information Technology*, 125–140. https://doi.org/10.5121/csit.2017.71011

Alnaggar, A., & Pitt, M. (2019a). Lifecycle exchange for asset data (LEAD) dataflow between building stakeholders using BIM open standards. *Journal of Facilities Management*, *17*(5), 385–411. https://doi.org/10.1108/JFM-06-2019-0030

Alnaggar, A., & Pitt, M. (2019b). Towards a conceptual framework to manage BIM/COBie asset data using a standard project management methodology. *Journal of Facilities Management*, *17*(2), 175–187. https://doi.org/10.1108/JFM-03-2018-0015

Alreshidi, E., Mourshed, M., & Rezgui, Y. (2017). Factors for effective BIM governance. *Journal of Building Engineering*, *10*, 89–101.

Amirebrahimi, S., Rajabifard, A., Mendis, P., & Ngo, T. (2015). A data model for integrating GIS and BIM for assessment and 3D visualisation of flood damage to building. *Locate*, *15*, 10–12.

Arayici, Y. (2015). *Building Information Modelling*. Bookboon Publisher. http://bookboon.com/en/building-information-modeling-ebook

Arayici, Y., Fernando, T., Munoz, V., & Bassanino, M. (2018). Interoperability specification development for integrated BIM use in performance based design. *Automation in Construction*, *85*, 167–181. https://doi.org/10.1016/j.autcon.2017.10.018

Arslan, M., Riaz, Z., & Munawar, S. (2017). Building information modeling (BIM) enabled facilities management using Hadoop architecture. In PICMET 2017 – Portland International Conference on Management of Engineering and Technology: Technology Management for the Interconnected World, Portland, OR, 2017-January, 1–6. https://doi.org/10.23919/PICMET.2017.8125462

Assunção, M. D., Calheiros, R. N., Bianchi, S., Netto, M. A. S., & Buyya, R. (2015). Big Data computing and clouds: Trends and future directions. *Journal of Parallel and Distributed Computing*, 79–80, 3–15. https://doi.org/10.1016/j.jpdc.2014.08.003

Beetz, J., van Berlo, L., de Laat, R., & van den Helm, P. (2010). Bimserver.org: An open source IFC model server. In Proceedings of the CIB W78 2010: 27th International Conference, Cairo, Egypt, November 16–18.

Bentley. (2020). *What are iModels ? Exchange Information for AEC Projects*. https://www.bentley.com/en/i-models/what-is-i-model/about-i-models.

BIMPLUS. (2020). *Features*. https://www.bimplus.net/index.php

BIMXML. (2020). *Building Information Model Extended Markup Language (BIMXML)*. http://bimxml.org/.

BuildingSMART. (2020). *IFC Schema Specifications: buildingSMART Technical*. https://technical.buildingsmart.org/standards/ifc/ifc-schema-specifications/.

Cabaleiro, M., Riveiro, B., Arias, P., Caamaño, J. C., & Vilán, J. A. (2014). Automatic 3D modelling of metal frame connections from LiDAR data for structural engineering purposes. *ISPRS Journal of Photogrammetry and Remote Sensing*, *96*, 47–56. https://doi.org/10.1016/j.isprsjprs.2014.07.006

Ceton, G. (2019). *OmniClass ® A Strategy for Classifying the Built Environment Introduction and User's Guide*.

Chen, H., Chang, K., & Lin, T. (2016). Automation in construction: A cloud-based system framework for performing online viewing, storage, and analysis on Big Data of massive BIMs. *Automation in Construction*, *71*, 34–48. https://doi.org/10.1016/j.autcon.2016.03.002

Choi, J., Shin, J., Kim, M., & Kim, I. (2016). Development of open BIM-based energy analysis software to improve the interoperability of energy performance assessment. *Automation in Construction*, *72*, 52–64. https://doi.org/10.1016/j.autcon.2016.07.004

Crawley, D. B., Hand, J. W., Kummert, M., & Griffith, B. T. (2008). Contrasting the capabilities of building energy performance simulation programs. *Building and Environment*, *43*(4), 661–673. https://doi.org/10.1016/j.buildenv.2006.10.027

Construction Specification Institute (CSI). (2020). *OmniClass-Background*. https://www.csiresources.org/standards/omniclass/standards-omniclass-background.

D'Amico, F., Calvi, A., Schiattarella, E., Prete, M., & Veraldi, V. (2019). BIM and GIS data integration: A novel approach of technical/environmental decision-making process in transport infrastructure design. In AIIT 2nd International Congress on Transport Infrastructure and Systems in a Changing World (TIS ROMA 2019), Rome, Italy, 803–810.

Ding, L., & Xu, X. (2014). Application of cloud storage on BIM life-cycle management. *International Journal of Advanced Robotic Systems*, *11*(1). https://doi.org/10.5772/58443

Dixit, M. K., Venkatraj, V., & Ostadalimakhmalbaf, M. (2019). Integration of facility management and building information modeling (BIM) A review of key issues and challenges. *37*(7), 455–483. https://doi.org/10.1108/F-03-2018-0043

Edirisinghe, R., London, K. A., Kalutara, P., & Aranda-Mena, G. (2017). Building information modelling for facility management: Are we there yet? *Engineering, Construction and Architectural Management*, *24*(6), 1119–1154. https://doi.org/10.1108/ECAM-06-2016-0139

El Asmi, E., Robert, S., Haas, B., & Zreik, K. (2015). A standardized approach to BIM and energy simulation connection. *International Journal of Design Sciences and Technology*, *21*(1), 59–82.

Farghaly, K., Abanda, H., Vidalakis, C., & Wood, G. (2017). BIM Big Data system architecture for asset management: A conceptual framework. *In*: Proceedings of the Joint Conference on Computing in Construction (JC3), Heraklion, Greece, July, 289–296. https://doi.org/10.24928/jc3-2017/0163

Farghaly, K., Wood, G., Fonbeyin, Abanda, H., & Vidalakis C. (2018). Taxonomy for BIM and asset management semantic interoperability. *Journal of Management in Engineering*, *34*(4), 1–13. https://doi.org/10.1061/(ASCE)ME.1943-5479.0000610

Gallaher, M. P., O'Connor, A. C., Dettbarn, J. L., & Gilday, L. T. (2004). Cost analysis of inadequate interoperability in the U.S. capital facilities industry. https://doi.org/10.6028/NIST.GCR.04-867

Gao, X., & Pishdad-Bozorgi, P. (2019). BIM-enabled facilities operation and maintenance: A review. In *Advanced Engineering Informatics*, *39*, 227–247. https://doi.org/10.1016/j.aei.2019.01.005

Garwood, T. L., Hughes, B. R., O'Connor, D., Calautit, J. K., Oates, M. R., & Hodgson, T. (2018). A framework for producing gbXML building geometry from point clouds for accurate and efficient building energy modelling. *Applied Energy, 224*, 527–537. https://doi.org/10.1016/j.apenergy.2018.04.046

Gerrish, T., Ruikar, K., Cook, M., Johnson, M., & Phillip, M. (2017). Using BIM capabilities to improve existing building energy modelling practices. *Engineering, Construction and Architectural Management, 24*(2), 190–208. https://doi.org/10.1108/ECAM-11-2015-0181

Gholami, E. (2015). Implementing building information modelling (BIM) in energy-efficient domestic retrofit : Quality checking of BIM model implementing building information modelling (BIM) in energy. In Proceedings of the 32nd CIB W78 conference, October, 11. https://www.researchgate.net/publication/284189992.

Goulding, J. S., Rahimian, F. P., & Wang, X. (2014). Virtual reality-based cloud BIM platform for integrated AEC projects. *Journal of Information Technology in Construction, 19* (August 2013), 308–325.

Graphisoft. (2017). *BIMcloud and BIM Server User Guide GRAPHISOFT® BIMcloud and BIM Server User Guide.* http://www.graphisoft.com

Graphisoft. (2020). *Overview.* https://www.graphisoft.com/bimcloud/overview/

Grilo, A., & Jardim-Goncalves, R. (2010). Value proposition on interoperability of BIM and collaborative working environments. *Automation in Construction.* https://doi.org/10.1016/j.autcon.2009.11.003

Ham, Y., & Golparvar-Fard, M. (2015). Mapping actual thermal properties to building elements in gbXML-based BIM for reliable building energy performance modeling. *Automation in Construction, 49*, 214–224. https://doi.org/10.1016/j.autcon.2014.07.009

Han, K. K., & Golparvar-Fard, M. (2017). Potential of big visual data and building information modeling for construction performance analytics: An exploratory study. *Automation in Construction.* https://doi.org/10.1016/j.autcon.2016.11.004

Hargaden, V., Papakostas, N., Newell, A., Khavia, A., & Scanlon, A. (2019). The role of blockchain technologies in construction engineering project management. Proceedings of the 2019 IEEE International Conference on Engineering, Technology and Innovation, ICE/ITMC, Valbonne Sophia-Antipolis, France. https://doi.org/10.1109/ICE.2019.8792582

Heaton, J., Parlikad, A. K., & Schooling, J. (2019). Design and development of BIM models to support operations and maintenance. *Computers in Industry, 111*, 172–186. https://doi.org/10.1016/j.compind.2019.08.001

ISO. (2018). *ISO - ISO 16739-1:2018 - Industry Foundation Classes (IFC) for Data Sharing in the Construction and Facility Management Industries: Part 1: Data Schema.* https://www.iso.org/standard/70303.html

Italianist, J. (2020). *EDMmodelServerTM (ifc).* https://jotneit.no/edmmodelserver-ifc

Jergeas, G., Williamson, E., Skulmoski, G., & Thomas, J. (1986). Stakeholder management on construction projects. *Preventing School Failure, 51*(3), 49–51.

Kambatla, K., Kollias, G., Kumar, V., & Grama, A. (2014). Trends in Big Data analytics. *Journal of Parallel and Distributed Computing, 74*(7), 2561–2573. https://doi.org/10.1016/j.jpdc.2014.01.003

Kamel, E., & Memari, A. M. (2019). Review of BIM's application in energy simulation: Tools, issues, and solutions. In *Automation in Construction, 97*, 164–180. https://doi.org/10.1016/j.autcon.2018.11.008

Kang, T. W., & Choi, H. S. (2018). BIM-based data mining method considering data integration and function extension. *KSCE Journal of Civil Engineering, 22*(5), 1523–1534. https://doi.org/10.1007/s12205-017-0561-6

Kang, T. W., & Hong, C. H. (2015). Automation in construction: A study on software architecture for effective BIM/GIS-based facility management data integration. *Automation in Construction, 54*, 25–38. https://doi.org/10.1016/j.autcon.2015.03.019

Kim, H., Shen, Z., Kim, I., Kim, K., Stumpf, A., & Yu, J. (2016). BIM IFC information mapping to building energy analysis (BEA) model with manually extended material information. *Automation in Construction, 68*, 183–193. https://doi.org/10.1016/j.autcon.2016.04.002

Kim, K., Kim, G., Yoo, D., & Yu, J. (2013). Semantic material name matching system for building energy analysis. *Automation in Construction, 30*, 242–255. https://doi.org/10.1016/j.autcon.2012.11.011

Koch, C., Neges, M., König, M., & Abramovici, M. (2014). Natural markers for augmented reality-based indoor navigation and facility maintenance. *Automation in Construction.* https://doi.org/10.1016/j.autcon.2014.08.009

Kučera, A., & Pitner, T. (2018). Semantic BMS: Allowing usage of building automation data in facility benchmarking. *Advanced Engineering Informatics.* https://doi.org/10.1016/j.aei.2018.01.002

Kučera, A., Glos, P., & Pitner, T. (2013). Fault detection in building management system networks. In IFAC Proceedings Volumes (IFAC-PapersOnline), Velke Karlovice, Czech Republic. https://doi.org/10.3182/20130925-3-CZ-3023.00027

Laakso, M., & Kiviniemi, A. (2012). The IFC standard: A review of history, development, and standardization. *Electronic Journal of Information Technology in Construction, 17*(May), 134–161.

Lee, P. C., Wang, Y., Lo, T. P., & Long, D. (2018). An integrated system framework of building information modelling and geographical information system for utility tunnel maintenance management. *Tunnelling and Underground Space Technology, 79*, 263–273. https://doi.org/10.1016/j.tust.2018.05.010

Liao, Y., Ramos, L. F. P., Saturno, M., Deschamps, F., de Freitas Rocha Loures, E., & Szejka, A. L. (2017). The role of interoperability in the fourth industrial revolution era. *IFAC-PapersOnLine, 50*(1), 12434–12439. https://doi.org/10.1016/j.ifacol.2017.08.1248

Lucas, J., & Thabet, W. (2018). Using a case-study approach to explore methods for transferring BIM-based asset data to facility management systems. In Proceeding of Construction Research Congress *2018, 1*, 148–157, New Orleans, Louisiana. https://doi.org/10.1213/01.ANE.0000149897.87025.A8

Microsoft Azure. (2020). *IaaS nedir? Hizmet Olarak Altyapı.* https://azure.microsoft.com/tr-tr/overview/what-is-iaas/

Murphy, S. N., Chueh, H. C., & Herrick, C. D. (2017). Information technology. In *Clinical and Translational Science: Principles of Human Research*: Second Edition (pp. 227–242). Elsevier. https://doi.org/10.1016/B978-0-12-802101-9.00013-2

The Open Geospatial Consortium (OGC). (2020). *CityGML.* https://www.ogc.org/standards/citygml

Oraee, M., Hosseini, M. R., Edwards, D. J., Li, H., Papadonikolaki, E., & Cao, D. (2019). Collaboration barriers in BIM-based construction networks: A conceptual model. *International Journal of Project Management, 37*(6), 839–854. https://doi.org/10.1016/j.ijproman.2019.05.004

Ozturk, G. B. (2020). Interoperability in building information modeling for AECO/FM industry. *Automation in Construction, 113.* https://doi.org/10.1016/j.autcon.2020.103122

Pärn, E. A., & Edwards, D. J. (2017). Conceptualising the FinDD API plug-in: A study of BIM–FM integration. *Automation in Construction, 80*, 11–21. https://doi.org/10.1016/j.autcon.2017.03.015

Pärn, E. A., Edwards, D. J., & Sing, M. C. P. (2017). The building information modelling trajectory in facilities management: A review. *Automation in Construction, 75*, 45–55. https://doi.org/10.1016/j.autcon.2016.12.003

Perera, S., Nanayakkara, S., Rodrigo, M. N. N., Senaratne, S., & Weinand, R. (2020). Blockchain technology: Is it hype or real in the construction industry? *Journal of Industrial Information Integration, 17*(August 2019), 100125. https://doi.org/10.1016/j.jii.2020.100125

Potkany, M., Vetrakova, M., & Babiakova, M. (2015). Facility management and its importance in the analysis of building life cycle. *Procedia Economics and Finance, 26*, 202–208. https://doi.org/10.1016/s2212-5671(15)00814-x

Puri, N., & Turkan, Y. (2020). Bridge construction progress monitoring using Lidar and 4D design models. *Automation in Construction, 109.* https://doi.org/10.1016/j.autcon.2019.102961

Rahmani Asl, M., Zarrinmehr, S., Bergin, M., & Yan, W. (2015). BPOpt: A framework for BIM-based performance optimization. *Energy and Buildings, 108*, 401–412. https://doi.org/10.1016/j.enbuild.2015.09.011

Ratajczak, J., Malacarne, G., Krause, D., & Matt, D. T. (2015). The BIM approach and stakeholders integration in the AEC sector: Benefits and obstacles in South Tyrolean context. *In*: Proceedings of the Fourth International Workshop on Design in Civil and Environmental Engineering, 30–31 October 2015, NTU, Taiwan, November 2016, 1–9.

Sacks, R., Eastman, C., Lee, G., & Teicholz, P. (2018). *BIM Handbook: A Guide to Building Information Modeling for Owners, Designers, Engineers, Contractors, and Facility Managers*. Wiley.

Sadeghi, M., Elliott, J. W., Porro, N., & Strong, K. (2019). Developing building information models (BIM) for building handover, operation and maintenance. *Journal of Facilities Management, 17*(3), 301–316. https://doi.org/10.1108/JFM-04-2018-0029

Sanhudo, L., Ramos, N. M. M., Martins, J. P., Almeida, R. M. S. F., Barreira, E., Simões, M. L., & Cardoso, V. (2020). A framework for in-situ geometric data acquisition using laser scanning for BIM modelling. *Journal of Building Engineering, 28*. https://doi.org/10.1016/j.jobe.2019.101073

Sani, M. J., & Rahman, A. A. (2018). GIS and BIM integration at data level: A review. *International Archives of the Photogrammetry, Remote Sensing and Spatial Information Sciences: ISPRS Archives, 42*(4/W9), 299–306. https://doi.org/10.5194/isprs-archives-XLII-4-W9-299-2018

Secom. (2020). *IFC Model Server - SECOM Intelligent Systems Laboratory*. https://www.secom.co.jp/isl/en/research/ifc-model-server/

Shalabi, F., & Turkan, Y. (2016). IFC BIM-based facility management approach to optimize data collection for corrective maintenance. *Journal of Performance of Constructed Facilities, 31*(1), 04016081. https://doi.org/10.1061/(asce)cf.1943-5509.0000941

Singh, V., Gu, N., & Wang, X. (2010). A theoretical framework of a BIM-based multi-disciplinary collaboration platform. *Automation in Construction, 20*(2), 134–144. https://doi.org/10.1016/j.autcon.2010.09.011

Stavropoulos, G., Krinidis, S., Ioannidis, D., Moustakas, K., & Tzovaras, D. (2015). A building performance evaluation & visualization system. *In*: Proceedings of the 2014 IEEE International Conference on Big Data, IEEE Big Data 2014, Washington, DC, 1077–1085. https://doi.org/10.1109/BigData.2014.7004342

Steel, J., Drogemuller, R., Toth, B., Tony Clark, C., Bettin Steel, J. J., Drogemuller, R., & Toth, B. (2012). Model interoperability in building information modelling. *Software & Systems Modeling, 11*, 99–109. https://doi.org/10.1007/s10270-010-0178-4

Stumpf, A., Kim, H., & Jenicek, E. (2009). Early design energy analysis using BIMs (building information models). In Construction Research Congress 2009, Seattle, Washington, 426–436. https://doi.org/10.1061/41020(339)44

Sustainability of Digital Formats: Planning for Library of Congress Collections. (2016). *EXPRESS Data Modeling Language*, ISO 10303-11. http://www.loc.gov/preservation/digital/formats/fdd/fdd000449.shtml

Sustainability of Digital Formats: Planning for Library of Congress Collections. (2017). *EXPRESS Data Modeling Language*, ISO 10303-11. http://www.loc.gov/preservation/digital/formats/fdd/fdd000449.shtml

Szeleczki, S. (2019). Interpreting the interoperability of the NATO's communication and information systems. *Scientific Bulletin, 24*(1), 95–107. https://doi.org/10.2478/bsaft-2019-0011

Taha, F. F., Hatem, W. A., & Jasim, N. A. (2020). Effectivity of BIM technology in using green energy strategies for construction projects. *Asian Journal of Civil Engineering, 21*(6), 995–1003. https://doi.org/10.1007/s42107-020-00256-w

Terreno, S., Asadi, S., & Anumba, C. (2019). An exploration of synergies between lean concepts and BIM in FM: A review and directions for future research. *Buildings, 9*(6), 1–28. https://doi.org/10.3390/BUILDINGS9060147

W3C. (2016). *Extensible Markup Language (XML)*. https://www.w3.org/XML/

Wang, C., & Cho, Y. K. (2015). Performance evaluation of automatically generated BIM from laser scanner data for sustainability analyses. *Procedia Engineering, 118*, 918–925. https://doi.org/10.1016/j.proeng.2015.08.531

Wang, J., Sun, W., Shou, W., Wang, X., Wu, C., Chong, H.-Y., Liu, Y., & Sun, C. (2015). Integrating BIM and Lidar for real-time construction quality control. *Journal of Intelligent & Robotic Systems, 79*, 417–432. https://doi.org/10.1007/s10846-014-0116-8

Wang, M., Deng, Y., Won, J., & Cheng, J. C. P. (2019). An integrated underground utility management and decision support based on BIM and GIS. *Automation in Construction, 107*. https://doi.org/10.1016/j.autcon.2019.102931

William East, E., Nisbet, N., & Liebich, T. (2012). Facility management handover model view. *Journal of Computing in Civil Engineering, 27*(1), 61–67. https://doi.org/10.1061/(asce)cp.1943-5487.0000196

Wong, A. K. D., Wong, F. K. . W., & Nadeem, A. (2013). Comparative roles of major stakeholders for the implementation of BIM in various countries. *Journal of Chemical Information and Modeling, 53*(9), 1689–1699. https://doi.org/10.1017/CBO9781107415324.004

Wu, W., Yang, X., & Fan, Q. (2014). GIS-BIM Based Virtual Facility Energy Assessment (VFEA)—Framework Development and Use Case of California State University, Fresno. *Computing in Civil and Building Engineering, 758*, 339–346. https://doi.org/10.1061/9780784413616.043

Wu, X., Zhu, X., Wu, G.-Q., & Ding, W. (2013). Data mining with Big Data. *IEEE Transactions on Knowledge and Data Engineering, 1*, 97–107.

Yalcinkaya, M., & Singh, V. (2019). VisualCOBie for facilities management: A BIM integrated, visual search and information management platform for COBie extension. *Facilities, 37*(7–8), 502–524. https://doi.org/10.1108/F-01-2018-0011

Zhang, J. P., Liu, Q., Yu, F. Q., Hu, Z. Z., & Zhao, W. Z. (2014). A framework of cloud-computing-based BIM ervice for building lifecycle. In 2014 International Conference on Computing in Civil and Building Engineering, Orlando, FL, June 23–25, 1514–1521.

Zheng, D. (2014). Cloud and open BIM-based building information interoperability re-search. *Journal of Service Science and Management, 7*, 47–56. https://doi.org/10.4236/jssm.2014.72005

13 BIM and ethics

Nicholas Nisbet

Introduction

BIM and ethics are not obviously commensurate. BIM is variously an outcome, a means, or a process of applying information technology to describe the built environment. More specifically, BIM is a tool for describing the management of change in the built environment. This can be referred to as the architecture, engineering, and construction and operation sector, or AECO for short. For some, BIM can be side-lined as merely a technology, often an evolution or complication of computer-aided draughting. As such, it may be restricted to aiding the delivery of contractually agreed documents and drawings. For others, it is the beginning of the urgent digitization of an industry that has a poor reputation for thrift, timeliness, and quality and a poor profile in terms of productivity, investment, and ambition.

Ethics may seem to be a polar opposite to BIM, affecting only personal relations and behaviour. Ethics involves "systematizing, defending, and recommending concepts of right and wrong behaviour" (IEP, 2020). Ethics naturally extends out from personal behaviour to describing the judgements to be made by larger entities such as practices, corporations, and state actors. The view that BIM and ethics have no relationship would have to rely on the position that corporate and individual entities' only obligation is to enhance shareholder or partner value through contractual relationships. For example, the UK RIBA Code of Conduct (see tables 13.1 and 13.2) proposes in its introduction an ethical scale, with the client (customer) interest midway between social and personal interests. The body of the text makes no further reference to the higher principles. This monochromatic view has not often been accepted in the built environment, and so it is at least probable that there is a pertinent relationship which needs examination.

Approaches

Ethics in the development of AECO professions

There's been an evolution of the relationship between society, the professions, and ethical expectations (Collins, 1971). In the eighteenth century, privileges were granted to some of the professions in recognition of the fact that they had a body of knowledge which had taken a long time to acquire, both collectively and as individuals. In return for these privileges, certain expectations were built into the charters and licences to practice, including to preserve the public good. Such charters were found in the practice of law, medicine, and architecture, and later in civil engineering and accountancy. This compact lasted until late in the twentieth century when the presumption became that professions were working as commercial

Table 13.1 An ethical ranking (Introduction clause 11, RIBA 2019)

The Code enshrines the following duties owed by Chartered Practices:
To the wider world Towards society and the end user Towards those commissioning services (i.e. clients – this may include professional clients, investors, and funders) Towards those in the workplace (i.e. colleagues, employees, employers) Towards the profession Towards oneself

actors. For example, fee scales were abolished and the rights to perform some functions were de-restricted. The body of the RIBA Code (RIBA 2019, section 3.2) has only three phrases referencing wider duties.

Interestingly, at the same time there was an increasing realization that corporate interest, and in particular shareholder interest, might not yield results that were in societies' best interests. The topic of corporate social responsibility and concerns around environmental impacts were reintroduced as behaviours that were to be expected from larger corporations in return for their freedom to operate on a national and global scale. Some aspects of this reintroduction were enforced by law and some through media pressure. The commercialization of the professions coincided with the introduction of tools such as word processing and computer-aided drafting, which promised advantages in efficiency. In many cases, these tools were adopted because the reduction of the typing pool and drawing office to conventional desks meant that there were reductions in the rented area needed. These tools did not change the role of documents. The core cultural assumption remained that mistakes can only be detected and evaluated by deploying similarly qualified individuals, or by awaiting disaster or confusion on site. All these relatively low-key developments were actually being adopted in parallel with the development of corporate databases and authoring tools that are now starting to fulfil the promise of the so-called IT revolution of 20 years ago.

The professions are now having to face the renewed challenge as to which of their services are commodities which can be carried out by less qualified personnel and which of those functions can be automated (Susskind, 2015). This is most clearly seen in professions such as accountancy, but the same hollowing out of the lower tiers of organizations has been noted in architecture and engineering and even in religious support networks. In each case, automated information processes and web services have been seen to reduce the need for apprentices and junior staff. The senior ranks of these professions have not yet had to answer the question as to where their replacements will come from.

Ethical aspects of three early BIM implementations

The social dimension of sustainability was present during BIM's foundational period in the late 1970s. Much of the theoretical discussion was being laid down in the United States by Chuck Eastman (Eastman and Henrion, 1977) and Nicholas Negroponte (Negroponte,

Table 13.2 Ethical factors taken within time, cost, and quality considerations. (RIBA 2019)

3.2 Chartered Practices should endeavour to deliver projects that:

(a) are safe;

(b) are cost-effective to use, maintain, and service; and

(c) minimize negative impacts on the environment during their anticipated life cycle.

1970). In the United Kingdom, there were three early implementations of BIM. Each was not only motivated by the intellectual and commercial curiosity to discover if such systems were possible and effective, but were also driven by an ethical intent to produce the largest amount of social goods at the least risk and the least expenditure. One example is the development of software for the Scottish Special Housing Association. This body was at the time responsible for all the social housing being built in Scotland and so had a substantial budget. The government, through a National Research and Development Council, sponsored the development of an early BIM system which was developed at Edinburgh University by the Edinburgh University Computer Aided Architectural Design Research Unit (Bijl, 1979). The software ran on the university mainframe and competed for resources with teams making an early investigation into artificial intelligence. One feature of the system was that it was not, in modern jargon, object-based, but treated the design of housing, with its complex roof forms, trusses, staircases, hallways, and other features, as a task involving the carving up of space. The best analogy and its nickname for this approach was the "soap-bubble" approach, capturing the way each individual bubble in soap-suds shares its boundaries up against all the other soap bubbles. The software helps the designer to maintain a single representation of this fragmentation of space. Each bubble was deemed to contain either air or a particular construction material or product. A direct outcome of this strategy was that no clash detection was ever required and no duplicate components could be double counted by mistake. Plans, sections, elevations, bills of quantities, schedules, U-value calculations, and compliance with the prevailing Parker Morris standards (Parker Morris, 1961) could all be derived automatically. Changes to major materials such as a switch from blockwork to no-fines concrete were effected instantaneously and competitive tenders obtained for both constructions. The application was saving 4% of construction costs on all the housing schemes. A second example of a BIM application was developed by Applied Research of Cambridge Ltd (Hoskins, 1977) for the Oxford Regional Health Authority, which was responsible for a substantial hospital building programme across the whole Thames Valley. In this case, hospital design was being rationalized to a system of building involving a framed structure on a repeating grid with a limited library of sized components. Again, the current object-based approach was rejected. Instead the grid system was represented internally by large (but sparse) matrices. Every space was accounted for in this matrix, and each cuboid cell was occupied by a construction component or by usable space. Considerable savings were effected. Various kinds of automated checks, automated design, and automated component selection were included in the system, exploiting the controlled vocabulary. A third system is less well documented. CEDAR (Webster 1976).) was a system developed by a UK Government agency, the Property Services Agency, supervised by the Office of Public Works, and latterly used by the Department of the Environment to manage the automated design of telephone exchanges, an almost forgotten building type that was at the time a crucial element in the modernization of the telephone network. Many of the publications relating to CEDAR focussed on the issues around training professionals to adapt to interfacing with "machines".

It is significant that all three of these systems were directed at socially important building types in which government or near-government bodies had a strong interest. Each was deployed with a strong sense of duty to obtain the best quality and the best value possible for the citizen.

A fourth early BIM implementation

There was a fourth system in development at that time called RUCAPS (Davison, 1978). This application was one of the first to adopt the purely object-orientated approach that has come to dominate all the major BIM applications since. In this paradigm, physical and spatial elements can be placed and orientated freely in 3D space with limited relationships between them. It was developed specifically to handle the production drawings for what was then the largest building in the world, Riyadh University. It was developed of necessity as at that time there were not enough architectural technicians in London to support the level of production needed. It was dedicated to the production drawing task related to the university departments. Conventional 2D draughting was used for the sports stadium and other prestige buildings. The requirement was to document for each departmental wing the concrete cladding elements, the partition layouts, and the room numbering of the teaching and office space. The output required was to produce plans, all elevations, and some schedules of the concrete panel types in an efficient and accurate manner. It was successful, given this limited ambition. Although heavily constrained by the size of minicomputer available and the fact that all programmes had to operate with 256K of memory, the application evolved to be used on some major office buildings in London, including King William Street in the City and the Royal Bank of Scotland building at Angel Islington in North London. As it grew, it incorporated 3D, and in its second incarnation, as Sonata, it included colour rendering and numerous other outputs. Sonata was used for the design and documentation of the British Library, then the largest building in the United Kingdom (CStWilson, 1998). This required the synchronization of the architectural and MEP designs at fortnightly intervals by the exchange of magnetic reel tapes. At the time of the handover to the client, no attention was given to preserving the object-orientated information. Instead, a number of large and degraded DXF files were delivered. Considerable effort was then spent on trying to rediscover the object basis so as to support facility and asset management activity (Wix and McLelland, 1986).

Subsequent incarnations such as Reflex and Revit have not yet recovered the power and sophistication of the three pioneer applications mentioned above. Much of the blame for this failure to achieve those higher aspirations must be laid at the feet of a failure of imagination on the part of both designers and clients to grasp the fuller potential. Instead the focus was on productivity and accuracy in the generation of conventional drawings and documentation.

Open BIM

In 1998, buildingSMART (then called the International Alliance for Interoperability) started to make the case for BIM or to deploy the approaches developed in academia (Tolman, 1999). Early demonstrations (BuildingSMART, 1998) focussed on the transfer of asset information between authoring platforms, and then on transferring information from those platforms to applications such as thermal and structural analysis and simulations. Recent releases of the IFC (BuildingSMART, 2020) have broadened the scope of the schema to include the spatial, physical, process, and criteria aspects of buildings and most infrastructure. IFC is based on standards that ensure that even in the absence of any previous application software, a new software application can self-generate from a small number of configuration files up to the point where the information in an IFC file can be reactivated. This feature of the standard was fundamental to the design of the underlying EXPRESS and STEP standards (ISO, 1980). More immediately important is that the information remains accessible, independent of the availability of specific versions and revisions of proprietary software and licences. The key point is that OpenBIM separates out the information about an asset and makes that information independent of the applications that produced it and independent of the applications that will use it. This opens the way to the ethical considerations around access to information.

Collaborative BIM

The first and second generations of BIM authoring tools inspired BAA (British Airports Authority) to expect BIM as part of the procurement of design and construction services for Heathrow Airport's Terminal 5. On appointment, many of the successful bidders demurred from this obligation. This led to BAA instigating an alternative strategy that maximized the use of more conventional computer-aided draughting (CAD) tools, but with 3D objects to be used. The key elements of the strategy were an easily accessible common data environment (CDE) and easily generated 3D or BIM models. These together allow information to be offered in a "SHARED" state with a collaborative compact between the parties to share asset information early and often, accurately and completely. The combination of CDE and BIM was formalized in British Standards (BSI, 2007) and is now being propagated worldwide (ISO, 2018) and mandated by governments and private sector contracts (CIC, 2018). The shift of emphasis is from cooperation, legally construed as not obstructing others, towards collaboration – actively pursuing support and mutual interests.

Open requirements

OpenBIM is only one aspect of a more systematic approach to AECO. A key moment in the development of BIM occurred around the turn of the century with the marketing of the early versions of the Solibri Model Checking application (Solibri, 2009) and the deployment of the Singapore ePlanCheck system (Liebich and Wix, 2002). Solibri demonstrated that model and data completeness can be checked for. ePlanCheck demonstrated that facilities can be checked for compliance to norms including zoning, planning, layout, and civil defence. These applications demonstrated automated checking of data, geometry, and regulations and that compliance requirements can be accurately represented with the computer, offering the prospect that the dependency on manual checking, inspection, and professional judgement could be broken and BIM could move from being a production solution to being an information solution.

There is increasing interest in the concept of a digital twin since NASA introduced the concept (Glaessgen 2012). Although the core meaning is being eroded by over-use, in essence the digital twin is required to be actively engaged with the real world. It must receive information from the physical world by means of IoT sensors and conventional monitoring and reporting. It must also control and configure the physical reality to protect and improve the performance of the asset. This control may be rules based or it may yet be based on artificial intelligence (AI) technology. However, there can be no digital twin unless there is a third sibling. This third sibling must represent the objectives that are set for the asset. Without something to set the objective to optimize against, perhaps in terms of benefits over costs or outcomes over investment, there is nothing for the digital twin to do. In order to document this digital triplet, alongside "open BIM" there is a need for "open Requirements". These may represent agreed standards, government regulations, client requirements, or recommendations coming from third parties like insurers and other influencers. Just as asset information needs to be visualized for humans and accessible for machine analysis, expectations need to be expressed in forms that are both human readable and machine operable. This need for a way of capturing the norms and expectations of our built environment is a prerequisite before the creation of digital twins. The maintenance of the ePlanCheck system in Singapore was undermined by the economics of the project and the economics of keeping the computer code up to date with the norms. AEC3 has demonstrated technologies that respect the normative text but make it accessible to the computer (Nisbet et al., 2008). The RASE approach to capturing the precise meaning of the original text of any document along with appropriate dictionaries enables automated assessments of openBIM information. The ability to share those requirements and expose them may be recognized as central to the requirements of an open society (Hackett, 2018). These requirements must have an expression that is computer compatible whilst retaining their basis in ordinary written language. One of the first demonstrations of the approach to capturing in an operable form the content of normative documents was to take the "Three Laws of Robotics" as proposed by Isaac Asimov in the 1940s (Asimov, 1950). Once captured as formal logic, it was also possible to generate the electrical circuit diagram that would have to be at the heart (or cortex) of a robot. The inputs to that circuit will of course be much more challenging to engineer. The real danger from AI is not that it will not work, as increasingly it does. The real danger is that there may be a change of mind about the objectives. Researchers in AI (Russell, 2019) are now starting to think in the same terms, saying that that the objectives provided to AI must not be built into AI. It must be possible to manage and change and contradict the criteria. The objectives must be defined and maintained outside of the digital twin. Objectives should be subject to the same scrutiny as new legislation and standards. The aim is to make sure that the ethics that are developed for society remain separate from the technology that we use to achieve them.

Two issues

Ethics and effort

Traditionally, labour and effort have been rewarded on the basis that the effort required is predictable, as expressed in the Biblical quotation "The labourer is worthy of his hire" (Bible, 1611). For example, the ploughing of a field of one acre is worth a day's subsistence. The introduction of a horse, oxen, or a tractor disrupts the equation by saying that the same work can be done quicker but with a prior investment. This represents a challenge to those labourers who have not invested in horses or oxen or tractors. Are they still worthy of a day's pay when others can do it quicker? This challenge is again raising its head: for example, many traditional design activities such as the preparation of drawings and cost estimates can be substantially accelerated by the use of BIM. The ethical question becomes, should any service be charged by the hour? Does charging by the hour act as not only a disincentive to automation and efficiency but also as a disincentive to the adoption of more accurate, trustworthy, and repeatable processes? Does charging by the hour embed old ways of working with their reliance on quality control – expecting mistakes to be discovered after the event often by others and often at the expense of the interests of the project or the asset owner or society as a whole? "Lean" thinking, as exemplified in lean production and lean construction, suggests that quality assurance during a process is worth more to the outcome than quality control afterwards. Automation and information management enable continuous quality assurance by the introduction of structured methods and tools. As the simplest example, using BIM, one would expect that the act of inserting a door into a wall will automatically ensure that any plans, sections, elevations, perspectives, walkthroughs, schedules, tables, and analysis will all be automatically "updated" and guaranteed to be consistent, not because of quality control after the event, but because the very process of using information can guarantee that those outcomes remain coordinated and correct.

Ethics and outcomes

There are specific outcomes that society in general expects from all work in the built environment. Facilities and infrastructure may still fail in spite of the knowledge and experience built up over three centuries of scientific, technical, and professional development. Society may not hesitate to challenge the current arrangements if the delivered outcomes are socially unacceptable. The tragic example of Grenfell (Grenfell, 2020) confirms this. It is also a challenge that society, corporations, and private individuals continue to be repeatedly disappointed by the outcomes that the AECO sector produces in terms of cost quality and timeliness. Companies and individuals fail to deliver and fail to sustain themselves economically. In addition, there are now increasingly pressing concerns for the environment. If the built environment continues to be responsible for up to 45% of our energy consumption (DEFRA, 2020) and therefore to be responsible for that percentage of our greenhouse gas emissions, then there will be a third crisis. What underlies these social, economic/functional, and environmental failures is the failure to create a learning environment: there is no knowledge base on which to evaluate our actions and refine our proposals. Not only is the sector not sustainable in social terms,

Table 13.3 BIM can underpin the ethical outcomes

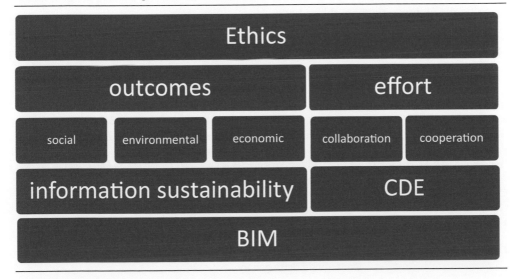

economic terms, and environmental terms but there is also an underlying weakness in the absence of sustainable information. There is not sufficient information about assets created 8, 10, or 15 years ago, let alone one or two centuries ago. Information created in the course of new projects is rapidly degraded into documents that serve only a short-term purpose such as contractual obligations. Traditional habits mean that most participants in the process are focussed on the delivery of documents to meet immediate delivery obligations without regard to the purpose of supporting the next stage of the project, whether those stages are the stages of a plan of work or the next trigger events in the in-use phase of the asset. There is little consideration of the role that the original information might have in the long-term future of the asset. Where was asbestos used in the new buildings of the 1970s? Where was high alumina cement used in the buildings of the 1960s? What materials were used in housing erected in the 1980s? Sustainable information will be crucial if the circularity or re-use of products and materials is to be achieved.

Conclusions

This chapter has illustrated the impact of BIM on ethics and of ethics on BIM. The ethical landscape is shifting so that robust criteria for social, environmental, and economic sustainability can be checked repeatedly and continuously through into the life of the asset. The qualities of the outcome that were previously "guaranteed" by professional bodies and best practice can be replaced by guarantees generated by new processes built around collaboration.

Current BIM technology has ethical content, encapsulating assumptions about the role of the users, managers, and owners. As this technology delivers a more connected information infrastructure, practitioners will need to steer the tools and their usage back towards ethical sustainability. The need is to preface discussions of social sustainability, of economic sustainability, and of environmental sustainability with a commitment to "information sustainability" – the creation and storage of information in ways that do not limit its use in the future. The prime consideration in any activity in design, engineering, construction, or occupancy should be to create information which is of lasting utility. The impact of ethics on BIM is to give urgency to using "open BIM" and "open Requirements" to create sustainable information and assets that together support both current and future society. The challenge remains to act on the ability to do more things better.

References

Asimov, Isaac (1950). "Runaround". I, Robot (The Isaac Asimov Collection ed.). New York City: Doubleday. p. 40. ISBN 0-385-42304-7

Bible, (1611). *Luke 10:7 King James Version.*

Bijl, A, (1979). Integrated CAAD systems. In A Bijl, D Stone, D Rosenthal (eds.) *EdCAAD Studies.*

BSI, (2007). *BS1192 Collaborative Production of Architectural, Engineering and Construction Information.* BSI.

BuildingSMART, (1998). *IFC1.0..* https://technical.buildingsmart.org/.

BuildingSMART, (2013) *ISO 16739: 2013 Industry Foundation Classes (IFC) for Data Sharing in the Construction and Facility Management Industries.* International Organization for Standardization.

CIC. (2018). *BIM PROTOCOL.* cic.org.uk.

Collins Peter. (1971). *Architectural Judgement. London: Faber & Faber; ISBN-10 : 0571094589.*

Davison JA, (1978). RUCAPS Cost-effective drafting for the building industry. In *CAD 78,* IPC Science and Technology Press.

Department for Environment, Food and Rural Affairs (DEFRA) (2020). *UK Government Department for Environment, Food and Rural Affairs.* randd.defra.gov.uk

DLR Group Inc. (2019). *Design Ethics: An Approach to Ethics and the Built Environment* (accessed Sept 2020). https://www.dlrgroup.com/media/733704/dlr-group-the-ethics-of-design.pdf.

Eastman C and Henrion M (1977) GLIDE: a language for design information systems. *ACM SIGGRAPH Computer Graphics*, 11, 24–33.

Glaessgen E and Stargel D. (2012). *The Digital Twin Paradigm for Future NASA and US Air Force Vehicles.* arc.aiaa.org.

Grenfell Tower. (2020). *Grenfell Tower Inquiry*. www.grenfelltowerinquiry.org.uk

Hackett, J (2018). *Independent Review of Building Regulations and Fire Safety.*

Hoskins EM (1977). The OXSYS system. *Computer Applications in Architecture.*

IEP (2020). https://www.iep.utm.edu/ethics/

ISO (1980) *ISO 10303-1(en): Industrial Automation Systems and Integration: Product Data Representation and Exchange: Part 1: Overview and Fundamental Principles.* ISO.

ISO (2018) *ISO 19650: Information Management Using Building Information Modelling.* ISO.

John Wilson C St (1998). *The Design and Construction of the British Library.* British Library.

Liebich, T. & Wix, J. & Qi, Z. (2004). Speeding-up the submission process: The Singapore e-plan checking project offers automatic plan checking based on IFC. In International Conference on Construction Information Technology (INCITE 2004). 245–252.

Morris SP. (1961). *Homes for Today and Tomorrow.* HM Stationery Office.

Negroponte N. (1970). *The Architecture Machine.* MIT Press.

Nisbet N, Wix J and Conover D. (2008). The future of virtual construction and regulation checking, In Brandon, P., Kocaturk, T. (Eds), *Virtual Futures for Design, Construction and Procurement.* Hoboken, NJ: Blackwell Publishing Ltd. ISBN:9781405170246

RIBA. (2019). *RIBA Code of Practice.* https://www.architecture.com/-/media/GatherContent/Test-resources-page/Additional-Documents/RIBA-Code-of-Practice--May-2019pdf.pdf

Russell S. (2019). *Human Compatible: Artificial Intelligence and the Problem of Control.* Penguin.

Solibri SMC. (2009). *Solibri Model Checker.* www.solibri.com

Susskind RE, Susskind D. (2015) *The Future of the Professions.* Oxford University Press.

Tolman FP. (1999). Product modeling standards for the building and construction industry: past, present and future. *Automation in Construction*, 8, 227–235.

Webster GJ and Johnson CW. (1976). The evaluation and selection of a computer system for interactive design. *Computer-Aided Design*, 8, 247–251.

Wix J and McLelland C. (1986). *Data Exchange between Computer Systems in the Construction Industry.* BSRIA.

Section 2

For educators and trainers

14 BIM teaching and learning frameworks in construction-related domains

What the literature says

Reza Taban, Mohsen Kalantari, and Elisa Lumantarna

Introduction

The rising demand for building information modelling (BIM) skills in the architecture, engineering, and construction (AEC) industry has increased the demand for BIM-enabled graduates from higher education courses in the field of AEC management. Higher education needs to adapt to these changing demands in two different ways: by utilizing BIM technology for teaching AEC concepts and by teaching to address the skill gap in the industry. Numerous studies have explored BIM capabilities, such as visual capabilities to enhance the experience of teaching the fundamentals of AEC (Jin et al., 2019), and the common suggestion is that maximizing the benefit of BIM relies on the adaptation of BIM implementation within industries. Hence, a number of studies have focused on teaching the skill competencies demanded by the immediate needs of the industry. Despite the positive efforts, BIM teaching and learning frameworks comprising overall strategy changes remain lacking due to the rate of change in BIM technology as a central challenge (Puolitaival and Forsythe, 2016).

For the past few years, academia has chosen different teaching approaches for BIM subjects at the vocational, undergraduate, and postgraduate levels. Some universities have introduced BIM-centered curricula at the undergraduate level only or as an elective subject (Brokbals and Čadež, 2017, Puolitaival and Forsythe, 2016). Other universities have planned to implement BIM in their course programmes. Naturally, restructuring a course programme takes a pedagogical study and significant effort. Those universities successful in restructuring their course programmes tend to be small institutions and are thus relatively flexible to conduct major programme changes.

In recent years, higher education institutions have taken the BIM transition seriously, and thus, the number of universities offering BIM-related subjects has increased. Meanwhile, many research studies have utilized BIM as an instrument for teaching and learning to oversee educational pedagogy (Ren and Zhang, 2014, Vimonsatit and Htut, 2016, Clevenger et al., 2015). However, the studies on pedagogical approaches to teaching and learning BIM for students in the AEC discipline are limited.

A framework allows us to organize information in the domain of knowledge and to understand and scrutinize BIM deliverables and higher education (Birmingham, 2015, Succar, 2009). Succar (2009) suggested an ontological representation of the knowledge area of the BIM framework in the industry. Zamora-Polo et al. (2019) proposed a conceptual framework of BIM education in a three-dimensional format of (1) pedagogical approach, (2) level of integration, and (3) competencies based on Succar's proposed ontological representation. This chapter explores the pedagogical approach to BIM education.

Previously introduced by Underwood et al. (2013), higher BIM education can be considered in progressive stages, namely, BIM-aware, BIM-focused, and BIM-enabled. BIM-aware education refers to subjects which ensure that graduates are aware of BIM. BIM-focused puts an emphasis on teaching students how to use BIM to perform specific tasks. BIM-enabled focuses on learning that is embedded in a virtual environment, in which BIM acts as a "vehicle" for learning. Underwood et al. argued for a progressive adaptation of BIM to education.

Witt and Kähkönen (2019) suggested that the existing BIM-enabled higher education cases can be arranged in two categories: (i) BIM acting as a learning tool, wherein traditional learning processes can be deployed through some aspects of BIM tools; and (ii) BIM acting as a learning environment (post-BIM transition), wherein learning takes place within a BIM context, such as a common platform and workflow process. The BIM role in the latter category is expected to result in a paradigm shift in the workflow of the AEC discipline. The current study analyzes and creates an integrated collection of the leading practices of BIM education pedagogy from peer-reviewed resources.

This chapter is organized into six sections. Section 1 gives a brief overview of BIM education. Section 2 presents the methodology used in this work, including a four-stage systematic review. Section 3 presents the findings of the research with a focus on the three key themes of BIM education, namely, teaching and learning BIM, curriculum, and the pedagogy of BIM education. Section 4 presents the state of BIM education in the AEC curriculum. Section 5 presents a critical review in three subsections: the first and second subsections map the proposed taxonomy and pedagogical practices and identify the current gaps and ongoing challenges of BIM in tertiary education, whilst the third subsection reviews the gaps in the theory of teaching and learning within the BIM education framework. Section 6 presents the conclusions outlining the analyzed state of the BIM education framework.

Methodology

This study focuses on the adapted theories of BIM education. The challenges and gaps outlined in the reviewed literature are investigated, and the teaching of BIM in higher education is explored. A content analysis of the selected articles and their reported case studies is conducted to understand the complexities of BIM education implementation in higher education according to the theoretical framework presented by Tight (2012).

On the basis of the works of Saunders et al. (2007) and Anfara (2008), which identified the collective and systematic approaches documented in peer-reviewed literature, this chapter presents the systematic review of the literature in four stages: 1) identifying studies discussing BIM education; 2) filtering studies discussing challenges; 3) filtering studies discussing BIM pedagogy in tertiary education; and 4) analyzing and synthesizing the studies in terms of curriculum, teaching and learning.

Stage 1 was conducted in January 2020. As shown in Figure 14.1, 302 peer-reviewed journal articles were obtained from the Scopus database. The script for searching published studies on the Scopus database is provided in Appendix A. Only articles published in the English language were considered in the literature search. The search results were filtered to identify the studies discussing the challenges in BIM education (n = 101). A total of 20 articles were identified for the final stage of the study in BIM education pedagogy. These papers were fully reviewed to establish the current BIM education pedagogy and frameworks.

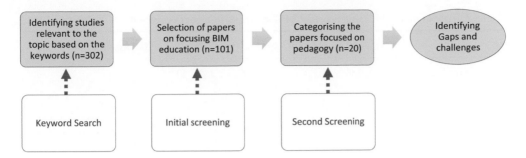

Figure 14.1 Systematic review of literature conducted in four stages.

Upon reviewing the selected papers, a qualitative analysis was conducted to identify the objectives of the studies on the basis of the following criteria:

* The study focuses on the pedagogy of BIM education;
* The study features a discussion of teaching and learning BIM;
* The study involves a discussion of the design of the BIM curriculum in higher education.

At the final stage, this study focuses on identifying the pedagogical elements, gaps, and challenges in BIM education with reference to effective education frameworks in the context of Australian higher education.

BIM education studies by teaching and learning strategies

Table 14.1 presents the approaches employed by the identified studies. These studies mostly discuss broad forms of teaching and learning strategies, such as the Bloom taxonomy and constructive pedagogy, rather than investigating specific ways of teaching and learning pedagogy. Some of the articles, such as those by Abdirad and Dossick (2016), involve a magnitude of collective analyses of previous case studies focusing on the BIM curriculum. Sacks and Pikas (2013) applied the Bloom taxonomy to their studies on undergraduate students (MacDonald and Mills, 2013, Pikas et al., 2013). Subsequently, the Bloom taxonomy has been found to be a dominantly chosen taxonomy in later BIM education studies. Only Zhang et al. (2019) explored a different learning method based on experiential learning theory. The exploration of innovative teaching and learning approaches to BIM education has been identified as an early trend of existing studies.

Figure 14.2 shows a map that outlines the discussions in BIM education. It reflects the recognized taxonomy and the different approaches to the curriculum.

BIM education in AEC courses involves tutorials and workshops in various modes: by conducting seminars or workshops (Becerik-Gerber et al., 2012, Russell et al., 2014), inviting industry professionals to lectures (Solnosky et al., 2014, Suwal et al., 2014), embedding BIM in existing courses (Huang, 2018), and integrating BIM to a single subject accompanied by a capstone subject (Ghosh et al., 2015).

BIM education fundamentally provides opportunities for interdisciplinary collaboration between disciplines. Many studies have suggested an integrated approach to teaching to shrink the gap between the AEC industry and the education sector (Forgues and Becerik-Gerber,

Table 14.1 Articles on teaching and learning BIM strategies in higher education

Study	Discipline	BT	CP	PbBL	PjBL	TBL	RBL	SL	Remark
(Abdirad and Dossick, 2016)	A,E,C	1		O	O	1	O	1	
(Ahn, Cho, and Lee, 2013)	C	1				O			Cognitive
(Asojo, 2012)	A			O					Adaptive control of thought – Rational
(Belayutham et al., 2018)	E			O	O				
(Benner and McArthur, 2019)	A	1	1	1	1	O	O	1	
(Ghosh et al., 2015)	C	1		O	1	O	O		
(Huang, 2018)	C	1			O				
(Jin et al., 2019)	A,E,C	1		1	1	1		O	
(Lee et al., 2019)	A,E,C	1			O	O	1		Tyler's curriculum model / Cognitive structure
(MacDonald and Mills, 2013)	A,E,C	1		O	O	O			IMAC
(Pikas et al., 2013)	CEM	1		1	1	O		O	
(Puolitaival and Kestle, 2018)	A, E, C	1		1	1	1	1		
(Sacks and Pikas, 2013)	CEM	1		O	1	1		O	
(Wang and Leite, 2014)	C			1	1	1		1	
(Wu and Luo, 2016)	C			1	1	1		1	
(Xu et al., 2018)	E			O	O	1	1	O	
(Zhang et al., 2016)	CEM	1		1	1	1		O	
(Zhang et al., 2018a)	C, EM			1	1	1	1	O	
(Zhang et al., 2018b)	CEM	1		1	O	O			
(Zhang et al., 2019)	CE, C			O	1	1		1	Kolb's experiential learning

1   Direct T/L approach   O   Indirect T/L approach

A | Architecture   BT | Bloom's taxonomy   PbBL | Problem-based learning
E | Engineering   CP | Constructive   PjBL | Project-based learning
EM | Engineering management   pedagogy   TBL | Team-based learning
C | Construction (management)   RBL | Role-based learning
CE | Civil engineering   SL | Self-learning
CEM | Civil engineering and
   management

2013). A number of studies have identified cases exhibiting interdisciplinary collaboration and integrated learning. However, Solnosky et al. (2015) suggested the need for BIM education for each discipline of AEC at the early stages of course programmes and the necessity of transforming existing knowledge into an interdisciplinary learning framework, such as a capstone multidisciplinary design subject.

Following the discussion of the trends in BIM education, the next section reviews the state of BIM education.

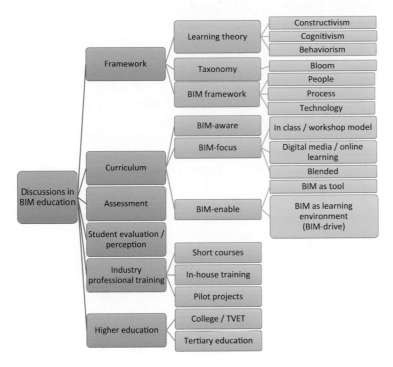

Figure 14.2 Conceptual map of pedagogical discussion in BIM education.

State of BIM education for the AEC curriculum

The AEC industry has undergone its most significant changes over the last 25 years. Consequently, the curriculum at universities has been in a constant state of change, adapting technological advancements and BIM (Clark and Shanna, 2012). Cocchiarella (2016) presented an epistemology which deals with filling the gap between traditional practices and BIM practices as a new dimension of knowledge.

Developing BIM subjects in a relatively new degree programme grants academics the opportunity to introduce BIM-integrated subjects with few restrictions. The option to offer advanced BIM subjects is thereby advocated (Wang and Leite, 2014, Puolitaival and Forsythe, 2016, Forsythe et al., 2013, Sacks and Pikas, 2013). Some academics suggest having interdisciplinary subjects to encourage multidisciplinary approaches. MacDonald and Mills (2013) indicated that some BIM topics should be taught as interdisciplinary topics or be infused across multiple subjects to encourage multidisciplinary approaches. Frameworks for BIM-related learning outcomes have been discussed for AEC education in some universities. However, BIM is currently only taught in a limited number of subjects, which are typically electives rather than core subjects.

Badrinath et al. (2016) differentiated BIM education in terms of national contexts, disciplines, and subject levels in tertiary education. Similarly, Abdirad and Dossick (2016) suggested the need for diverse research designs and settings to bridge the gaps identified in BIM curriculum research. The study highlighted the lack of case studies on the challenges of graduate-level BIM subjects in AEC course programmes which have diverse students with different prior education, skills, and experience.

Addressing the BIM pedagogy, Jin et al. (2019) discussed technical interoperability issues. They introduced the integrated design and construction delivery approach for BIM implementation under the UK BIM Level 2/3 framework. Brokbals and Čadež (2017) itemized the gap between the target and the current state of BIM education, especially in BIM modules with interdisciplinary BIM projects, in which students need to increase their experience in the new work style and communication processes.

In practice, several studies have reported the integration of BIM as a part of existing curricula. For example, Forsythe et al. (2013) suggested a programme-wide implementation strategy in which individual subjects tap into BIM according to the specific learning outcomes of subjects. This approach proposes to address the areas of knowledge with a small number of BIM models as representative of building types throughout an undergraduate degree programme. Furthermore, Forsythe et al. asserted that the utilization of different problems in BIM models accelerates learning potential across a variety of learning specialization. Lee et al. (2019) argued that curricula need to impart the necessary BIM skills to accommodate industry requirements.

The perceived BIM-enabled graduates of vocational and higher education institutions may vary depending on whether they are fresh graduates or BIM-trained industry professionals. According to the findings of Clevenger et al. (2010), three strategies are used to adapt BIM to AEC education:

- Developing a single subject in BIM covering the generic BIM principles;
- Incorporating BIM adaptation procedures into existing core topics to enhance BIM specialization;
- Enabling capstone subjects to provide students with multidisciplinary specialization experience.

BIM as a single subject has been offered as a core or elective subject in many universities. This offering has been noted as insufficient because of the scaffolding principle of students' knowledge during the study. Wu and Issa (2014) suggested that a single subject can be disruptive because of the potential mixed messages on BIM topics associated with other subjects. Ahn et al. (2013) suggested inputs from industry professionals for developing learning objectives for a new subject on the basis of the lessons learned from the limitations of the current objectives in BIM subjects in construction programmes.

Puolitaival and Forsythe (2016) reported some cases wherein entire undergraduate programmes utilizing BIM have been developed with a focus on up to two BIM training subjects, with the remaining subjects using BIM as a visualization tool. The benefit of scaffolding can be utilized if training opportunities are provided in multiple subjects. The diversity of students' prior learning, including their technical knowledge and digital literacy, affects the curriculum design and learning outcomes. For example, the design of a BIM subject at a basic level may exhaust some students. An option is to have a mid-level curriculum and provide students with supporting materials in addition to the allocation of workshops to satisfy the needs of students who require knowledge of the basic principles.

Sacks and Pikas (2013) categorized the perceived intended learning objectives of BIM education into process, technology, and application. The intended learning objectives need to include the people-related areas of BIM integration, such as collaboration, and interpersonal skills in the BIM process, including open-mindedness, ability to adapt, self-learning ability, communication, leadership, teamwork, and engagement suitable to the BIM context. Additionally, Zhang et al. (2018a) presented the key competencies and abilities required in

AEC management by outlining the skills and knowledge related to each discipline. This study focused on the preparation of the knowledge from fundamental to the core of each discipline and the people abilities outlined in the category of teamwork in the intended learning objectives.

Wu and Issa (2014) performed a comparative analysis of stakeholder perceptions in educational and professional communities. The survey findings suggested substantial gains in BIM adoption and implementation in both communities and identified a gap between the rapid growth of the BIM-related job market and the incentives in place to encourage students to commit to a BIM-oriented career path. Sacks and Pikas (2013) analyzed the university education gap leading to the compilation of a framework for the development of BIM content for undergraduate and graduate construction engineering and management degree programmes. The requirements and framework are intended to provide educators with essential knowledge as they develop and implement BIM content in their programmes.

From the assessment perspective, Rubenstone (2007) mentioned the difficulties in BIM adaptation in engineering and the lack of proper textbooks on BIM such that students seek results from vendor-based and free web-based materials. Reports have identified that the lack of resource standardization, information security, BIM contracts, and change management stop universities from offering these areas. The lack of BIM management in the curriculum has also been reported; the education limitation seems to prevent students from learning beyond the understanding of BIM management. This section reviews the state of BIM education in AEC. The theoretical and pedagogical frameworks of BIM education are discussed subsequently.

Theoretical and pedagogical frameworks and gaps

Theoretical frameworks

Many BIM educators refer to the Bloom taxonomy as the principle of BIM education. Sacks and Pikas (2013) utilized the principles of Bloom's taxonomy to establish a teaching and learning framework to meet the competency levels for construction engineering and management graduates and professionals.

Bloom's taxonomy (Bloom et al., 1984) identifies three domains of educational activities: (1) cognitive, i.e. mental skills (knowledge); (2) affective, i.e. growth in feeling or emotional areas (attitude); and (3) psychomotor, i.e. manual or physical skills (skills). This taxonomy is defined in six categories of the hierarchical framework of achievement, in which a complex skill or ability is built on prior knowledge (Krathwohl, 2002). The original Bloom taxonomy categorized cognitive domains according to the competency levels of "knowing, understanding, applying, analysing, synthesizing and evaluating" (defined based on Bloom's taxonomy in Bloom et al., 1956). Categorizing BIM competency principles in three groups, Sacks and Pikas (2013) introduced BIM processes, BIM technologies, and BIM applications based on the original Bloom taxonomy. The recent studies have revised the hierarchy of Bloom's taxonomy to "remember, understand, apply, analyse, evaluate and create" (Anderson and Krathwohl, 2001).

BIM education studies lack a pedagogical analysis approach. For example, Zhang et al. (2016) adopted the revised Bloom taxonomy in the context of BIM education. In their studies, "analyse, evaluate and create" are located parallel at the highest hierarchy level over "apply, understand and remember", respectively. The Bloom taxonomy has also been incorporated into a four-year construction management (CM) undergraduate programme (Huang, 2018).

A three-level competency framework is introduced, and it consists of the fundamental level of BIM contents in CM courses, an application level of BIM for solving real-world problems, and an advanced level of BIM topics in CM courses.

However, some BIM education studies do not establish the relative importance of competency levels. Most of the studies focus on identifying the competencies which have been considered to be most important in the industry and AEC degree programmes. For example, Gerber et al. (2015) reported that BIM authoring, energy simulation, BIM-based collaboration, model-based estimation, and 4D simulation are the most taught BIM uses and competencies. Ku and Taiebat (2011) found that constructability and visualization are currently required BIM uses, model-based estimation and cost control are short-term BIM uses, and facility management and energy analysis are long-term priorities. Wong et al. (2011) suggested three BIM educational approaches for the Hong Kong Polytechnic's nominating student-centered curriculum and emphasized a collaborative design and an integrated curriculum. Wong's model presents BIM applications in productivity analysis, plan reading skills, and the transformative context.

Abdirad and Dossick (2016) reported that industry professionals use varying rankings of several BIM uses and competencies, which are considered the most important subjects in BIM curricula, in different studies and disciplines. On the basis of these studies, educators can prioritize subjects for BIM implementation when there is insufficient room for a comprehensive integration of BIM into the AEC curricula. Naturally, the identified competencies deem to follow the local industry demand for the required competencies.

MacDonald and Mills (2013) suggested the incorporation of the benchmarking component of the IMAC (illustration, manipulation, application, collaboration) framework into existing subjects to plot targets for future curriculum developments. The IMAC framework considers a teaching delivery method at four stages of illustration, manipulation, application, and collaboration to achieve profound levels of learning as students progress through their education. The framework recognizes the different collaboration targets for different disciplines. For example, architecture graduates are expected to create BIM models from scratch, whereas engineers and construction managers are expected to manipulate and review the existing models for their own analysis purposes. The study also suggested how BIM education can incorporate students from different AEC disciplines into study programmes of varying levels, lengths, and skills. This framework focuses on the practical application of BIM tools as the core of education and seeks collaboration as the highest stage in the framework.

Pedagogical frameworks

The current BIM education uses a variety of approaches, including team-based learning (TBL), role-based learning, project-based learning, problem-based learning, and self-learning. The subsection shows that project-based learning, TBL, and problem-based learning are the most utilized teaching and learning methods for simulating industry cases. These methods are driven by the BIM-based communication environment wherein BIM tools are utilized as the communication environment.

TBL is discussed in 50% of the selected studies. Half of the discussions regard TBL as an effective structure of teaching and learning in BIM education, whilst the rest seems to emphasize the value of the concept. For example, team assignments in BIM subjects are described as a way to enable students to experience and learn collaboration, integration, and teamwork. Mathews (2013) reported discussion groups as a success factor in BIM implementation.

Parfitt et al. (2013) found that comprehensive learning will not be achieved on capstone subjects because each student may focus only on one aspect of an assignment.

Most TBL studies in the context of BIM education use an experimental approach rather than a theoretical approach. In an innovative experimental study, Ghosh et al. (2015) applied a vertical integration of collaboration in a BIM-enabled CM course. In this course, first-year students had to collect and prepare information as BIM inputs for senior-level students who had to use BIM for a site logistic planning assignment. Although the first-year students did not use BIM software, they developed an appreciation for collaborative information management and BIM implementation before using the application. Zhao et al. (2015) focused on theoretical implications and suggested that TBL has a standpoint of fulfilling knowledge and competencies, such as discipline integration, collaboration, the use of different BIM processes and roles and collaborative learning for students with different skills and knowledge levels.

Teamwork skills are one of the skills required in advanced professional workplaces that use BIM. Universities have thus established collaborative BIM subjects which offer collaborative tasks between local students or two or more campuses in virtual settings; this approach is gaining momentum in AEC education (Soibelman et al., 2011, Pikas et al., 2013). However, the level of success has not been recorded; nevertheless, instructors believe that the subjects are successful in terms of collaborative BIM and implementing and dealing with challenges associated with distributed teams. The problems introduced to the students in these subjects are intended to be similar to those in industry practice, such as time zone differences, task coordination, interoperability and data-sharing issues, and the development of interpersonal relationships in the workplace and virtual worlds.

TBL and project-based learning are delivered simultaneously in AEC education. Zhang et al. (2018b) remarked that the practical cognitive process mapping in information exchange assists students' learning evaluation and that a lack of preparation of prior knowledge can be common. AEC programmes have experimented on global collaborations between students from different countries. This form of collaboration has some advantages in teamwork settings in the contexts of international construction and globalization, in which BIM development varies across countries. It has been recorded as a challenging procedure across different schools, along with the need to maintain the scope of work and manage communication technologies and language barriers (Liu and Kramer, 2011).

Self-learning can be considered a learning objective in the digital era because of the characteristics of the BIM process. Becerik-Gerber et al. (2012) reported the advantages of the flexibility of self-learning for students who have to self-learn different BIM tools whilst being taught a few tools by instructors. Gier (2008) suggested that instructors in BIM-enabled subjects need to help students engage in experiencing and constructing the knowledge of BIM implementation workflows. Students should initiate their active participation in BIM subjects, and instructors should carefully design subject activities and materials to enable the self-construction of knowledge. Meeting these needs requires instructors to guide and facilitate students rather than demonstrate knowledge (Abdirad and Dossick, 2016). The following subsection reviews the gaps in the teaching and learning frameworks for BIM education.

Gaps in teaching and learning frameworks for BIM education

Theoretical approaches to teaching and learning can be explored fully in different aspects of BIM education. Most studies on BIM education in the AEC field only explore cognitivism approaches; others focus on constructivism, whilst a few explore workplace learning theories. Interestingly, only a handful of studies have investigated the learning theories behind the approaches. For example, prior knowledge and skills in the constructivism approach, especially within the postgraduate level of BIM education, have yet to be fully identified.

Most BIM studies mention the Bloom taxonomy, whilst higher educators generally focus on cognitivism, constructivism, and behaviourism (Nagowah and Nagowah, 2009). Nonetheless, cognitive structure is the most widely used teaching theory in the pedagogical approach in BIM education. For example, Zhang et al. (2016) suggested the cognitive structure of BIM in education whilst following the Bloom taxonomy.

Teaching and learning theories have broadly developed, thus granting the opportunity to revisit successful approaches. MacDonald and Mills (2013) proposed an integrated project delivery approach to construction education to reflect the collaborative approach successfully used in the industry. They introduced the IMAC framework to integrate BIM into the existing curriculum. However, BIM advancement requires the development of BIM education for the fulfilment of other advanced levels of competency and the adaptation of the revised Bloom taxonomy, which introduces the revised cognitive process of "create" as a new process level.

Puolitaival and Forsythe (2016) argued that students' model development shows the benefits of understanding complex geometries but still suggested that difficulties exist in teaching a single BIM in the context of construction and project management practices. Advanced BIM education, such as a developed BIM curriculum at the higher level of competency required in the industry, and a comprehensive integration of advanced BIM curricula into existing programme degrees remain lacking.

Conclusion

The successful implementation of BIM in AEC management curricula has been a challenge for universities. The organizational structure of big universities is not considered to be the best hierarchical model to respond to the rapidly changing BIM literacy affecting the industry.

Many studies focus on BIM education and identify competency skills, whilst a few explore the educational approach to achieve an educational framework. A large number of articles focus on the skills required for BIM literacy; however, BIM adaptation requires new teamwork skillsets for BIM-enabled workplaces relative to traditional workplaces. A number of BIM educators use their subjects as a simulation of the workplace. This method has its limitations, such as the availability of materials and scenarios and the restrictions in the time frame, resources, and assessment criteria.

Several successful methodologies have been mentioned in previous studies, and they include Kolb's experiential learning and cross-disciplinary teamwork design, which is most suitable for capstone subjects. Other teaching and learning methods, such as TBL, project-based learning, and problem-based learning, may be ideal for a single BIM subject, subjects with BIM content within the curriculum, and capstone subjects. The review of the educational principles in the area of BIM reveals that project-based learning, TBL, and problem-based learning are the most common teaching methods. These methods are encouraged by the BIM-based communication environment. The advancement of TBL depends on the availability of BIM tools and the delivering discipline.

Studies also suggest self-learning as a substitute for face-to-face learning in the context of a shortage of resources. The popularity of project-based learning among BIM educators reveals the suitability of this pedagogy for the development of the required skill sets. Many of the studied articles reveal the importance of project-based learning that utilizes real-world exercises.

The continuous changes in the BIM landscape also present a challenge for the education sector. This challenge intensifies in the accreditation process with professional associations. Focusing on BIM tertiary education, this work identifies the following challenges associated with teaching and learning strategies in BIM education:

- Further investigations into the suitability of advanced pedagogical strategies implemented in BIM education are necessary.
- The current curriculum has not been able to impart the BIM skills necessary to accommodate the requirements of the industry.
- The standards which can be used as a reference in BIM education and as a baseline and specialization in different disciplines are not enough.
- Further research should explore the AEC disciplines in tertiary education to develop a framework that allows a multidisciplinary approach in BIM education to fill the gaps in the technical and managerial skill sets of AEC graduates.

This chapter explains that BIM educators are looking for a structured way to establish structured learning outcomes on the basis of the cognitive domain of the Bloom taxonomy with a focus on knowledge, whilst disregarding the affective and psychomotor domains. Structured learning outcomes do not prepare current students for the evolving industry. There is a need to investigate the adaptive framework for BIM education to satisfy the future needs of the industry.

References

Abdirad, H. & Dossick, C. S. 2016. BIM curriculum design in architecture, engineering, and construction education: A systematic review. *Journal of Information Technology in Construction*, 21, 250–271.

Ahn, Y. H., Cho, C. S. & Lee, N. 2013. Building information modeling: Systematic course development for undergraduate construction students. *Journal of Professional Issues in Engineering Education and Practice*, 139, 290–300.

Anderson, L. & Krathwohl, D. 2001. *A Taxonomy for Learning, Teaching, and Assessing: A Revision of Bloom's Taxonomy of Educational Objectives*, New York: Addison Wesley Longman.

Anfara, V. A. 2008. Theoretical frameworks. In *The SAGE Encyclopedia of Qualitative Research Methods*. California: SAGE, 870–874.

Asojo, A. O. 2012. An instructional design for building information modeling (BIM) and revit in interior design curriculum. *Art, Design and Communication in Higher Education*, 11, 143–154.

Badrinath, A. C., Chang, Y. T. & Hsieh, S. H. 2016. A review of tertiary BIM education for advanced engineering communication with visualisation. *Visualization in Engineering*, 4, 1–17.

Becerik-Gerber, B., Ku, K. & Jazizadeh, F. 2012. BIM-enabled virtual and collaborative construction engineering and management. *Journal of Professional Issues in Engineering Education and Practice*, 138, 234–245.

Belayutham, S., Zabidin, N. S. & Ibrahim, C. K. I. C. 2018. Dynamic representation of barriers for adopting building information modelling in Malaysian tertiary education. *Construction Economics and Building*, 18, 24–44.

Benner, J. & Mcarthur, J. J. 2019. Data-driven design as a vehicle for BIM and sustainability education. *Buildings*, 9, 103.

Birmingham, S. 2015. *Higher Education Standards Framework (Threshold Standards) 2015*, Canberra, Australian Capital Territory: Commonwealth of Australia.

Bloom, B. S., Engelhart, M. D., Furst, E. J., Hill, W. H. & Krathwohl, D. R. 1956. *Taxonomy of Educational Objectives: The Classification of Educational Goals*, London: Longmans.

Bloom, B. S., Krathwohl, D. R. & Masia, B. B. 1984. Bloom taxonomy of educational objectives. In *Allyn and Bacon*, London: Pearson Education.

Brokbals, S. & Čadež, I. 2017. Academic teaching of BIM – development – status quo – demand for action. *Bautechnik*, 94, 851–856.

Clark, C. & Shanna, S.-M. 2012. Applying BIM in design curriculum. In *Computational Design Methods and Technologies: Applications in CAD, CAM and CAE Education*, Hershey, PA: IGI Global.

Clevenger, C., Lopez Del Puerto, C. & Glick, S. 2015. Interactive BIM-enabled safety training piloted in construction education. *Advances in Engineering Education*, 4, n3.

Clevenger, C. M., Ozbek, M., Glick, S. & Porter, D. 2010. Integrating BIM into construction management education. In EcoBuild Proceedings of the BIM-Related Academic Workshop, Washington, D.C.

Cocchiarella, L. 2016. Bim: Dimensions of space, thought, and education. *Disegnarecon*, 9, 3.1–3.5.

Forgues, D. & Becerik-Gerber, B. 2013. Integrated project delivery and building information modeling: Redefining the relationship between education and practice. *International Journal of Design Education*, 6, 47–56.

Forsythe, P., Jupp, J. & Sawhney, A. 2013. Building information modelling in tertiary construction project management education: A programme-wide implementation strategy. *Journal for Education in the Built Environment*, 8, 16–34.

Gerber, D. J., Khashe, S. & Smith, I. F. C. 2015. Surveying the evolution of computing in architecture, engineering, and construction education. *Journal of Computing in Civil Engineering*, 29, 04014060.

Ghosh, A., Parrish, K. & Chasey, A. D. 2015. Implementing a vertically integrated BIM curriculum in an undergraduate construction management program. *International Journal of Construction Education and Research*, 11, 121–139.

Huang, Y. 2018. Developing a three-level framework for building information modeling education in construction management. *Universal Journal of Educational Research*, 6, 1991–2000.

Jin, R., Zou, Y., Gidado, K., Ashton, P. & Painting, N. 2019. Scientometric analysis of BIM-based research in construction engineering and management. *Engineering, Construction and Architectural Management*, 26, 1750–1776.

Krathwohl, D. R. 2002. A revision of Bloom's taxonomy: An overview. *Theory into practice*, 41, 212–218.

Ku, K. & Taiebat, M. 2011. BIM experiences and expectations: The constructors' perspective. *International Journal of Construction Education and Research*, 7, 175–197.

Lee, S., Lee, J. & Ahn, Y. 2019. Sustainable BIM-based construction engineering education curriculum for practice-oriented training. *Sustainability*, 11.

Liu, J. & Kramer, S. 2011. East meets west: Teaching BIM in a study abroad class with Chinese & American university students. In 47th Associated Schools of Construction Annual International Conference, University of Nebraska-Lincoln, Omaha, NE.

Macdonald, J. & Mills, J. 2013. An IPD approach to construction education. *Construction Economics and Building*, 13, 93–103.

Mathews, M. 2013. BIM collaboration in student architectural technologist learning. *Journal of Engineering, Design and Technology*, 11, 190–206.

Nagowah, L. & Nagowah, S. 2009. A reflection on the dominant learning theories: Behaviourism, cognitivism and constructivism. *International Journal of Learning*, 16, 279–285.

Parfitt, M. K., Holland, R. J. & Solnosky, R. L. 2013. Results of a pilot multi-disciplinary BIM-enhanced integrated project delivery capstone engineering design course in architectural engineering. In AEI 2013: Building Solutions for Architectural Engineering – Proceedings of the 2013 Architectural Engineering National Conference, State College, PA, 43–52.

Pikas, E., Sacks, R. & Hazzan, O. E. 2013. Building information modeling education for construction engineering and management. II: Procedures and implementation case study. *Journal of Construction Engineering and Management*, 139, 05013002.

Puolitaival, T. & Forsythe, P. 2016. Practical challenges of BIM education. *Structural Survey*, 34, 351–366.

Puolitaival, T. & Kestle, L. 2018. Teaching and learning in AEC education-the building information modelling factor. *Journal of Information Technology in Construction*, 23, 195–214.

Ren, S. & Zhang, W. 2014. Application of BIM software in construction design education. *World Transactions on Engineering and Technology Education*, 12, 432–436.

Rubenstone, J. 2007. As new tool lands on more campuses, students seek 'A' in BIM. *ENR (Engineering News-Record)*, 259, 40–43.

Russell, D., Cho, Y. K. & Cylwik, E. 2014. Learning opportunities and career implications of experience with BIM/VDC. *Practice Periodical on Structural Design and Construction*, 19, 111–121.

Sacks, R. & Pikas, E. 2013. Building information modeling education for construction engineering and management. I: Industry requirements, state of the art, and gap analysis. *Journal of Construction Engineering and Management*, 139, 04013016.

Saunders, M., Lewis, P. & Thornhill, A. 2007. Research methods. In *Business Students*, 4th edition, London: Pearson Education Limited.

Soibelman, L., Sacks, R., Akinci, B., Dikmen, I., Birgonul, M. T. & Eybpoosh, M. 2011. Preparing civil engineers for international collaboration in construction management. *Journal of Professional Issues in Engineering Education and Practice*, 137, 141–150.

Solnosky, R., Parfitt, M. K. & Holland, R. 2015. Delivery methods for a multi-disciplinary architectural engineering capstone design course. *Architectural Engineering and Design Management*, 11, 305–324.

Solnosky, R., Parfitt, M. K. & Holland, R. J. 2014. IPD and BIM-focused capstone course based on AEC industry needs and involvement. *Journal of Professional Issues in Engineering Education and Practice*, 140, A4013001.

Succar, B. 2009. Building information modelling framework: A research and delivery foundation for industry stakeholders. *Automation in Construction*, 18, 357–375.

Suwal, S., Jäväjä, P. & Salin, J. 2014. BIM Education: Implementing and reviewing "OpeBIM": BIM for teachers. *Computing in Civil and Building Engineering*, 2014, 2151–2158.

Tight, M. 2012. *Researching Higher Education*, Maidenhead, UK: McGraw-Hill Education.

Underwood, J., Khosrowshahi, F., Pittard, S., Greenwood, D. & Platts, T. 2013. Embedding building information modelling (BIM) within the taught curriculum: Supporting BIM implementation and adoption through the development of learning outcomes within the UK academic context for built environment programmes, Higher Education Academy. <https://www.heacademy.ac.uk/system/files/bim_june2013.pdf>

Vimonsatit, V. & Htut, T. 2016. Civil engineering students' response to visualisation learning experience with building information model. *Australasian Journal of Engineering Education*, 21, 27–38.

Wang, L. & Leite, F. 2014. Process-oriented approach of teaching building information modeling in construction management. *Journal of Professional Issues in Engineering Education and Practice*, 140, 04014004.

Witt, E. & Kähkönen, K. 2019. BIM-enabled education: A systematic literature review. In 10th Nordic Conference on Construction Economics and Organization (Emerald Reach Proceedings Series, Volume 2). Emerald Publishing Limited, Bingley, 261–269.

Wong, K.-D. A., Wong, F. K. & Nadeem, A. 2011. Building information modelling for tertiary construction education in Hong Kong. *Journal of Information Technology in Construction*, 16, 467–476.

Wu, W. & Issa, R. R. A. 2014. BIM education and recruiting: Survey-based comparative analysis of issues, perceptions, and collaboration opportunities. *Journal of Professional Issues in Engineering Education and Practice*, 140, 04013014.

Wu, W. & Luo, Y. 2016. Pedagogy and assessment of student learning in BIM and sustainable design and construction. *Journal of Information Technology in Construction*, 21, 218–232.

Xu, J., Li, B. K. & Luo, S. M. 2018. Practice and exploration on teaching reform of engineering project management course in universities based on bim simulation technology. *Eurasia Journal of Mathematics, Science and Technology Education*, 14, 1827–1835.

Zamora-Polo, F., Luque-Sendra, A., Sánchez-Martín, J. & Aguayo-González, F. 2019. Conceptual framework for the use of building information modeling in engineering education. *International Journal of Engineering Education*, 35, 744–755.

Zhang, J., Wu, W. & Li, H. 2018a. Enhancing building information modeling competency among civil engineering and management students with team-based learning. *Journal of Professional Issues in Engineering Education and Practice*, 144, 05018001.

Zhang, J., Xie, H. & Li, H. 2016. Exploring the cognitive structure and quality elements: Building information modeling education in civil engineering and management. *International Journal of Engineering Education*, 32, 1679–1690.

Zhang, J., Xie, H. & Li, H. 2018b. Project based learning with implementation planning for student engagement in BIM classes. *International Journal of Engineering Education*, 35, 310–322.

Zhang, J., Xie, H., Schmidt, K., Xia, B., Li, H. & Skitmore, M. 2019. Integrated experiential learning-based framework to facilitate project planning in civil engineering and construction management courses. *Journal of Professional Issues in Engineering Education and Practice*, 145, 05019005.

Zhao, D., Mccoy, A. P., Bulbul, T., Fiori, C. & Nikkhoo, P. 2015. Building collaborative construction skills through BIM-integrated learning environment. *International Journal of Construction Education and Research*, 11, 97–120.

15 Educating the "T-shaped" BIM professional

Lessons from academia

Igor Martek, Wei Wu, Mehran Oraee, and M. Reza Hosseini

The need for "T-shaped" BIM professionals

The knowledge economy calls for T-shaped professionals

Professional services are a fundamental driver of developed economies. The unit of market exchange for such services is "knowledge". Knowledge is traded for profit and created to improve service innovation and preserve a competitive advantage. The trade, exchange and manufacture of knowledge run so deep in modern economies that terms such as "knowledge economy" or "knowledge-based society" are commonplace. Barile et al. (2015) have defined the knowledge economy as one displaying four trends: 1) dissolution of the value chain, 2) globalization of competition, 3) convergence of technologies, and 4) integration of knowledge. A firm's competitive success rests increasingly on its ability to leverage value from intangible knowledge rather than physical resources. Productive capital seeks out cognitive flexibility rather than material efficiency. In such a scenario, self-contained linear value-chains give way to open-ended strategic networks. This, in turn, leads to a conflating of commercial boundaries. Industries such as construction, manufacturing, advanced materials, transport and logistics, telecommunications, electronics, and biotechnology become merged, integrated, and hybridized. In the global economy, economies of scale remain important, but economies of flexibility much more so.

Paralleling the shift from a resource-based view of the economy to a knowledge-based view has been a shift in our understanding of "expertise". Three waves are evident from the "history of science" studies. A century ago, with the dramatic rise of industrial and scientific achievement, the world was in awe at the power of science to dramatically improve the lives of people: delivering hygiene and medicine, radio and telephone, rail and automotive travel, and bridges, high-rise buildings, and other engineering achievements. The emphasis lies in understanding and duplicating recipes for success across these accomplishments (Woolgar and Latour, 1986, Collins and Evans, 2002). An expert was someone who knew that recipe. In the 1960s, Post-Modernist ideologies took hold, viewing scientific knowledge as no different from other forms of knowledge – that it was subjective and contextual. To "social constructivists", science was a platform for imposing one worldview over others that were equally valid (Beck et al., 1994). Think of alternate healing therapies or aboriginal "dream-time" explanations for natural phenomena. In the domain of building construction, primitive engineering feats, such as an Inuit igloo or traditional Khmer bamboo hut, might compete for admiration with a modern mega-structure, such as Dubai's Burj Khalifa.

In the 2000s, the third iteration of expertise emerged. "Knowledge science" is a term that considers not only expert knowledge but also experience. In this understanding, there are many kinds of knowledge derived from the layered application of subject expertise over many and varied situations. As one applies a received skill set, interaction with the context of application generates nuances that refine that skill set – experience. Put simply, most people recognize that years of applying knowledge to many and varied problems generate "wisdom". From a theoretical perspective, the key notion is that the object of knowledge application does not remain inert but rather engages with the expert to refine and develop their expertise. From a practical perspective, no two equivalently trained experts will, over time and with different practical experiences, share identical "expertise" (Laitila, 2009).

Shotter (2010) identified three kinds of knowledge: Knowing "what", "how", and "from". The first two types of knowledge are familiar. "What" refers to understanding the inherent qualities of an object, such as a steel girder – its engineering properties. "How" refers to processes – the steps required to bring it to the site, lift it into place, and fix it in position. Both these kinds of knowledge can be codified in texts and taught to novices through traditional classroom formats (Gibbons et al., 1994). The third kind of knowledge – "From" – cannot be codified (written down). It is therefore described as "tacit knowledge". It arises from interaction with the object of expert knowledge. Furthering the analogy it emerges from working with and installing steel girders across multiple projects and purposes, under varying conditions and demands. Tacit knowledge is an introspective insight arising from the application of expert knowledge in specific contexts. It is highly personal and unique (Nonaka and Takeuchi, 1995). In mapping these three forms of knowledge, codified knowledge – the knowledge that can be taught and learnt passively ("What" and "How") – is represented vertically. Tacit knowledge – the knowledge that comes from experience and insight ("From") – is represented horizontally. Greater depth of knowledge speaks to greater formal training. A greater breadth of knowledge speaks to a broader scope of experience.

Given that industries, including and especially construction, compete in knowledge economies, professionals must operate with appropriate knowledge proficiencies. Specifically, this includes both the vertical (technical) dimension of knowledge, as well as the horizontal (application) dimension. This gives rise to the concept of the "T-shaped" knowledge worker. The term was first conceived by David Guest in 1991 and popularized by Tim Brown, CEO and president of IDEO, a prominent product design firm in California (Rogers and Freuler, 2015). The proficient T-shaped professional is endowed with a specific disciplinary knowledge skill set, while also adapting, innovating, and applying that skill set across traditional boundaries in new applications. The concept was developed in the late 1990s from management studies of effective problem-solvers and change managers operating in large organizations. It was observed that successful organizations evolve at a pace faster than the rate of change occurring within their competitive environment. This adaptive change was facilitated invariably by people endowed with deep specialized disciplinary or functional knowledge but who were able to find fresh applications of that knowledge to capture new and uncharted markets. These persons are the epitome of what is now termed the T-shaped professional. They possess vertical "competencies" that also reach horizontally to generate new "capacities". It is this ability to apply deep knowledge in wholly new ways that ignites competitive advantage (McIntosh and Taylor, 2013). See Figure 15.1.

The T-shaped BIM professional

In construction, the "building information modelling" (BIM) platform is the core tool of knowledge management. According to ISO 19650, BIM uses a shared digital representation of a built asset to facilitate design, construction, and operational processes, and represent these activities as a reliable basis for decision-making. It is about extracting benefits

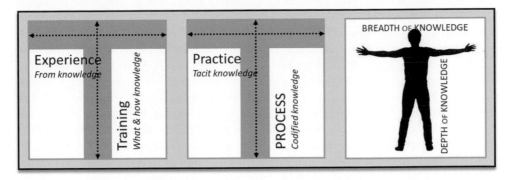

Figure 15.1 Knowledge paradigms defining the "T-shaped" professional. Source: Compiled by authors.

through better specification and delivery of just the right amount of information concerning the design, construction, operation, and maintenance of buildings and infrastructure, using appropriate technologies (ISO, 2018). At its core, BIM is a life-cycle approach to better design, procurement, assembly, and operation capital facilities, in the process of which a broad spectrum of stakeholders are involved, generating and consuming data and information to an unprecedented extent (Eastman et al., 2011). Historically, the construction industry has evolved into silos of thought and expertise. To successfully implement BIM, however, industry professionals must transform into skilled practitioners who can engage collaboratively with other disciplines at multiple levels. Those silos, while vital, need to become more permeable. This transition will herald a need for significant domain-specific knowledge, but in conjunction with closer collaboration between disciplines. Prior research literature, e.g., Sacks and Pikas (2013), Succar and Sher (2014), Wu and Issa (2014), and Hardin and McCool (2015), acknowledges the unique skill-set requirements of BIM to encompass not only technical dimensions but also interpersonal and teamwork-based competencies that are essential to an integrated project delivery (IPD) common with BIM.

The BIM manager – who represents a new career path that has emerged as a consequence of the global momentum in BIM – usually champions the application of BIM across construction projects within the organization (Kiker, 2009). BIM managers typically come from a strong "computer-aided-drafting" (CAD) or IT background. They may also have strong construction or project management knowledge. This defines their vertical expertise. But the effective BIM professional is much more than that. They require a range of tacit competencies spanning numerous interconnected domains. According to "The practical guide for BIM project management" (Baldwin, 2019), there are six distinct functional requirements of the BIM professional: (1) professional competence, (2) construction experience, (3) technical software competence, (4) knowledge of norms and standards, (5) management competence, and (6) sales and networking ability. More could be added to this list, ranging from aptitudes as distinct as promoting organizational values, through negotiation and interpersonal capacities, to highly refined problem-solving skills. Indeed, given that construction projects are necessarily bespoke, with every project precisely tailored to the requirements of a unique site, client requirements, and delivery conditions, construction project management survives on effective knowledge utilization. In that regard, the competent BIM professional stands out as the industry's iconic "T-shaped" professional.

Another driver for the rising demand for T-shaped BIM professionals is the proliferation of BIM use (Wu et al., 2018a). In the last decade, BIM adoption and implementation has been on the rise, as exemplified by increased utilization rates along with governmental mandates (McGraw-Hill Construction, 2012, 2014, Smith, 2014, Dodge Data & Analytics, 2015, Fenby-Taylor et al., 2016, Harun et al., 2016, NBS, 2017). BIM-driven project successes and discipline-specific BIM innovations are commonplace. A broad spectrum of best practices has been documented in a series of BIM SmartMarket Reports, e.g., Dodge Data & Analytics (2017) and McGraw-Hill Construction (2014). A total of 25 typical BIM use cases along a facility's life cycle were summarized in the BIM Project Execution Planning (PXP) Guide (CICRP, 2010), which was later integrated into the NBIMS-US Version 3, Section 5.9 (NIBS, 2015). Internationally, buildingSMART has also proposed a roadmap for life-cycle BIM, which classifies a total of 40 BIM use cases in accordance with a project's *Design*, *Procure*, *Assemble*, and *Operate* phases (Sjøgren, 2012). The diversification and specialization of BIM applications has generated substantial incentive for companies to develop a competent workforce able to perform the day-to-day tasks within the BIM domain. The challenge faced by corporate human resource managers is to identify and recruit competent BIM professionals in the absence of an established standardization of BIM job titles, responsibilities, and qualifications. Most companies have no guidelines to follow, nor do they have any benchmark or instruments to evaluate and validate their candidates' BIM competency declarations (Sacks and Pikas, 2013, Wu and Issa, 2013, 2014). In parallel with other initiatives such as the BIM Excellence by Succar et al. (2013), O*NET by Uhm et al. (2017) and BIM Body of Knowledge by Wu et al. (2018a, b), the concept of the T-shaped BIM professional offers a convenient and immediate solution to defining and cultivating the desired BIM workforce the market is looking for.

Horizontal and vertical (H&V) education – The promise of a better pedagogy

Traditional education aims at instilling graduates with requisite disciplinary knowledge and training – the "What" and "How": the vertical "I" of the "T". A strong traditional education programme is one where all the information and procedures the student will need to operate within their chosen profession, once they leave the classroom, is provided. Its purpose is not to offer experiences of real-world application, but rather to prepare students with sufficient knowledge sets to begin engaging with real-world challenges. The paradigm is that classrooms ensure preparation for practising in a particular profession, while it is only once the graduate takes up their profession that they will begin accumulating experiences from which they will further learn how to better apply their knowledge. The outside world – the "university of hard knocks", so to speak – will push back against their theoretical and abstract education and reveal nuances of what works best in practice. Thus, the novice entering their chosen profession begins to expand the "crossbar" horizontal dimension of the "T". Within this traditional view, a "T-shaped" professional is one who is well educated but steeped in years of practical experience across multiple projects.

But not all education is traditional. Certain professions – here healthcare stands out – recognize the value, indeed the necessity of situational experience, and have attempted to bring to their academic training a contextual focus. Indeed, healthcare education pedagogy is replete with reference to horizontal and vertical (H&V) learning. The advances made by such forward-looking programmes offer important insights into how the more traditional education of the BIM professional may be improved. Before going there, however, it needs to be pointed out that the terminology of *horizontal* and *vertical* is applied to education with a mixture of meanings. It is therefore useful to quickly review the tenets and various applications of the H&V terminology, dismissing what is not useful to this discussion and settling on what is.

Tenets of H&V education

First and foremost, in terms of education policy, vertical alignment refers to the appropriately graded, staged sequencing of increasingly demanding subjects, while horizontal alignment refers to the matching of subject content with assessments. In other words, at every level of an academic course, assessments should examine students' mastery of the subject, as demanded by that level – not above nor below (Case and Zucker, 2005). This hierarchical framework of defining learning outcomes of a field of study resonates with Bloom's taxonomy to a great extent (Bloom et al., 1956), which offers six orders of learning to empower educators to convey intended learning objectives, provide insights into students' progress towards a planned learning trajectory, and foster continuous cognitive development. Increasingly, Bloom's taxonomy has been used by state accreditors and university administrations as a standard practice by faculty to ensure desired educational outcomes.

Another understanding of H&V terminology occurs in the appreciation that the truly educated graduate offers more to society than expertise in a narrow field of endeavour. The study conducted by Wu and Issa (2014) briefly touched upon the discussion on whether a specialist or generalist should be the desired college BIM educational outcome, with participation from both educators and industry professionals. The agreement was that BIM as a transformative trend was calling for strong communication and teamwork skills, a capacity to work efficiently within co-located teams, and abilities to apply fundamental engineering, management, and computer skills in real-world scenarios (Becerik-Gerber et al., 2012). In other words, it was highly expected that in addition to discipline-specific knowledge, skills, and abilities (KSAs), a foundational understanding of BIM as a collaborative, interoperable, and integrated solution for a highly specialized and fragmented industry should be instilled in students. In producing the "well-rounded" individual, schools may require students to take elective subjects unrelated to their discipline. The scope of inclusion of non-core units strictly considered unnecessary to the training of the professional is a measure of the course's horizontal integration

A third understanding of H&V education relates to the interaction between students at various stages of their education. In this interpretation, vertical integration refers to collaborative work undertaken by students in different years, engaged on the same project, This has been applied in engineering education, with senior students directing junior students. The junior students benefit from the concentrated instruction provided by a cohort of seniors, while the seniors benefit by learning through providing instruction (Solnosky et al., 2015). Collectively, the approach instils interpersonal and collaborative skills essential to real-world engineering problem-solving. Conversely, horizontal learning is simply where students of the same year learn from each other – through group work (Orkwis et al., 1997). The stratification of BIM competency or expertise levels again echoes the need to align educational outcomes with assessment frameworks such as Bloom's taxonomy. It also reminds us to recognize the need and importance to accept the coexistence and coherence of BIM professionals at their different career stages, as exemplified in the *Entry*, *Middle*, and *Full-Performance* BIM professionals categorized in the BIM BOK initiative (Wu et al., 2018a). When students graduate with college degrees, the depth of their technical skills and empirical knowledge is limited to the entry/novice level. As they start to participate in real BIM projects and play roles and take on responsibilities in a project team, the vertical integration of the H&V education continues to apply, and paves a pathway via which novices will become seasoned professionals and eventually BIM experts. Noticeably, in addition to the *cognitive domain* of knowledge, the journey to becoming BIM experts also involves learning of the two other domains of Bloom's taxonomy, which are referred to as the *affective domain* (Krathwohl et al., 1964) and the *kinesthetic/psychomotor domain* (Dave, 1970, Harrow, 1972). Unlike the cognitive domain that has a strong focus on the technical dimension of BIM, for the affective domain, we expect that students' educational background and participation in practical experience will promote their understanding of values in BIM professional practice, enhance their awareness of BIM professionals' identity, and develop a genuine career interest in BIM. For the kinesthetic/psychomotor domain, we expect that the intellectual and emotional engagement in BIM practices will motivate students to invest time and passion in persistent practice to drive desirable behavioural change in order to perform relevant job-specific tasks better and faster, and facilitate their transition from novices to experts.

A fourth perspective is that apart from discipline-specific skills that students must learn to gain accreditation within their chosen profession, students ought also to be endowed with generic attributes that all graduates will share in common. In Deakin University, Australia, these are known as "graduate learning outcomes", and there are eight. Besides course-specific discipline knowledge and capabilities are competencies in communication, digital literacy, critical thinking, problem solving, self-management, teamwork, and global citizenship. Each subject within a programme will address one or more of these "outcomes", with the idea that by the time a student graduates, they will have sufficiently acquired these generic competences. Other universities have taken the approach further. The University of Rochester, for example, has adopted "developmental vectors", as first proposed by (Chickering and Reisser, 1993). Here, verticality refers to learning objectives, and at Rochester there are seven: intellectual and social competence, emotional management, interdependence, interpersonal maturity, self-identity, purposefulness, and personal integrity. Rochester, however, has also precisely articulated the degree of development a student should attain at every year of their course. For example, with respect to, say, developing purpose, in first year, students need to know how to access counselling services, but by final year, they must have crafted a substantive post-graduation career plan. This progression represents the horizontality of learning objectives.

Pedagogical approaches to H&V BIM education

BIM pedagogy, including the design and delivery of learning activities are the next logical steps following the discussion on H&V education and its application in cultivating the T-shaped BIM talent. The term *pedagogy* broadly refers to "any conscious activity by one person designed to enhance learning by another" (Watkins and Mortimore, 1999). Good pedagogical design needs to express the congruence between the content, teaching strategies, learning environment, assessment, and feedback, and to reflect underlying theories of learning and value (Kalantzis and Cope, 2010, Hudson, 2011). In other words, pedagogy is a systematic mechanism that facilitates the realization of learning outcomes with predictable and controlled results. Pedagogy design heavily relies on theoretical foundations to avoid inconsistencies and undesirable results. Globally, most popular learning theories, including *behaviourism, information processing, constructivism* and *social cognitive* share common principles to enhance learning from instruction, but differ in their individual focuses on how learning occurs and the role of memory and motivation, as well as self-regulation and implications for instruction (Schunk, 2012).

Do any of these educational pedagogies offer insight into a better approach to training the BIM professional? Referring back to the pervasiveness of the "knowledge economy", we need to remain cognizant of two prevailing trends that graduates will face entering the workplace. Firstly, knowledge is growing at a pace that can only overwhelm even the mightiest of intellects. This means that specialization in any field is an ever-deeper but narrowing band of expertise within the full gamut of capabilities required to deliver complex projects. This leads to fragmentation of knowledge, and these fragments must be brought together. In this regard, more than ever, professionals must work as teams of complementary specialists. This leads to a second trend: teams are replacing individuals. No longer can the responsibility of a project rest in the hands of one person, as was traditionally the case, but rather must reside with the organization (Ball and Shannon, 1984). In this respect, the BIM professional requires deep expertise, but also the deftness to assemble, coordinate, and leverage the complementary expertise of a multi-skilled team. These two demands typify the "T-shaped" BIM professional – the ability to lead teams in complex problem-solving.

The inadequacy of traditional education in shaping professionals fit for a knowledge economy is evident. In engineering schools, modes of instruction have hardly changed over the decades. Additionally, construction and engineering faculty are seldom exposed to teaching training, neither do we set our foot in exploring what learning theories are and what scientific implications they have for learning in general. As highly application-oriented as the construction industry is, teaching and training usually take place on a trial-and-error basis. Of course, there are important exceptions, but for the most part, construction management and engineering instruction proceeds somewhat as follows. Courses are compartmentalized into packages of subjects; each is composed of topics relevant to the subject. Students digest these subjects, attending lectures and tutorials and completing assignments and exams. Over time, the student completes a number of subjects, but with little understanding of how these evidently related themes are to be brought together in dealing with a real-world scenario; they remain siloed and abstract. Then, at the end of the course, as if in recognition of this deficiency, the student is confronted with a "capstone" unit or "senior project". Here, the student is meant to synthesize all they have learnt in preceding subjects to deliver a comprehensive solution to some problem brief at a level of proficiency that demonstrates they are graduation ready. It is often at this point that student performance breaks down. Rather than showcase student proficiency, what is revealed is the program's emphasis on base analysis over integrated design and problem-solving (Orkwis et al., 1997).

Since the BIM professional's core function will be problem-solving (Becerik-Gerber et al., 2012), with a reliance on team contribution, these parameters must guide the design of BIM education programmes. The nature of problem-solving corresponds to the *constructivism* learning theory, which advocates the reflection of "self" in learning: we learn as we adjust our mental models to accommodate new experiences in search for meaning. As experience cannot be taught and understanding originates from the perception from individual perspectives and lenses employed to assess the experience, constructivism discounts the value of standardized and universal tests. Instead, it calls for individualized, customized, and experience-driven learning, with scaffolding and assistance from instructors. A few widely used pedagogical approaches, such as problem-based learning and project-based learning, can find their roots in constructivism.

Small group, problem-led learning ought to be introduced at the earliest stages of the curriculum. There are two variations of the problem-based pedagogies in this regard: "problem-based learning" and "problem-oriented learning" (Snyman and Kroon, 2005). Either approach delivers learning, integrating knowledge with skill. Problem-based learning begins with a stimulus in the form of a defined problem for the class to solve. Problem-oriented learning centres on case studies in which a situation is presented without an explicit question being evident. Thus, a problem must first be identified before a solution can be proposed to remedy it. From there, students proceed to explore what it is they need to know to engage with the case scenario and deliver the desired outcome (Vidic and Weitlauf, 2002). Variations exist – case-based learning, team-based learning, inquiry-based learning – but they all share a common objective: to engage students collaboratively in solving problems. Beginning with a problem set, and from there progressing to identifying the required cognitive tools to solve the problem, turns the traditional education paradigm on its head. However, this is what educating the "T-shaped" BIM professional demands.

Project-based learning (PBL) is another proven effective student-centred pedagogical approach (Baş, 2011) that focuses on real-world issues (Chinowsky et al., 2006). Empirical evidence found in the adoption and implementation of BIM has suggested that integration, instead of specialization, sets new skill-set requirements for the workforce in the construction industry featured with a brand-new technological infrastructure. PBL holds the promise of cultivating the desired "integrative" competency with breadth and depth (Goedert et al., 2013), which constitute the foundational tenets of the H&V education. PBL allows students to gain knowledge (Liu et al., 2010) and develop critical thinking, creativity (Kubiatko and Vaculova, 2011), and many soft skills, e.g., leadership and communication (Walters and Sirotiak, 2011). These conducive outcomes align with the mission of cultivating T-shaped BIM professionals. Aside from the student learning process, PBL helps redefine and transform the roles of instructors. In traditional education mode, instructors are the point of authority and function as the source of solutions. In contrast, instructors in PBL will work as mentors and/or expert consultants who help students formulate their own strategies towards the accomplishment of learning objectives by offering open-ended, heuristic suggestions instead of spoon-feeding a set of answer keys. PBL underscores the development of students' metacognition and self-monitoring skills in handling, analyzing, and resolving complex problems in real-world scenarios (Chinowsky et al., 2006). In the past two decades, interest in PBL has been increasing significantly in BIM education (Goedert et al., 2013, Lee et al., 2013, Abdirad and Dossick, 2016, Anderson et al., 2020).

Plotting a pathway forward

The concept of H&V education, however understood and however beneficial in principle, represents a shift from traditional education modes. The question arises as to whether traditional institutions are suited to delivering H&V education. Indeed, shifts in requirements would be few, but some observations on the nature of traditional institutions are warranted. Universities are vertically organized. Faculties and schools are "siloed", with little interaction between them. Schools compete for scarce resources and promote vested interests above that of the broader mission of the institution. Professors within a discipline will tend to have a closer relationship with those academics around the world who share the same research interest, rather than with colleagues within their own school. The fragmentation of academic work, too, has an impact. The emergence of mass education has led to larger class cohorts, reduced opportunities for mentorship, and increased impersonalization. Rising administrative burdens, programmatic course delivery, fragmentation, and outsourcing of content deliverers all combine to ossify the traditional emphasis on vertical course delivery. Disengagement among academics from their work is a further concern. Work satisfaction is falling, with scores for "limited or above" work satisfaction being for the United States 64%, Australia 55%, and the UK 46% (Jones, 2011). There is a further caution; Australian engineering graduates who are perceived to have a mismatched education – studying units outside the requirements of their chosen profession – appear to suffer a 3.4% earnings penalty, while for architecture graduates, the penalty is 5.1%.

The foregoing discussion spotlights the rise of the knowledge economy and highlights both the need for the development of "T-shaped" BIM professionals and the potential of H&V education to better prepare such professionals. It also points out that there might be some systemic resistance from institutions with respect to delivering H&V education, while the employment marketplace may not fully appreciate the benefits of a laterally trained graduate. With these constraints in mind, it appears that a rich, effective, educational curriculum able to prepare competent "T-shaped" BIM professionals must deliver the following: (1) rich, and relevant conceptual BIM content; (2) content themes that interface with others; (3) deep skill in using BIM; (4) exposure to broader and peripheral issues impacting BIM and construction management and engineering; (5) problem-based and project-based teaching format; (6) team building, team management, and interpersonal skills; (7) peer-driven mentee and mentorship; (8) collaborative learning; and (9) real-world focus. These nine focus areas could be easily converted to measurable metrics and adopted by BIM educators to evaluate a given BIM curriculum against a set of H&V BIM educational outcomes.

H&V BIM curriculum proposal

A proposed BIM curriculum for cultivating T-shaped BIM professionals is yet to be fully developed. Nevertheless, the roadmap to achieve this goal is established as common curriculum development guidelines have laid out its key elements and procedures, which include the *needs assessment* (i.e., learning objectives or outcomes), *learning design and implementation (delivery)*, and *outcomes assessment* (Jenkins, 2007). A sample BIM curriculum that embodies the H&V BIM education criteria is discussed below as a test case into how critical curriculum design decisions can be made to address and deliver the specific competency expectations for T-shaped BIM professionals.

Integrated social and technical BIM curriculum: Homeless shelter case study

The core component of this sample BIM curriculum includes a learner-centred development approach and integrative technical and socio-emotional learning outcomes. Homelessness represents a significant and ubiquitous social issue faced by modern society. Exploring the interaction between technology solutions such as BIM with homeless shelter design provided an ideal testbed for problem-based, project-based, and experiential learning pedagogies, with the potential to cultivate a spectrum of technical BIM skills, such as design authoring, visualization, model-based quantity take-off, and cost estimating. More importantly, this curriculum was positioned to explore and foster construction management and engineering students' development of empathy, professional identity, teamwork, and collaborative learning. Examined through the lens of the nine focus areas suggested by the H&V BIM education framework proposed above, the *Design for the Homeless (DfH) BIM curriculum* met the majority, if not all, of these criteria.

Needs assessment: The curriculum was developed as an intervention to stimulate student engagement and professional identity among undergraduate construction management and engineering students at an early stage (i.e., sophomore and junior years). An incentive for this intervention was the considerably lower first-year retention rate in construction management and engineering programmes than in other degree programmes. The expectation was that participation in projects like *DfH* could students develop a sense of belonging. The experience in exploring the technical strength of BIM and a team-based learning environment facilitated by BIM with a common goal for improving the lives of the disadvantaged, homeless population could significantly promote students' professional awareness and self-efficacy.

Learning design and implementation: The DfH curriculum tasked students, assigned to teams with three or four members, to explore and produce a design that featured extreme low cost ($300, materials only) and technical viability, with elaborations on the design concepts, design exhibits, quantity surveying, and cost estimating. A poster showcase, written reflection, and oral presentation were additional learning activities embedded to boost microlearning opportunities and self-efficacy reinforcement. The curriculum delivery adopted a coupled experiential and project-based learning (E+PBL) pedagogy (Wu and Luo, 2018), as illustrated in Figure 15.2.

Outcome assessment: A portfolio of assessment measures, including student artefacts evaluation, peer-evaluation, progress report, summative assessment survey, and external subject matter expert review, were utilized for learning outcome assessment. Both quantitative and qualitative data were collected and analyzed to understand the technical and socio-emotional learning outcomes. Developing appropriate outcome assessment instrument was informed

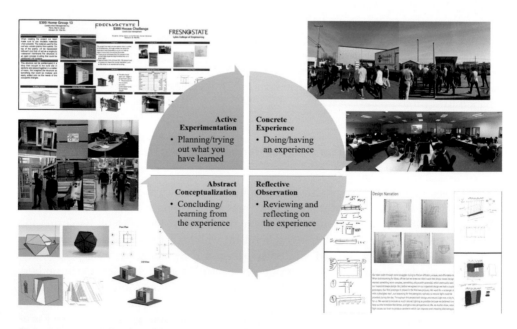

Figure 15.2 The delivery and implementation of the *DfH* BIM curriculum.

by various theoretical frameworks, including George D. Kuh's student engagement and high impact practices (HIPs) framework (Kuh, 2008), Bandura's Self-efficacy theory (Bandura, 1977), the Social Cognitive Career Theory (SCCT) (Lent et al., 2002), and the professional identity theory (Eliot and Turns, 2011).

Conclusion

This chapter critically reviewed the context of the emerging demand for T-shaped BIM professionals, which featured the parallel development of the knowledge economy and the technological infrastructure for more collaborative and integrative new business paradigms. This rapidly changing business environment has also imposed brand-new skill-set requirements on knowledge workers, who will play a critical role in the ongoing digital transformation and industry 4.0 revolution. Traditional educational standards and curriculum for developing and classifying individual professionals' expertise and qualifications have fundamentally changed as the boundaries between disciplines and domains become blurry. Nevertheless, neither the traditional notion of *generalist* or *specialist* fully embodies the essence of what is deemed necessary to future BIM professionals, as both the depth and breadth of knowledge, skills, and abilities have been redefined for executing and delivering capital projects via their life-cycle phases under the efforts of a large integrative project team that relies on a *common data environment* for collaboration, communication, and coordination. This chapter went ahead and proposed the concept of T-shaped BIM professionals via the H&V education framework. After delineating T-shaped BIM professionals' characteristics, the authors challenged the educational community to redesign the existing BIM curriculum and suggested a series of benchmarks that reflected the tenets of H&V learning to ensure measurable success in cultivating T-shaped BIM professionals. A sample *Design for the Homeless* BIM curriculum that advocated an integrated technical and socio-emotional learning was discussed to provide preliminary details of the proposed H&V BIM curriculum development.

Some extended discussion is dedicated to potential future trends in BIM education. First, the proliferation of BIM innovations and practices in field operations in return enhances our understanding of how and when learning occurs, in what application contexts, and via what experiences. Secondly, synergies between BIM and other ICT advancements in business operations may also help adapt the pedagogical approaches in classroom BIM teaching and job-site BIM training. The rising of cloud computing and web-based collaboration platforms and recent booms in virtual learning and training environments enabled by virtual reality/augmented reality/mixed reality (VR/AR/MR) have presented learners with much more interactive, complex, and higher-order learning opportunities that simulate real-world scenario BIM applications, and expand the boundaries of BIM learning and training to be more robust, independent of constraints from locations, facilities, and time. Data analytics, the science of deriving insights from data, and sensor technology are yet another trend in learning and development across industry sectors that could revolutionize our understanding of how people learn and assess how well they learn. Last but not least, many of us have also speculated that the COVID-19 pandemic may fundamentally change the notion of higher education and revolutionize how learning should be designed and delivered in virtual environments due to social distancing constraints. Regardless, T-shaped BIM professionals must overcome uncertainties, embrace new opportunities, stay nimble and adaptive to provide quality service, and foster the transformation of the construction and engineering industry to a digital knowledge economy.

References

Abdirad, H. and Dossick, C. S. 2016. BIM curriculum design in architecture, engineering, and construction education: A systematic review. *Journal of Information Technology in Construction*, Vol. 21. pp. 250–271.

Anderson, A. Dossick, C. S. and Osburn, L. 2020. Curriculum to prepare AEC students for BIM-enabled globally distributed projects. *International Journal of Construction Education and Research*, Vol. 16 No. 4. pp. 270–289.

Baldwin, M. 2019. *The BIM-Manager: A Practical Guide for BIM Project Management*, Beuth Verlag GmbH.

Ball, M. J. and Shannon, R. H. 1984. Vertical and horizontal curricula: How they can work together in the integration of medical computer science and the classic medical sciences. *Medical Informatics*, Vol. 9 No. 3–4. pp. 281–288.

Bandura, A. 1977. Self-efficacy: Toward a unifying theory of behavioral change. *Psychological Review*, Vol. 84 No. 2. pp. 191–215.

Barile, S. Saviano, M. and Simone, C. 2015. Service economy, knowledge, and the need for T-shaped innovators. *World Wide Web*, Vol. 18 No. 4. pp. 1177–1197.

Baş, G. 2011. Investigating the effects of project-based learning on students' academic achievement and attitudes toward English lesson. *The Online Journal of New Horizons in Education*, Vol. 1 No. 4. pp. 1–15.

Becerik-Gerber, B. Ku, K. H. and Jazizadeh, F. 2012. BIM-enabled virtual and collaborative construction engineering and management. *Journal of Professional Issues in Engineering Education and Practice*, Vol. 138 No. 3. pp. 234–245.

Beck, U. Giddens, A. and Lash, S. 1994. *Reflexive Modernization: Politics, Tradition and Aesthetics in the Modern Social Order*, Stanford University Press, Stanford, CA.

Bloom, B. S. Engelhart, M. D. Hill, W. H. Furst, E. J. and Krathwohl, D. R. 1956. *Taxonomy of Educational Objectives: The Classification of Educational Goals, Handbook I: Cognitive Domain*, David McKay, New York.

Case, B. and Zucker, S. 2005. Methodologies for alignment of standards and assessments. *In*: China–US Conference on Alignment of Assessments and Instruction, Beijing, China. 1–5.

Chickering, A. W. and Reisser, L. 1993. *Education and Identity. The Jossey-Bass Higher and Adult Education Series*, ERIC.

Chinowsky, P. S. Brown, H. Szajnman, A. and Realph, A. 2006. Developing knowledge landscapes through project-based learning. *Journal of Professional Issues in Engineering Education and Practice*, Vol. 132 No. 2. pp. 118–124.

Computer Integrated Construction Research Program (CICRP) 2010. *BIM Project Execution Planning Guide: Version 2.0.*, 2nd ed, Pennsylvania State University, University Park, PA.

Collins, H. M. and Evans, R. 2002. The third wave of science studies: Studies of expertise and experience. *Social Studies of Science*, Vol. 32 No. 2. pp. 235–296.

Dave, R. H. 1970. Psychomotor levels. *In*: Armstrong, R. J. (ed.) *Developing and Writing Behavioral Objectives*, Educational Innovators Press, Tucson, AZ.

Dodge Data & Analytics 2015. The business value of BIM in China. *In*: Jones, S. A., Bernstein, H. M., Russo, M. A., Laquidara-Carr, D. & Taylor, W. (eds.) *SmartMarket Report*, Dodge Data & Analytics, Bedford, MA.

Dodge Data & Analytics 2017. *The Business Value of BIM for Infrastructure 2017*, Dodge Data & Analytics, Bedford, MA.

Eastman, C. M. Teicholz, P. Sacks, R. and Liston, K. 2011. *BIM Handbook: A Guide to Building Information Modeling for Owners, Managers, Designers, Engineers and Contractors*, John Wiley and Sons, Hoboken, NJ.

Eliot, M. and Turns, J. (2011), Constructing professional portfolios: Sense-making and professional identity development for engineering undergraduates. *Journal of Engineering Education*, Vol. 100 No. 4. pp. 630–654.

Fenby-Taylor, H. Thompson, N. Philp, D. MacLaren, A. Rossiter, D. and Bartley, T. 2016. *Scotland Global BIM Study*, Heriot-Watt University, Edinburgh, Scotland.

Gibbons, M. Limoges, C. Nowotny, H. Schwartzman, S.Scott, P. and Trow, M. 1994. *The New Production of Knowledge: The Dynamics of Science and Research in Contemporary Societies*, sage.

Goedert, J. D. Pawloski, R. Rokooeisadabad, S. and Subramaniam, M. 2013. Project-oriented pedagogical model for construction engineering education using cyberinfrastructure tools. *Journal of Professional Issues in Engineering Education and Practice*, Vol. 139 No. 4. pp. 301–309.

Hardin, B. and McCool, D. 2015. *BIM and Construction Management: Proven Tools, Methods, and Workflows*, Wiley, Indianapolis, IN.

Harrow, A. J. 1972. *A Taxonomy of the Psychomotor Domain: A Guide for Developing Behavioral Objectives*, David McKay, New York.

Harun, A. N. Samad, S. A. Nawi, M. N. M. and Haron, N. A. 2016. Existing practices of building information modeling (BIM) implementation in the public sector. *International Journal of Supply Chain Management*, Vol. 5 No. 4. pp. 166–177.

Hudson, B. 2011. Didactical design for technology enhanced learning. *In*: Hudson, B. & Meyer, M. A. (eds.) *Beyond Fragmentation: Didactics, Learning and Teaching in Europe*, Verlag Barbara Budrich, Opladen.

ISO 2018. *Organization and Digitization of Information about Buildings and Civil Engineering Works, Including Building Information Modelling (BIM): Information Management Using Building Information Modelling: Part 1: Concepts and Principles*. ISO 19650-1:2018, International Organization for Standardization, Geneva, Switzerland.

Jenkins, D. E. P. 2007. Curriculum development for future engineers. *European Journal of Engineering Education*, Vol. 8 No. 1. pp. 1–4.

Jones, N. 2011. *Structural Impact*, Cambridge University Press, Cambridge, UK.

Kalantzis, M. and Cope, B. 2010. The teacher as designer: Pedagogy in the New Media age. *E-Learning and Digital Media*, Vol. 7 No. 3. pp. 200–222.

Kiker, M. W. 2009. BIM manager: The newest position. *In*: *Autodesk University 2009*, Autodesk, Inc., Las Vegas, NV.

Krathwohl, D. R. Bloom, B. S. and Masia, B. B. 1964. *Taxonomy of Educational Objectives, Handbook II: Affective Domain (The Classification of Educational Goals)*, David McKay Company, Inc., New York.

Kubiatko, M. and Vaculova, I. 2011. Project-based learning: Characteristic and the experiences with application in the science subjects. *Energy Education Science and Technology Part B: Social and Educational Studies*, Vol. 3 No. 1–2. pp. 65–74.

Kuh, G. D. 2008. *High-Impact Educational Practices: What They Are, Who Has Access to Them, and Why They Matter*, Association of American Colleges and Universities, Washington, DC.

Laitila, A. 2009, The expertise question revisited: Horizontal and vertical expertise. *Contemporary Family Therapy*, Vol. 31 No. 4. pp. 239–250.

Lee, N. Dossick, C. S. and Foley, S. P. 2013. Guideline for building information modeling in construction engineering and management education. *Journal of Professional Issues in Engineering Education and Practice*, Vol. 139 No. 4. pp. 266–274.

Lent, R. W. Brown, S. D. and Hackett, G. 2002. "Social cognitive career theory". *Career Choice and Development*, Vol. 4. pp. 255–311.

Liu, Y.-H. Lou, S.-J. Shih, R.-C. Meng, H.-J. and Lee, C.-P. 2010. A case study of online project-based learning: The Beer King project. *International Journal of Technology in Teaching & Learning*, Vol. 6 No. 1. pp. 43–57.

McGraw-Hill Construction 2012. *The Business Value of BIM in North America: Multi-Year Trend Analysis and User Ratings (2007–2012)*, McGraw-Hill Construction, Bedford, MA.

McGraw-Hill Construction 2014. The business value of BIM for construction in major global markets: How contractors around the world are driving innovation with building information modeling. *In*: Bernstein, H. M., Jones, S. A., Russo, M. A., Laquidara-Carr, D. & Taylor, W. (eds.) *SmartMarket Report*, McGraw-Hill Construction, Bedford, MA.

McIntosh, B. S. and Taylor, A. 2013. Developing T-shaped water professionals: Building capacity in collaboration, learning, and leadership to drive innovation. *Journal of Contemporary Water Research & Education*, Vol. 150 No. 1. pp. 6–17.

National Building Specification (NBS) 2017. *National BIM Report 2017*, National Building Specification, Newcastle, UK.

National Institute of Building Sciences (NIBS) 2015. *National BIM Standard: United States Version 3: Transforming the Building Supply Chain Through Open and Interoperable Information Exchanges*, National Institute of Building Sciences, Washington, DC.

Nonaka, I. and Takeuchi, H. 1995. *The Knowledge-Creating Company: How Japanese Companies Create the Dynamics of Innovation*, Oxford University Press, Oxford, UK.

Orkwis, P. Walker, B. Jeng, S. Khosla, P. Slater, G. and Simitses, G. 1997. Horizontal and vertical integration of design: An approach to curriculum revision. *International Journal of Engineering Education*, Vol. 13. pp. 220–226.

Rogers, P. and Freuler, R. J. 2015. The "T-Shaped" engineer. *In*: 122nd ASEE Annual Conference & Expo, Seattle, WA. American Society for Engineering Education.

Sacks, R. and Pikas, E. 2013. Building information modeling education for construction engineering and management. I: Industry requirements, state of the art, and gap analysis. *Journal of Construction Engineering and Management*, Vol. 139 No. 11. pp. 04013016.

Schunk, D. H. 2012. *Learning Theories: An Educational Perspective*, Pearson, Boston, MA.

Shotter, J. 2010. *Social Construction on the Edge*, Taos Institute, Chagrin Falls, OH.

Sjøgren, J. 2012. Product & process room business propotion. *buildingSMART UserCOM*, buildingSMART, Tokyo, Japan.

Smith, P. 2014. BIM implementation: Global initiatives & creative approaches. *In*: Hajdu, M. & Skibniewski, M. J. (eds.) Creative Construction Conference 2014, June 21–24, Prague, Czech Republic. Diamond Congress Ltd., Budapest, Hungary, 605–612.

Snyman, W. and Kroon, J. 2005. Vertical and horizontal integration of knowledge and skills – a working model. *European Journal of Dental Education*, Vol. 9 No. 1. pp. 26–31.

Solnosky, R. Parfitt, M. K. and Holland, R. 2015. Delivery methods for a multi-disciplinary architectural engineering capstone design course. *Architectural Engineering and Design Management*, Vol. 11 No. 4. pp. 305–324.

Succar, B. and Sher, W. 2014. A competency knowledge-base for BIM learning. *Australasian Journal of Construction Economics and Building: Conference Series*, Vol. 2 No. 2. pp. 1–10.

Succar, B. Sher, W. and Williams, A. 2013. An integrated approach to BIM competency assessment, acquisition and application. *Automation in Construction*, Vol. 35. pp. 174–189.

Uhm, M. Lee, G. and Jeon, B. 2017. An analysis of BIM jobs and competencies based on the use of terms in the industry. *Automation in Construction*, Vol. 81. pp. 67–98.

Vidic, B. and Weitlauf, H. M. 2002. Horizontal and vertical integration of academic disciplines in the medical school curriculum. *Clinical Anatomy: The Official Journal of the American Association of Clinical Anatomists and the British Association of Clinical Anatomists*, Vol. 15 No. 3. pp. 233–235.

Walters, R. C. and Sirotiak, T. 2011. Assessing the effect of project based learning on leadership abilities and communication skills. *In*: Sulbaran, T. (ed.) 47th ASC Annual International Conference, April 6–9, 2011, Omaha, NE. Associated Schools of Construction.

Watkins, C. and Mortimore, P. 1999. *Understanding Pedagogy: And Its Impact on Learning*, SAGE Publications.

Woolgar, S. and Latour, B. 1986. *Laboratory Life: The Construction of Scientific Facts*, Princeton University Press.

Wu, W. and Issa, R. 2013. Impacts of BIM on talent acquisition in the construction industry. *In*: Smith, S. D. & Ahiaga-Dagbui, D. D. (eds.) 29th Annual ARCOM Conference, Reading, UK. Association of Researchers in Construction Management.

Wu, W. and Issa, R. R. A. 2014. BIM education and recruiting: Survey-based comparative analysis of issues, perceptions, and collaboration opportunities. *Journal of Professional Issues in Engineering Education and Practice*, Vol. 140 No. 2. pp. 04013014.

Wu, W. and Luo, Y. 2018. Technological and social dimensions of engaging lower division undergraduate construction management and engineering students with experiential and project based learning. *In*: Wang, C., Harper, C., Lee, Y., Harris, R. & Berryman, C. (eds.) Construction Research Congress 2018, New Orleans, LA. American Society of Civil Engineers, Reston, VA, 97–107.

Wu, W. Mayo, G. McCuen, T. L. Issa, R. R. A. and Smith, D. K. 2018a. Building information modeling body of knowledge. I: Background, framework, and initial development. *Journal of Construction Engineering and Management*, Vol. 144 No. 8, 04018065.

Wu, W. Mayo, G. McCuen, T. L. Issa, R. R. A. and Smith, D. K. 2018b. Building information modeling body of knowledge. II: Consensus building and use cases. *Journal of Construction Engineering and Management*, Vol. 144 No. 8, 04018066.

16 Developing digerati leaders

Education beyond the building information modelling (BIM) ecosystem

Eleni Papadonikolaki

The building information modelling (BIM) ecosystem

Origins of building information modelling (BIM) for data management

The built environment is a largely project-based sector (Morris, 2004). As projects are the main organizational vessel, teams move from project to project, and knowledge is lost among projects. Ackoff (1989) categorized data, information, knowledge, and wisdom into a pyramid; data is the foundation to eliciting information, information is the basis for knowledge, and, at the top of the pyramid, wisdom is developed after the accumulation of knowledge. Hence, data and information are crucial for managing the built environment. Historical advancements in hardware and software engineering created new information technology (IT) capabilities to be useful in built-environment projects and megaprojects (Whyte and Levitt, 2011) to help manage project data and information. The first applications of IT in construction were focused on managing construction processes, drawing upon relevant developments in the manufacturing industry. This chapter starts with a non-typical review of BIM focusing on its relation to data and information, and explains how the BIM ecosystem and pedagogy are transformed in the current digital economy.

At a data level, during the 1970s and mid-1980s, one of the most predominant lines of thought in construction was about structuring product data to represent knowledge about facts and artefacts (Eastman, 1999). This shift naturally followed similar changes in other industries, such as the manufacturing and aerospace industries. Product definitions were developed around the mid-1980s to support "the direct and complete exchange or sharing of a product model amongst computer applications, without human intervention" (Dado et al., 2010). Product modelling in construction contributed to the computerization of construction (Eastman, 1999). Dado et al. (2010) defined product models as integrated representations of information and data about an activity over its product life cycle. Later, the term building product model (BPM) was used to denote information about a building component embedded in a product model (Eastman, 1999). The origins of building information modelling (BIM) were found in the above approaches for object-oriented building product modelling that took place in the 1990s (Eastman et al., 2008) – essentially BIM authoring applications. Data is useful to the extent that it is interpretable; hence, understanding how to extract information from data is central to BIM practices.

At an information level, in the last decades, the built environment has been transformed by "wakes" of innovation across project networks (Boland et al., 2007). From digital three-dimensional (3D) representations of built assets to automated design and construction processes using BIM and various realities (Whyte et al., 2000), the sector witnessed changes in technologies, work practices, and knowledge across multiple communities (Boland et al., 2007). Presently, BIM is considered the most representative digital technology and information aggregator in construction, globally. As a digital technology, BIM is at the forefront of a digital built environment and works as a digital platform, allowing other technologies to work with it (Morgan and Papadonikolaki, 2018), and it is highly pervasive by transforming business as usual in the built environment. To this end, BIM is not an isolated technology, but instead a central technology that is becoming increasingly democratized and integrated with other digital technologies.

Defining BIM ecosystem in the context of data and information management

Until very recently, computer-aided architectural design (CAAD) software solutions were standard tools of computerization in architecture, and these not only portrayed contemporary architectural processes but also supported automated, semi-automatic, and standardized processes (Aouad et al., 2012). BIM as a technology-driven approach that includes integrated software solutions has "parametric intelligence" (Eastman et al., 2008). Since it generates, collects represents and manages building data and information, it relates to many other digital artefacts, such as data repositories, visualization tools, and associated processual functionalities. Although there is a plethora of BIM definitions and interpretations, in this chapter, BIM is considered as a set of technologies for generating, sharing, and managing building data and information in the built environment.

Whereas currently, the term *building information modelling* and, its acronym, BIM, have been made synonymous with various commercial software solutions used in the built environment, BIM is essentially an "umbrella" term. This umbrella term denotes, apart from its merely commercial instances, e.g. software applications such as Autodesk Revit, Bentley MicroStation, Graphisoft ArchiCAD, and numerous others, the process of modelling building products, also known as "BIM-ing", the BIM-based processes in design and construction (execution) of projects, as well as the outcome of these processes, i.e. the building information model. To understand the future directions that BIM-related education will take in the coming years, this chapter will focus on BIM through a digital technology lens.

Chapter aim

As explained, the BIM ecosystem has evolved over the last three decades to represent a hybrid field that is interconnected and crucial for managing data and information related to the built environment. The BIM ecosystem, through its continuously evolving nature, remains an innovation in construction. *Innovation* refers to a new product, service, or process (Abernathy and Clark, 1985). Innovations are highly context-dependent and interconnected with their environment relevant to a structuration view based on Giddens' (1984) theory about the constitution of society. Hence, the BIM ecosystem shapes and is continuously shaped by the built-environment industry and practice, as well as the institutions forming it. Friedland and Alford (1991) viewed society as an inter-institutional system where institutional logics such as market, professions, and communities influenced organizations and individuals. Higher education is an important institution with a crucial role to play in the democratization of innovations and in particular, within the context of this book chapter, how it can responsibly shape a digital built environment.

Higher education institutions (HEIs) are key drivers in the knowledge economy, and they have always been encouraged to develop links with industry and businesses to train novices for the market (Olssen and Peters, 2005). To this end, HEIs have an important role to play among its inter-related market, economy, and professional community institutions, and closer relations between industry and government are encouraged (Olssen and Peters, 2005). Based on the 1996 OECD (Organization for Economic Co-operation and Development) report, Foray and Lundvall (1998) recognize that future workers need to acquire a range of skills and continuously adapt those in a "learning economy". Hence, for technological innovations such as BIM, the importance of knowledge and technology diffusion needs a deeper understanding of cross-institutional collaboration and "national innovation systems" (Lundvall, 1998, Lundvall et al., 2002). BIM and digital technologies are highly pervasive and systemic (Egyedi and Sherif, 2008). This chapter aims to provide an alternative view of BIM education in higher education and the evolution of the BIM concept towards digital engineering (DE), and how leaders of digerati – that is, digital literates – with both soft competences and technical skills can be developed through cross-institutional collaboration, recognizing that the BIM ecosystem (including data and information management) as an innovation requires contextual sensitivities to thrive. The ensuing section will discuss the historical background and the evolution of BIM.

Historical perspectives and evolution of BIM

Digital solutions in the built environment

With the advent of technology, general-purpose IT solutions started receiving traction in managing construction projects. With IT, built-environment professionals could use informational, analytical, and decision-making tools to assist the development of construction projects. As projects are key organizational vessels in the built environment, and a project is a nexus for processing information (Winch, 2005), managing information flows is an inherent aspect of project management (Turner, 2006). The graphical and visualization capabilities of computer-aided architectural design (CAAD), which were developed in the 1950s, supported visualization, communication, and process modelling (Aouad et al., 2012). BIM has been touted as a new type of IT, whose modelling capabilities combine benefits from both CAAD and process modelling (Aouad et al., 2012), featuring built-in features of generating and managing building information, e.g. visualization, automated drawings, and quantity take-off.

IT applications have focused either on design or management capabilities in construction (Forbes and Ahmed, 2010). Past IT solutions for management in construction focused primarily on exchanging information to manage invoices, quantities, and construction crews. Until recently, for communication, most organizations used extranets, enterprise resource planning (ERP), and online project databases for information management (Ajam et al., 2010). Through these massive information systems, different parties could exchange on demand various project documents, orders, and invoices in the form of electronic data interchange (EDI). Building information, such as printed documents, from CAAD or 3D models were not as frequently exchanged. However, none of these technologies became globally accepted by all construction disciplines either at a file type or the type of information systems (Demian and Walters, 2014, Samuelson and Björk, 2013, Samuelson and Björk, 2014).

Since its introduction, BIM was clearly deemed as appropriate for offering both design and management capabilities and thus being inclusive and integrative in nature. BIM not only offers information artefacts, such as CAAD and quantity take-offs, but also contributes to information management through online platforms called common data environments (CDE), where heterogeneous information can be integrated and presented. Because of these features, BIM is more inclusive and accessible as a technology across the construction supply chain and all built-environment stakeholders. Hence, BIM could be described as a digital platform, because it allows other technologies to connect with it, for example virtual reality, augmented reality, and data from the Internet of Things (IoT) (Morgan and Papadonikolaki, 2018). These digital solutions offered by BIM put it at the centre of digital transformation in the built environment, as it has applicability and linkages across the whole life cycle. The above imply that BIM is a digital technology that is continuously evolving in nature.

Evolution of BIM towards digital engineering

BIM is a relatively old concept, having its origins in the product modelling efforts of the 1970s and mid-1980s. However, BIM could still be considered an innovation for construction, as despite having old technological origins, implementing it across projects in the built environment at a wide scale involving all actors is something entirely new. From a practical perspective, the need for aligning BIM with numerous existing processes, old and new standards, protocols, and workflows is novel and thus, innovative per se. Nevertheless, BIM is an evolving , and scholars and practitioners have moved towards more broad descriptions of BIM, such as "Building Information Management" (Becerik-Gerber and Kensek, 2009), "digitally-enabled working" (Dainty et al., 2017), "digitization" (Morgan, 2017), and "digital engineering" (Golizadeh et al., 2018).

Digital engineering (DE) is considered an advanced computerization of systems engineering practices and is an industry agnostic term originating in the manufacturing industry. Although the origins of BIM can be found in product modelling to support the representation of data, the origins of DE are found in data science. At the same time, DE is more oriented towards life-cycle considerations and systems integration than BIM, and BIM can be considered as a subset of DE. Table 16.1 summarizes key studies and policy initiatives that have contributed to the evolving nature of BIM towards digital engineering.

In a recent study of how digital innovations have evolved in the built environment, it was deduced that various technologies preceded and co-occurred with BIM and shape the landscape of digital innovations (Papadonikolaki et al., 2020). Since the 2000s, there has been a proliferation of scholarly production of research on digital transformation. After the 2000s, a steady increase in research on CAD, digital prototyping, internet applications, algorithms for generating design, and general-purpose IT solutions was observed. A sharp increase in research on BIM applications is dominant, followed by studies on digital prototyping, robotics, Big Data analytics, and applications for developing smart cities. Figure 16.1 illustrates how often digital technologies, such as BIM, appear across scientific literature since 2000.

From the above, it is understood that BIM is a hybrid concept that is continuously evolving and leads the digital transformation of the built environment (Figure 16.1). Various institutions and standardization efforts have contributed to BIM becoming an embedded concept (Table 16.1), as explained in the next section.

The institutional setting and standardization of BIM

According to the duality of structure and agency in structurational theory, digital innovations such as BIM not only shape their environment but also get shaped by it (Giddens, 1984). As

Table 16.1 Key studies and milestones in the evolution of the concept of building information modeling

Year	Milestone	Source
1992	Introduction of term *building information modeling*	van Nederveen and Tolman (1992)
1994	*International Alliance for Interoperability* was founded	Bazjanac and Crawley (1997)
1995	Start of *Industry Foundation Classes (IFC)* initiatives	Bazjanac and Crawley (1997)
1999	*Building Product Models* book was published	Eastman (1999)
2005	*International Alliance for Interoperability* was renamed *BuildingSMART*	www.buildingsmart.org
2007	*National BIM Standards* was founded in the United States	NIBS (2007)
2008	*Building Information Modelling (BIM) Handbook* was published	Eastman et al. (2008)
2009	Introduction of *building information "management"* concept	Becerik-Gerber and Kensek (2009)
2011	The *UK BIM strategy* was announced	GCCG (2011)
2015	The *Digital Built Britain* strategic plan was published	HMG (2015)
2017	*European Handbook of BIM* for public sector was published	EUBIM (2017)
2018	Europe publishes *ISO 19650 standards*	www.iso.org
2019	The Institution of Engineering and Technology launches the *Digital Twins* for the Built Environment report	www.theiet.org
2019	UK launches the *UK BIM Framework*	UKBIMFramework (2020)

adapted from Papadonikolaki (2018).

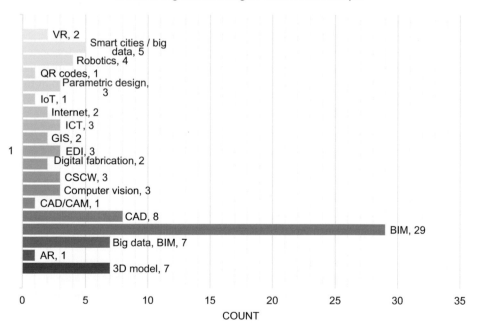

Count of digital technologies across data sample

Figure 16.1 Count of digital technologies in scientific research – adapted from Papadonikolaki et al. VR: Virtual Reality; QR: Quick Response; IoT: Internet of Things; ICT: Information and Communication Technologies; GIS: Geographic Information System; EDI: Electronic Data Interchange; CSCW: Computer-Supported Cooperative Work; CAD/CAM: Computer-Aided Design/Computer-Aided Manufacturing; CAD: Computer-Aided Design; BIM: Building Information Modelling; AR: Augmented Reality; 3D: Three-Dimensional.

context is important for understanding the setting of digital innovations, considering institutional logics stresses the importance of the relations between agency (behaviour, values, intentions) and context (individuals, organizations, institutions) (Friedland and Alford, 1991). The heterogeneous and evolving context of digital innovations is crucial for understanding how they are diffused and democratized (Rogers et al., 2005). Digital innovations in construction are shaped by actors and forces beyond the traditional demand and supply chain, including clients, developers, policy-makers, and users. Instead, institutional players such as the government, industry bodies, standardization authorities, professional institutions, and megaprojects are often vessels, drivers, and agents of digital innovation at a macro level. Therefore, with the involvement of different institutions in its development, the concept of BIM became embedded in its context.

To ensure successful roll out and diffusion of its digital technologies, sponsors of new standards engage in standardization processes to develop, legitimize, and implement them (Narayanan and Chen, 2012). Among other key institutions, the government is one of the most important standardization actors (Gao et al., 2014). This is because the government typically promotes technology development, implementation, and diffusion through research and development, technology investment, and forming state-led standard-setting consortia (Funk and Methe, 2001). Especially in the United Kingdom (UK), a complex cross-institutional system has been mobilized to support the development and democratization of digital innovations such as BIM. With the support of industry bodies, standardization authorities, professional institutions, and megaprojects, which are seen as institutional projects, the UK government has recently spearheaded the mandate of BIM in public projects and standardization efforts pivoted towards developing a UK BIM framework (see also Table 16.1). Figure 16.2 illustrates the long effort towards the democratization, legitimization, and diffusion of digital innovations such as BIM in UK construction. Although Figure 16.2 is UK-centric, similar patterns of interaction between institutions and the democratization of BIM have been discovered in other countries around the world.

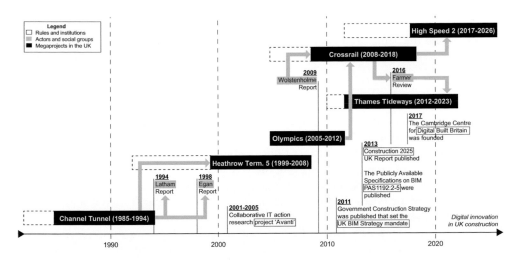

Figure 16.2 Timeline of digital innovations in the UK construction industry influencing and being influenced by institutions, actors, and megaprojects – adapted by Papadonikolaki et al. (2018).

The human-derived need to trust exchanged data and information as to accuracy and conciseness, and the machine-based necessity for interoperability, motivated the efforts for standardization. Standards in IT are solutions to satisfy and balance the needs and requirements of a diverse group of users in a seamless manner for electronic communication within and between computers (Laakso and Kiviniemi, 2012). Standards are expected to be used "during a certain period, by a substantial number of the parties for whom they are meant" (Vries, 2005). Various efforts took place aiming at creating a standardized model for building product information in construction. Those were organized by industrial consortia and cross-institutional bodies in the 1970s (Björk and Laakso, 2010).

Standards could refer to various functions, such as information exchange among applications, collaboration processes, and protocols and other data specifications. For example, for transferring concise information among various BIM applications, the Industry Foundation Classes (IFC) standard has been the most long-lived (Björk and Laakso, 2010), developed by the international non-profit consortium International Alliance for Interoperability, now named BuildingSMART. However, standards also come with drawbacks, and in particular with regards to enabling the software vendor lock-in and hindering the quality of information during conversion. Standards are typically created by reaching a consensus among heterogeneous institutions that negotiate appropriate common definitions and accepted processes.

State of the art in BIM pedagogy

Key institutions in global BIM education

Within the afore-described heterogeneous and complex institutional setting, the democratization of digital innovations such as BIM in the construction sector is becoming increasingly challenging. Across the world, there have been numerous efforts in designing BIM-related educational frameworks, curricula, and courses. These efforts currently take place for two main reasons. On the one hand, BIM education is paramount for providing the much-needed skills and competences to novices in the sector. On the other hand, BIM education is responding to the transformation of construction's institutional setting as HEIs are key institutions in driving the knowledge economy. To this end, HEIs are not only responding but also driving the demand for BIM education to provide relevant skills and competences to the industry. Across the world, a complex inter-institutional setting shapes BIM-related education.

The key institutions involved in global BIM education vary depending on the specific context. In this chapter, the UK setting has been used as an example to illustrate relationships among key institutions. In 2011, the UK government intervened in the market by mandating BIM in public procurement from 2016, and it set up the BIM Task Group, a government-funded group managed by the Cabinet Office, which consisted of practitioners seconded by their employers to support the success of the UK BIM Level 2 mandate. This BIM Task Group worked in close collaboration with an appended BIM Academic Forum to support a strong need for HEI-led BIM education. The BIM Academic Forum has worked on identifying requirements for BIM pedagogy, but no concrete rules for HEIs have emerged. Most recently, the Centre for Digital Built Britain, a partnership between the Department of Business, Energy & Industrial Strategy and the University of Cambridge, was formed to understand how the construction and infrastructure sectors could use a digital approach to better design, build, operate, and integrate the built environment. Although primarily dedicated in research and policy, the association of the Centre for Digital Built Britain with other

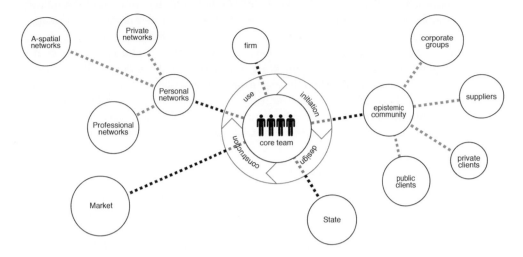

Figure 16.3 Project ecologies related to construction setting – adapted by Papadonikolaki (2017).

HEIs informs education and industry. According to Grabher and Ibert (2012), in project-based environments, such institutions are organized in a complex inter-related system of project ecologies, as illustrated in Figure 16.3.

BIM education across learners' levels

A recent study by Badrinath et al. (2016) identified BIM education efforts in HEIs mainly across Europe, the United States, Brazil, South-East Asia, and Australia. Most of the existing BIM-related curricula have been developed at undergraduate levels and significantly less so at postgraduate levels. Abdirad and Dossick (2016) conducted a critical and systematic review of BIM-related curricula and found 38 undergraduate as opposed to only 7 graduate relevant courses. In the early years of BIM proliferation, mainly undergraduate courses had introduced it to their curricula (Becerik-Gerber et al., 2011); however, later, Gerber et al. (2015) found that this demographic had shifted towards more postgraduate offerings of BIM-related courses.

After graduation, there is a wealth of offerings for lifelong learning, primarily offered by software vendors and focusing on technical skills. Other institutions involved in professional education are professional institutions such as the Royal Institution of British Architects (RIBA), the Institution of Civil Engineers (ICE), the Royal Institution of Chartered Surveyors (RICS), and the Chartered Institute of Building (CIOB), which offer short courses for continuing professional development (CPD). These offerings, too, are usually delivered face-to-face and offer opportunities for skills development. Similarly, the former government-funded Building Research Establishment (BRE) was established to improve the housing stock in the United Kingdom. Currently, BRE is privatized and engaged not only in research but also in certifications and regulations in construction. Its education arm, the BRE Academy, provides training and certification of BIM compliance to the industry, either directly to individuals or to organizations.

Apart from face-to-face offerings, other educational offerings are delivered online by a variety of providers; these focus on technical skills and do not carry a direct link to the scholarly community and research developments. Online offerings draw upon primarily industry developments and less to scholarly research but ensure a quick update of knowledge for professionals already in the workplace. This suggests that as BIM education is a shared responsibility of industry and academia (Succar and Sher, 2014), "industry–academia alliances" are pragmatic approaches for educating and upskilling the industry (Badrinath et al., 2016). Essentially, technical BIM offerings from industry players can complement the offerings of HEIs and provide up-to-date industry knowledge (Abdirad and Dossick, 2016) to support mature learners who are already in the workplace. Naturally, there exist training differences among the various professions in the built environment, as discussed next.

BIM education across professional boundaries

The disciplines which have mostly embraced BIM-related curricula are architectural engineering, building engineering, civil engineering, quantity surveying, and construction engineering management. Various efforts have been focused on either developing new and often optional modules on BIM or integrating BIM into existing offerings and modules. Especially at undergraduate levels, there is a trend towards multi-disciplinary BIM-related curricula which can break the disciplinary silos in built-environment collaboration and emulate the multi-disciplinary nature of construction industry practice (Adamu et al., 2015, Wong et al., 2011). Similarly, Solnosky et al. (2015), aiming to develop "T-shaped" individuals whose knowledge goes "deep in at least one field of study while being knowledgeable in many others", deployed collaborative multi-disciplinary BIM-related curriculum design. Hence, multi-disciplinarity in BIM education is commonly sought.

An emphasis on multi-disciplinary BIM-related education is seen as the pinnacle of good practice, and there have been many novel and noteworthy approaches in pushing the boundaries of professional practice within the curricula to reduce the traditional fragmentation across the construction supply chain (Dederichs et al., 2011). Zhao et al. (2015) identified that a majority of BIM-related offerings are primarily vocational and focus mainly on technology training without providing the much-needed industry context of collaboration around setting common goals, communicating, coordinating work, and cooperating.

Project-based learning (PBL) approaches have been well-integrated in BIM education to enhance student learning outcomes through a student-centred pedagogical approach focusing on practical issues and allowing students to learn and develop critical thinking, creativity, leadership, and communication skills (Wu and Luo, 2015). These approaches work better at undergraduate levels, as the students can learn through PBL incrementally over a number of years and gradually build up their knowledge. These approaches are similar to on-the-job training and echo the professional learning experiences of recent graduates in the workplace environment. Therefore, multi-disciplinary offerings focusing on technical skills through PBL approaches have been the norm in BIM pedagogy. The ensuing section discusses new ideas entering this field.

New debates in BIM pedagogy

Pluralism of BIM and digital artefacts

Although BIM is considered to be a successor of 3D CAD, it provides a variety of digital artefacts and instrumentalities that promise to revolutionize construction work. Looking at its complex institutional setting, the numerous stakeholders involved, and the complex required standards, BIM is a highly interconnected IT system and is not an off-the-shelf solution to be simply added to construction processes (Papadonikolaki et al., 2019). Therefore, it requires continuous translation, coordination, and governance from various multi-disciplinary actors involved in the built environment. BIM offers a "multifunctional set of instrumentalities" for different purposes that will be increasingly becoming integrated (Miettinen and Paavola, 2014), and its "ready packed" commercial solutions show immediate improvements in construction processes (Jacobsson and Linderoth, 2010). There are various digital objects nested in a BIM process that form a digital infrastructure for project delivery (Whyte and Lobo, 2010). To this end, BIM is considered an interconnected set of digital artefacts including the following (Papadonikolaki et al., 2016):

- 3D models produced by digital tools, including BIM;
- generated two-dimensional (2D) documentation;
- web-based information management platforms, e.g. CDE;
- specialized sessions for kick-off and clash detections;
- BIM execution plans and protocols;
- information exchange and collaboration standards;
- decision-making instruments, such as contract addendums.

Because all these above interconnected artefacts and instrumentalities require very good coordination and alignment with existing construction processes, such as coordinated design, they cannot be seen as add-ons, because this would have adverse effects and hinder any improvements from BIM (Plesner and Horst, 2013). In BIM-using projects, complex socio-technical processes emerge to align actors with these information artefacts (Sackey et al., 2014). Therefore, the additional effort to coordinate the alignment of these instrumentalities and artefacts has cascading effects on the built-environment professionals involved. Existing roles are adjusted or new functions emerge, such as BIM managers and BIM coordinators (Badi and Diamantidou, 2017), to manage the various artefacts and instrumentalities of BIM. At the same time, BIM affects collaborative processes by transforming the information exchange and inciting denser and highly interdependent interactions among actors (Jaradat et al., 2013). All the above-described diverse digital artefacts and instrumentalities imply numerous novel and transformative ways in which the various professionals in the built environment interact with one another.

Changing roles and professionalization

With the numerous artefacts and instrumentalities of BIM, digital technologies imply new ways to organize work that is increasingly becoming socio-technical in nature. Papadonikolaki et al. (2019) found that identifying a BIM-working environment as a socio-technical and not solely technology-laden setting helped with collaborating and creating conditions for technology acceptance and greater performance. Built-environment professionals as social actors enacting specific roles are called on to adapt to the changing workplace setting of BIM. The BIM-using actors need not only their domain-related knowledge but also technical and digital skills and competences to address relational issues and soft collaboration skills (Liu et al., 2016). Dossick and Neff (2010), studying interactions among building design engineers, found that BIM enhanced transparency in their work by showing the connections among the actors, whether interferences or clashes. Hence, BIM changes business as usual and how disciplines work together.

The changing nature of the digital deliverables and integration of activities and interdependences across professional roles has implications for construction actors who engage in roles beyond the disciplines in which they were originally trained (Jaradat et al., 2013). Davies et al. (2015) stressed that a "combination of personality, experience, and training or education" is necessary to develop social competences for collaboration, communication, conflict management, negotiation, and teamwork in the BIM workplace. Developing social competences could, thus, support these emerging BIM-related roles. These social competences revolve mainly around communication and complement the existing technical skills, including the technical skills that BIM use requires. These competences are *soft* competences to accompany BIM-based work and are skills that do not require domain – discipline-related – expertise or BIM-related technical knowledge, unlike *hard* skills that are crucial for working with digital artefacts of BIM, such as 3D models, documentation, and CDEs. Table 16.2 includes key communication competences for novel roles in a digital economy and practices for leading BIM-centric work. These communication and leadership competences are with regards to mediating in negotiations or conflicts among team members, translating meaning among different professionals, and facilitating knowledge transfer within a firm.

Nature of leaders in the BIM ecosystem

From the above-stated argument, it is understood not only that new roles emerge in a digital economy, but also that these roles become more central than expected in projects and not simply auxiliary. Undoubtedly, project managers are central in teams and broker across different communities because they are members of multiple groups (Koskinen, 2008). At the

Table 16.2 Communication competences needed for novel roles in digital economy

Communication competences in digital economy

Mediation in negotiations/conflicts	*Translation of meaning across different professional domains*	*Facilitation of knowledge transfer within firm*
• *Mediating* between architect and senior technicians • *Connecting* experienced people and recent graduates (reverse mentoring) • *Resolving* resistance from external stakeholders • *Facilitating* and supporting the transition of senior designers and engineers; dealing with model accountability issues	• *Updating* project team about client requirements and mandates • *Delegating* work among project team and pushing people outside their comfort zone • *Engaging* with external stakeholders and then the internal team • *Consolidating* team's understanding of various datasets and file formats • *Answering* questions the project team asks about BIM models	• *Testing* new solutions before sharing them in the firm • *Sharing* centrally knowledge of good practices across the firm through repositories, meetings, and social media • *Developing* knowledge by selecting appropriate people for digital teams • *Capturing* knowledge across projects; making digital business-as-usual • *Upskilling* people; transferring knowledge from experienced people to the whole project team

Source: Adapted by Azzouz and Papadonikolaki (2020).

same time, project managers can display leadership qualities while they try to balance concern for task and concern for people (Jacques et al., 2008, Mäkilouko, 2004). Müller and Turner (2010) found that in project settings, leaders should develop communication competences tailored across different audiences and thus communication is a key competence for leaders. Project managers are centrally positioned in project teams; however, they might not have all the necessary digital and BIM-related knowledge to support digital innovation. These new BIM-related and digital roles are not simply project managers but show additional qualities as they care for people and focus on communication. From Table 16.2, it can be concluded that these new BIM-related roles are called on to use communication competences, which are key for project managers enhanced with leadership qualities.

Existing roles are adjusted or new functions emerge, such as BIM managers and BIM coordinators (Badi and Diamantidou, 2017), to manage digital innovations. A plethora of new terminologies describe BIM roles (Akintola et al., 2017), project-based or organizational (Davies et al., 2017, Papadonikolaki and Azzouz, 2018). Hosseini et al. (2018) claim that the new BIM-related roles are the same as project managers', apart from digital skills. The new individuals guide teams to improve processes by ensuring the implementation of digital technologies and manage resistance to change (Azzouz and Papadonikolaki, 2020). These agents were brokering knowledge about BIM and digital innovation at four distinct levels within and outside their organization, such as

- internal project teams;
- external project stakeholders and clients;
- senior management within their firm;
- an intra-firm network of BIM-related roles.

As the BIM ecosystem is highly volatile, it is continuously informed by its context through policies, markets, institutions, and the advent of new digital technologies. Accordingly, the new BIM-related roles change and adjust to fit this evolving context. Akintola et al. (2017) discussed the transitory nature of newly created BIM-related roles and argued that these would only stay relevant as long as industry is learning. These individuals are key organizational roles to drive innovation and support their organizations until it reaches an industry-wide digital maturity and project managers develop these digital competences instead (Hosseini et al., 2018). These ephemeral roles, such as BIM-related roles, remain relevant to project managers (Hosseini et al., 2018, Akintola et al., 2017), but also illustrate the future of digitally enhanced project managers, who are centrally positioned, digitally savvy, and client-facing individuals equipped with emotional intelligence and leadership skills. This new generation of leaders will lead digerati (digital literates) in leveraging the digital economy for the built environment. To this end, these new leaders of built-environment digerati need to demonstrate both communication and leadership competences as well as versatile digital skills.

From BIM skills to digital engineering management competences

Matching supply and demand

Apart from empirically informed scholarly work, targeted insights into industry in the built environment are needed to understand how to align BIM education with practice. A number of studies aimed at identifying the demand of digerati in the industry, helping understand how the supply of digital leaders in the built environment should be streamlined. Such studies are focused on analysing the contents of BIM-related job descriptions. As early in the development of the BIM ecosystem as 2009–2010, Barison and Santos (2011) analysed a relatively large sample – for its time – of 31 job descriptions for BIM-related positions across the built environment. They found that job titles such as BIM Manager, BIM Modeler, BIM Trainer, Director of BIM Technologies, BIM Consultant, and Manager of BIM were frequent across all disciplines of the built environment and primarily of general contractors (Barison and Santos, 2011). Apart from technical and digital skills, that is, by being able to work with multiple BIM applications, a number of transferable and soft competences were key for these jobs, such as oral and written communication, organization skills, providing training, presentation skills, leadership, interpersonal skills, and mentoring.

A more recent study with a larger sample of 242 job postings identified the emergence of numerous new job titles and categorized them into eight main categories (Uhm et al., 2017). In increasing order of required leadership competences, these posts were for directors, BIM project managers, BIM managers, BIM designers/engineers, BIM coordinators, and BIM technicians (Uhm et al., 2017), where leadership, communication, creativity, cooperation, and task management were of paramount importance, more than simply technical/digital skills. With a sample of 191 job listings in English-speaking countries, Hosseini et al. (2018) further identify that despite the terminology variations in BIM-related roles, the required knowledge, skills, and abilities were identical. Furthermore, the BIM manager role was no different from that of the project manager, apart from BIM knowledge (Hosseini et al., 2018), which implies that project managers will need to assimilate digital skills in the future, as well as develop leadership capabilities needed to communicate, develop, and mentor their teams. This demand for project managers with digital skills and leadership qualities is shaping BIM education initiatives. Therefore, these studies also confirmed the shift from BIM to digital engineering (DE) as a more comprehensive discipline that requires a broad set of technical skills and social competences beyond simply BIM-related technical expertise.

Learning from engineering management

The BIM ecosystem is firmly rooted in engineering, including architectural engineering, civil engineering, building services engineering, and even software engineering, given its relevance to data and information management. The shift to DE has made the link between BIM and engineering clearer. The DE ecosystem has been slowly expanded to include concepts such as digital twins (Bolton et al., 2018) and the need to create cyber-physical systems, as well as data science approaches to analyse and understand the data-intensive built environment and make robust decisions. All these relate to the need to align with the fourth industrial revolution – or industry 4.0 – an industry supported by digital technologies and automation that leverages the power of cyber-physical systems, the Internet of Things (IoT), and cloud and cognitive computing (Lasi et al., 2014) through a digital thread.

Engineering management is an innately multi-disciplinary field situated between industrial engineering and public policy (Kocaoglu, 1990). By virtue of its focus on life-cycle approach, technology, and processes, engineering management has relevance to project management. The engineering management field has a main focus on life-cycle thinking and systems (of humans, organizations, projects, resource, technology, and strategy) thinking and is supported by technology throughout (Kocaoglu, 1990). What differentiates engineering managers from project managers is that they possess the skills and competences to apply engineering principles and at the same time organize people and projects (Lannes, 2001).

Hence, in the DE ecosystem, the aforementioned unique skills of digerati leaders, who possess project management skills, digital skills, and leadership competences, relate to the hybrid field of engineering management.

Historically, the project management discipline has shared roots with engineering management in Taylorism through the need to organize the design, execution, and operation of large-scale engineering works (Kotnour and Farr, 2005). In a sense, project management and engineering management are intertwined. As projects are inseparable and essentially embedded into their issue, organizational and institutional contexts are quintessential for understanding and managing them (Blomquist and Packendorff, 1998). Not only should projects' relational context be continuously managed, but their wider institutional environment also merits equal management focus (Blomquist and Packendorff, 1998). Similarly, Söderlund (2004) acknowledged that whereas project management discipline has its "intellectual *roots*" in process planning and a Taylorist approach to workflows, it has transformed into a hybrid field which incorporates many strands of social science and organizational management. Projects are typically delivered through temporary project-based organizations involving various firms from various disciplines of the built environment (Hobday, 2000), collocated or geographically dispersed. Hence, projects behave more and more as organizations, and organizational management fields such as human resources, leadership, and strategy are intertwined with project management.

From the above argument, it appears that the areas of organizational management, project management, and engineering are shaping a framework for developing leaders who will work in digitally enabled projects and develop digeratis in their teams. Accordingly, the education of digerati leaders requires a multi-disciplinary and socio-technical approach that considers organizational mindset, project management skillset, and engineering skills. Developing leaders with such boundary-spanning capabilities requires a non-traditional and multi-disciplinary pedagogy, as explained next.

Inductive pedagogy for leadership development

The need to consolidate the teaching of digital and technical concepts with managerial ideas implies the need for novel teaching methodologies in a digital economy and DE. There are two main ways of teaching: teacher-led and student-led. These are also referred to as deductive and inductive modes, respectively. Similar to research the former is organized deductively – that is, from principles leading to phenomena – or inductively – that is, phenomena leading to principles. Teacher-led approaches or deductive teaching methods are very good for explaining complex concepts and ideas to students and start from fundamentals and proceed to applications (Felder and Silverman, 1988). Contrarily, student-led approaches or inductive teaching focus on discovery and inquiry first and then showing the underlying theories. For example, PBL is an inductive method.

Inductive teaching has its roots in Aristotle's Peripatetic School, which was not just a venue for philosophical dialogues, but an early form of higher education (Lynch, 1972). The Athenian Schools were voluntary, aimed at mature learners and less formal than our HEIs. But because of the fluid and flexible system of advanced studies in the Peripatetic School, independent thinking could flourish (Lynch, 1972). Armstrong (2004) stated that the analytic thinking that teachers express can impact students' learning immensely. Inductive teaching promotes effective learning through enhanced abstract reasoning skills, longer retention, improved ability to apply principles, confidence in problem-solving abilities, and increased capability for inventive thought (Felder and Silverman, 1988).

From reviewing curricula on teaching digital and technical skills, Abdirad and Dossick (2016) found that teacher-led tutoring by giving step-by-step instructions to students was the most common BIM tutoring approach in AEC degree programmes. However, they also found that learning new digital or technical skills was very frustrating for students (Abdirad and Dossick, 2016). Indeed, induction is the natural human learning style, and especially mature learners from the workplace have this learning style (Felder and Silverman, 1988). However, deduction is the natural human teaching style (Felder and Silverman, 1988), and instructors should make a larger effort to act more as coaches and facilitators of knowledge and learning rather than as keepers of it all (Abdirad and Dossick, 2016). It is commonly believed that inductive teaching can make teachers better at their work (Anderson et al., 2014).

Additionally, in teaching management concepts and ideas, the inductive teaching methodology has been shown to be extremely fruitful for teaching mature learners about decision-making. This aligns with the need for undergraduates and postgraduates to be offered practitioner-oriented education. Case teaching helps students recognize patterns from real-world scenarios and develop experience in simulated business environments (Anderson et al., 2014). This student-led teaching approach transfers experience from teacher, research, and workplace directly to the students to eventually help them develop their judgement (Anderson et al., 2014). At the same time, as it is student-led and the teacher is more a facilitator and coach and less a knowledge guru, students learn from one another as in PBL teaching method.

This pedagogy helps students not only understand management theories in their context but also develop their communication, negotiation, and leadership competences by helping teachers and students converse in the classroom. The case teaching method essentially simulates boardrooms and workplace meetings. Additionally, through the simulated environment, inductive teaching allows for interactions among students and teachers and relates to formative assessment, which undoubtedly enhances pedagogic practice (Yorke, 2003). Such inductive teaching approaches can also simulate conditions of decision-making in uncertainty, for instance when data might be incomplete or not all information is available. In our current digital economy, the challenge of eliciting information from a wealth of data is usually overlooked, and it requires leadership qualities to overcome it.

Outlook for a digital engineering ecosystem

As BIM can be considered a subset of DE, and as the digital technologies in the built environment continuously evolve, there exists no comprehensive digital platform to capture all needs and functionalities for the multi-disciplinary built environment (Eastman et al., 2008). Working with DE requires various actors to collaborate and move from data representation to data analysis. Such collaboration is hardly real-time and could become concurrent only in highly homogeneous environments (Cerovsek, 2011). Therefore, information management across the DE ecosystem becomes a laborious effort for continuously achieving interoperability through alignment with standards. Whereas digital and BIM support collaboration via digital artefacts, they generate additional needs for coordination and collaboration because of their continuous development.

Collaboration among actors is challenging even through digital technologies, and this becomes even more challenging as emerging technologies such as IoT, artificial intelligence (AI), and blockchain technologies enter the built-environment digitization space. Simultaneously, the multi-disciplinary actors in the built environment exchange information in non-central, largely asymmetric (due to contractual limitations), and asynchronous configurations that require managerial competences to coordinate, such as communication, negotiation, and leadership. To this end, future leaders in the digital built environment need to develop not only digital and technical skills, but also leadership competences, such as communication, negotiation, mentoring, and developing other colleagues. Similarly, as the built environment is project-based and organized through projects, knowledge of briefing, front-end management, collaboration, and hand-over are crucial skills for leading positive change in this setting. Hence, emphasis is needed on educating novices in the areas of organization, projects, and engineering.

Through this dynamically evolving, pluralistic and multi-disciplinary setting, education about DE and the role of HEIs in it becomes increasingly important. The sectoral systems of innovation view (Lundvall, 1998) offer important insights into how HEIs contribute to economic development. A cross-institutional approach places learning – in organizations, HEIs and training, or intermediary organizations – at the heart of economic development (Kruss et al., 2015). HEIs are part of this same DE ecosystem that helps organizations learn and develop market solutions in the digital economy. As education in DE is closely connected to engineering and practice, inductive teaching approaches used heavily in medicine, engineering, and management fields can help develop a new generation of future leaders in the built environment who will understand the digital economy. Moving towards DE helps the built environment converse and learn from other industries where digital technologies are more developed and allow knowledge externalities to shape its digital future.

References

Abdirad, H. & Dossick, C. S. 2016. BIM curriculum design in architecture, engineering, and construction education: A systematic review. *Journal of Information Technology in Construction*, 21, 250–271.

Abernathy, W. J. & Clark, K. B. 1985. Innovation: Mapping the winds of creative destruction. *Research Policy*, 14, 3–22.

Ackoff, R. L. 1989. From data to wisdom. *Journal of Applied Systems Analysis*, 16, 3–9.

Adamu, Z. A., Emmitt, S. & Soetanto, R. 2015. Social BIM: Co-creation with shared situational awareness. *Journal of Information Technology in Construction*, 20, 230–252.

Ajam, M., Alshawi, M. & Mezher, T. 2010. Augmented process model for e-tendering: Towards integrating object models with document management systems. *Automation in Construction*, 19, 762–778.

Akintola, A., Venkatachalam, S. & Root, D. 2017. New BIM roles' legitimacy and changing power dynamics on BIM-enabled projects. *Journal of Construction Engineering and Management*, 143, 04017066.

Anderson, E., Schiano, W. T. & Schiano, B. 2014. *Teaching with Cases: A Practical Guide*, Brighton, MA, Harvard Business Press.

Aouad, G., Wu, S., Lee, A. & Onyenobi, T. 2012. *Computer Aided Design Guide for Architecture, Engineering and Construction*, London, Routledge.

Armstrong, S. J. 2004. The impact of supervisors' cognitive styles on the quality of research supervision in management education. *British Journal of Educational Psychology*, 74, 599–616.

Azzouz, A. & Papadonikolaki, E. 2020. Boundary-spanning for managing digital innovation in the AEC sector. *Architectural Engineering and Design Management*, 16(5), 356–373.

Badi, S. & Diamantidou, D. 2017. A social network perspective of building information modelling in Greek construction projects. *Architectural Engineering and Design Management*, 13, 406–422.

Badrinath, A. C., Chang, Y. T. & Hsieh, S. H. 2016. A review of tertiary BIM education for advanced engineering communication with visualization. *Visualization in Engineering*, 4, 1–17.

Barison, M. B. & Santos, E. T. 2011. The competencies of BIM specialists: A comparative analysis of the literature review and job ad descriptions. *Journal of Computing in Civil Engineering*, 2011, 594–602.

Bazjanac, V. & Crawley, D. B. 1997. *The Implementation of Industry Foundation Classes in Simulation Tools for the Building Industry*, Berkeley, CA, Lawrence Berkeley National Laboratory.

Becerik-Gerber, B. & Kensek, K. 2009. Building information modeling in architecture, engineering, and construction: Emerging research directions and trends. *Journal of Professional Issues in Engineering Education and Practice*, 136, 139–147.

Becerik-Gerber B., Gerber D. J., & Ku K. 2011. The pace of technological innovation in architecture, engineering, and construction education: Integrating recent trends into the curricula. *Electronic Journal of Information Technology in Construction*, 16, 411–432.

Björk, B.-C. & Laakso, M. 2010. CAD standardization in the construction industry: A process view. *Automation in Construction*, 19, 398–406.

Blomquist, T. & Packendorff, J. 1998. Learning from renewal projects: Content, context and embeddedness. *In: Projects as Arenas for Renewal and Learning Process*, New York, Springer.

Boland, R. J., Lyytinen, K.. & Yoo, Y.. 2007. Wakes of innovation in project networks: The case of digital 3-D representations in architecture, engineering, and construction. *Organization Science*, 18, 631–647.

Bolton, A., Butler, L., Dabson, I., Enzer, M., Evans, M., Fenemore, T., Harradence, F., Keaney, E., Kemp, A. & Luck, A. 2018. *Gemini Principles. CDBB_REP_006*, Cambridge, UK, University of Cambridge.

Cerovsek, T. 2011. A review and outlook for a 'Building Information Model'(BIM): A multi-standpoint framework for technological development. *Advanced engineering Informatics*, 25, 224–244.

Dado, E., Beheshti, R. & Van de Ruitenbeek, M. 2010. Product modelling in the building and construction industry: A history and perspectives. *In*: Underwood, J. & Isikdag, U. (eds.) *Handbook of Research on Building Information Modelling and Construction Informatics: Concepts and Technologies*, Hershey, PA, IGI Global Publishing.

Dainty, A., Leiringer, R., Fernie, S. & Harty, C. 2017. BIM and the small construction firm: A critical perspective. *Building Research & Information*, 1–14.

Davies, K., Mcmeel, D. & Wilkinson, S. 2015. Soft skill requirements in a BIM project team. *In*: Beetz, J., Van Berlo, L., Hartmann, T. & Amor, R. (eds.) 32nd CIB W78 Information Technology for Construction Conference (CIB W78 2015), Eindhoven, Netherlands.

Davies, K., Wilkinson, S. & Mcmeel, D. 2017. A review of specialist role definitions in BIM guides and standards. *Journal of Information Technology in Construction*, 22, 185–203.

Dederichs, A. S., Karlshøj, J. & Hertz, K. 2011. Multidisciplinary teaching: Engineering course in advanced building design. *Journal of Professional Issues in Engineering Education and Practice*, 137, 12–19.

Demian, P. & Walters, D. 2014. The advantages of information management through building information modelling. *Construction Management and Economics*, 32, 1153–1165.

Dossick, C. S. & Neff, G. 2010. Organizational divisions in BIM-enabled commercial construction. *Journal of Construction Engineering and Management*, 136, 459–467.

Eastman, C. 1999. *Building Product Models: Computer Environments, Supporting Design and Construction*, Boca Raton, FL, CRC Press.

Eastman, C., Teicholz, P., Sacks, R. & Liston, K. 2008. *BIM Handbook: A Guide to Building Information Modeling for Owners, Managers, Designers, Engineers, and Contractors*, Hoboken, NJ, Wiley.

Egyedi, T. M. & Sherif, M. H. 2008. Standards' dynamics through an innovation lens: Next generation ethernet networks. *In*: K-INGN 2008: First ITU-T Kaleidoscope Academic Conference-Innovations in NGN: Future Network and Services. IEEE, 127–134, Geneva, 12–13 May 2008.

EUBIM. 2017. *Handbook for the Introduction of Building Information Modelling by the European Public Sector* [Online]. European BIM (EUBIM) Taskgroup. Available: http://www.eubim.eu/wp-content/uplo ads/2017/07/EUBIM_Handbook_Web_Optimized.pdf [Accessed 2020].

Felder, R. M. & Silverman, L. K. 1988. Learning and teaching styles in engineering education. *Engineering Education*, 78, 674–681.

Foray, D. & Lundvall, B.-Ä. 1998. The knowledge-based economy: From the economics of knowledge to the learning economy. *The Economic Impact of Knowledge*, 115–121.

Forbes, L. H. & Ahmed, S. M. 2010. Information and communication technology/building information modeling. *Modern Construction: Lean Project Delivery and Integrated Practices*, Boca Raton, FL, CRC Press.

Friedland, R. & Alford, R. R. 1991. Bringing society back in: Symbols, practices and institutional contradictions. *The New Institutionalism in Organizational Analysis*, 232–263.

Funk, J. L. & Methe, D. T. 2001. Market-and committee-based mechanisms in the creation and diffusion of global industry standards: The case of mobile communication. *Research Policy*, 30, 589–610.

Gao, P., YU, J. & Lyytinen, K. 2014. Government in standardization in the catching-up context: Case of China's mobile system. *Telecommunications Policy*, 38, 200–209.

Gerber, D. J., Khashe, S. & Smith, I. F. 2015. Surveying the evolution of computing in architecture, engineering, and construction education. *Journal of Computing in Civil Engineering*, 29, 04014060.

Giddens, A. 1984. *The Constitution of Society: An Outline of the Theory of Structuration*, Cambridge, MA, Polity Press.

Golizadeh, H., Hon, C. K., Drogemuller, R. & Hosseini, M. R. 2018. Digital engineering potential in addressing causes of construction accidents. *Automation in Construction*, 95, 284–295.

Government Construction Client Group (GCCG) 2011. *Government Construction Client Group: BIM Working Party Strategy Paper*, GCCG.

Grabher, G. & Ibert, O. 2012. Project ecologies: A contextual view on temporary organizations. *In*: Morris, P. W. G., Pinto, J. & Söderlund, J. (eds.) *The Oxford Handbook of Project Management*, Oxford, UK, Oxford University Press.

HM Government (HMG). 2015. *Digital Built Britain, Level 3 BIM Strategic Plan* [Online]. HM Government. Available: https://www.cdbb.cam.ac.uk/news/2015DBBStrategy [Accessed 13/04/2020].

Hobday, M. 2000. The project-based organization: An ideal form for managing complex products and systems? *Research Policy*, 29, 871–893.

Hosseini, M. R., Martek, I., Papadonikolaki, E., Sheikhkhoshkar, M., Banihashemi, S. & Arashpour, M. 2018. Viability of the BIM manager enduring as a distinct role: Association rule mining of job advertisements. *Journal of Construction Engineering and Management*, 144, 04018085.

Jacobsson, M. & Linderoth, H. C. 2010. The influence of contextual elements, actors' frames of reference, and technology on the adoption and use of ICT in construction projects: A Swedish case study. *Construction Management and Economics*, 28, 13–23.

Jacques, P. H., Garger, J. & Thomas, M. 2008. Assessing leader behaviors in project managers. *Management Research News*, 31(1), 4–11.

Jaradat, S., Whyte, J. & Luck, R. 2013. Professionalism in digitally mediated project work. *Building Research and Information*, 41, 51–59.

Kocaoglu, D. F. 1990. Research and educational characteristics of the engineering management discipline. *IEEE Transactions on Engineering Management*, 37, 172–176.

Koskinen, K. U. 2008. Boundary brokering as a promoting factor in competence sharing in a project work context. *International Journal of Project Organization and Management*, 1, 119–132.

Kotnour, T. & Farr, J. V. 2005. Engineering management: Past, present, and future. *Engineering Management Journal*, 17, 15–26.

Kruss, G., Mcgrath, S., Petersen, I.-H. & Gastrow, M. 2015. Higher education and economic development: The importance of building technological capabilities. *International Journal of Educational Development*, 43, 22–31.

Laakso, M. & Kiviniemi, A. 2012. The IFC standard: A review of history, development, and standardization. *Journal of Information Technology in Construction*, 17, 134–161.

Lannes, W. J. 2001. What is engineering management? *IEEE Transactions on Engineering Management*, 48, 107–115.

Lasi, H., Fettke, P., Kemper, H.-G., Feld, T. & Hoffmann, M. 2014. Industry 4.0. *Business & Information Systems Engineering*, 6, 239–242.

Liu, Y., Van Nederveen, S. & Hertogh, M. 2016. Understanding effects of BIM on collaborative design and construction: An empirical study in China. *International Journal of Project Management*, 35, 686–698.

Lundvall, B.-Å. 1998. Why study national systems and national styles of innovation? *Technology Analysis & Strategic Management*, 10, 403–422.

Lundvall, B.-Å., Johnson, B., Andersen, E. S. & Dalum, B. 2002. National systems of production, innovation and competence building. *Research Policy*, 31, 213–231.

Lynch, J. P. 1972. *Aristotle's School: A Study of a Greek Educational Institution*, University of California Press.

Mäkilouko, M. 2004. Coping with multicultural projects: The leadership styles of Finnish project managers. *International Journal of Project Management*, 22, 387–396.

Miettinen, R. & Paavola, S. 2014. Beyond the BIM utopia: Approaches to the development and implementation of building information modeling. *Automation in construction*, 43, 84–91.

Morgan, B. 2017. Organizing for digitization in firms: A multiple level perspective. *In*: Chan, P. W. & Neilson, C. J., eds. Proceedings of the 33rd Annual Association of Researchers in Construction Management Conference (ARCOM 2017), September 4–6 2017, Cambridge, UK. Association of Researchers in Construction Management.

Morgan, B. & Papadonikolaki, E. 2018. Organizing for digitization: A balancing act. Academy of Management Big Data, University of Surrey. Academy of Management, Vol. Surrey No. 2018.

Morris, P. W. G. 2004. Project management in the construction industry. *In*: Morris, P. W. G. & Pinto, J. K. (eds.) *The Wiley Guide to Managing Projects*, Hoboken, NJ, Wiley.

Müller, R. & Turner, R. 2010. Leadership competency profiles of successful project managers. *International Journal of Project Management*, 28, 437–448.

Narayanan, V. K. & Chen, T. 2012. Research on technology standards: Accomplishment and challenges. *Research Policy*, 41, 1375–1406.

National Institute of Building Sciences (NIBS). 2007. *United States National Building Information Modeling Standard*: Version 1-Part : Overview, Principles, and Methodologies [Online]. National Institute of Building Sciences. Available: https://buildinginformationmanagement.files.wordpress.com/2011/06/nbimsv1_p1.pdf [Accessed].

Olssen, M. & Peters, M. A. 2005. Neoliberalism, higher education and the knowledge economy: From the free market to knowledge capitalism. *Journal of Education Policy*, 20, 313–345.

Papadonikolaki, E. 2017. Unravelling project ecologies of innovation: A review of BIM policy and diffusion. *In*: *IRNOP (International Research Network on Organizing by Projects)*, Boston, MA, International Research Network on Organizing by Projects (IRNOP).

Papadonikolaki, E. 2018. Loosely coupled systems of innovation: Aligning BIM adoption with implementation in Dutch construction. *Journal of Management in Engineering*, 34, 05018009.

Papadonikolaki, E. & Azzouz, A. 2018. Boundary spanning and knowledge brokering for digital innovation. *In*: 16th Engineering Project Organization Conference (EPOC), Brijuni, Croatia. Engineering Project Organization Society (EPOS).

Papadonikolaki, E., Krystallis, I. & Morgan, B. 2020. Digital transformation in construction: Systematic literature review of evolving concepts. *In*: Engineering Project Organization Conference, 2020/10/05, Boulder, CO. Engineering Project Organization Society.

Papadonikolaki, E., Morgan, B. & Davies, A. 2018. Paving the way to digital innovation: Megaprojects, institutions and agency. *In*: *European Group for Organizational Studies (EGOS)*, Brussels, Belgium, European Group for Organizational Studies (EGOS).

Papadonikolaki, E., Van Oel, C. & Kagioglou, M. 2019. Organising and managing boundaries: A structurational view of collaboration with building information modelling (BIM). *International Journal of Project Management*, 37, 378–394.

Papadonikolaki, E., Vrijhoef, R. & Wamelink, H. 2016. The interdependences of BIM and supply chain partnering: Empirical explorations. *Architectural Engineering and Design Management*, 12, 476–494.

Plesner, U. & Horst, M. 2013. Before stabilization. *Information, Communication & Society*, 16, 1115–1138.

Rogers, E. M., Medina, U. E., Rivera, M. A. & Wiley, C. J. 2005. Complex adaptive systems and the diffusion of innovations. *The Innovation Journal: The Public Sector Innovation Journal*, 10, 1–26.

Sackey, E., Tuuli, M. & Dainty, A. 2014. Sociotechnical systems approach to BIM implementation in a multidisciplinary construction context. *Journal of Management in Engineering*, 31, A4014005.

Samuelson, O. & Björk, B.-C. 2013. Adoption processes for EDM, EDI and BIM technologies in the construction industry. *Journal of Civil Engineering and Management*, 19, S172–S187.

Samuelson, O. & Björk, B.-C. 2014. A longitudinal study of the adoption of IT technology in the Swedish building sector. *Automation in Construction*, 37, 182–190.

Söderlund, J. 2004. Building theories of project management: Past research, questions for the future. *International Journal of Project Management*, 22, 183–191.

Solnosky, R., Parfitt, M. K. & Holland, R. 2015. Delivery methods for a multi-disciplinary architectural engineering capstone design course. *Architectural Engineering and Design Management*, 11, 305–324.

Succar, B. & Sher, W. A competency knowledge-base for BIM learning. *Australasian Journal of Construction Economics and Building: Conference Series*, 2014. 1–10.

Turner, R. J. 2006. Towards a theory of project management: The nature of the project. *International Journal of Project Management*, 24, 1–3.

Uhm, M., Lee, G. & Jeon, B. 2017. An analysis of BIM jobs and competencies based on the use of terms in the industry. *Automation in Construction*, 81, 67–98.

UKBIMFRAMEWORK. 2020. *About us* [Online]. Available: https://ukbimframework.org/about/ [Accessed].

Van Nederveen, G. & Tolman, F. 1992. Modelling multiple views on buildings. *Automation in Construction*, 1, 215–224.

Vries, H. J., DE 2005. IT standards typology. *In*: Jakobs, K. (ed.) *Advanced Topics in Information Technology Standards and Standardization Research*, Hershey, PA, IGI Global.

Whyte, J. & Levitt, R. 2011. Information management and the management of projects. *The Oxford Handbook of Project Management*, Oxford, UK, Oxford University Press.

Whyte, J. & Lobo, S. 2010. Coordination and control in project-based work: Digital objects and infrastructures for delivery. *Construction Management and Economics*, 28, 557–567.

Whyte, J., Bouchlaghem, N., Thorpe, A. & Mccaffer, R. 2000. From CAD to virtual reality: Modelling approaches, data exchange and interactive 3D building design tools. *Automation in Construction*, 10, 43–55.

Winch, G. M. 2005. Rethinking project management: Project organizations as information processing systems. *In*: Slevin, D. P., Cleland, D. I. & Pinto, J. K. (eds.) *Innovations: Project Management Research 2004*. Newton Square, PA, Project Management Institute.

Wong, K. D. A. W., Wong, K. W. F. & Nadeem, A. 2011. Building information modelling for tertiary construction education in Hong Kong. *Electronic Journal of Information Technology in Construction*, 16, 467–476.

Wu, W. & Luo, Y. 2015. Project-based learning for enhanced BIM implementation in the sustainability domain. *In*: Proceedings of 9th BIM Academic Symposium and Job Task Analysis Review, Washington, DC. 2–9.

Yorke, M. 2003. Formative assessment in higher education: Moves towards theory and the enhancement of pedagogic practice. *Higher Education*, 45, 477–501.

Zhao, D., Mccoy, A. P., Bulbul, T., Fiori, C. & Nikkhoo, P. 2015. Building collaborative construction skills through BIM-integrated learning environment. *International Journal of Construction Education and Research*, 11, 97–120.

17 Incorporating collaborative problem solving (CPS) principles in BIM education

Abbas Mehrabi Boshrabadi, Mehran Oraee, Igor Martek, and M. Reza Hosseini

Introduction

The importance of collaborative problem solving (CPS) skills in developing highly skilled students to successfully meet the demands of increasingly complex and interactive tasks in the workplace is now a well-established fact within various contexts, particularly engineering education (Zou and Mickleborough, 2015) (Azizan et al., 2018). While there might be several reasons for this prevalence, much of the attention has focused on the nature of problems facing engineering graduates in real-life projects, being mostly ill structured and requiring collaboration. As Oraee et al. (2019) note, engineers nowadays face a myriad of complex, poorly defined, and multi-dimensional challenges. In real workplace contexts, addressing design problems relies on a coalescence of knowledge from various domains, and consequently tasks associated with finding solutions are usually distributed among a variety of people from various disciplines. In a company, for example, engineers might engage in teamwork and collaborate with different personnel from different departments, including engineers who may be discipline experts, technical professionals (e.g., draftsperson, survey designers), or administrators, many of whom might work for other organizations and companies (Mignone et al., 2016). In recent years, thanks to the capabilities of online collaboration platforms, engineering jobs are outsourced to companies and groups based internationally, often with huge time-zone differences (Hosseini et al., 2018b).

To cope with these challenges, students need to apply effective cognitive and social skills to analyse and solve ill-structured problems while collaborating with diverse groups of people who have different views and communication strategies. These people might have no history of collaboration, use various technical standards and codes, come from various cultural backgrounds, and been trained based on diverse working cultures. The problem is compounded when people put their commitment to their own companies above that of a team – one that is temporarily formed for handling tasks in a project (interested readers are referred to Hosseini et al. [2015, 2018b] for details). In this regard, Cunningham and Kelly (2017) also note that engineering projects are simply too complex and interdisciplinary for any individual to handle alone. Advances in such complex projects depend on team members' collaboration, with many minds working together to communicate and share ideas over time (Dossick and Neff, 2011, Emmitt, 2014). Collaborative problem-solving skills are thus of immense value, as engineering possesses a unique set of epistemic practices, including envisioning multiple solutions and reliance on teamwork.

When it comes to BIM implementation, evidence shows that successful implementation of BIM in the industry relies on a fundamental change of project delivery modus operandi (Mignone et al., 2016). Much of this change entails enhancing collaborative problem solving among project participants. This shift towards early and continued engagement of all key parties involved in a project is in fact the central development in new methods of project delivery, like integrated project delivery (IPD) and alliancing (Durdyev et al., 2019), which are the most suitable methods of delivery for BIM-enabled projects (Elghaish et al., 2020).

In light of this background, it is argued in this chapter that, though it has remained an untapped area, learning activities in BIM education should be designed with the development of graduates' CPS skills in mind. That is, fostering students' proficiency in working effectively with others in engaging with problem-solving tasks must become the mainstay of designing activities and developing learning content for BIM-related training programmes and subjects. This is a still-nascent area in BIM teaching and training, where educators face many challenges. Chief among all these is that the concept of CPS must be borrowed from the field of education, to be adapted for the setting of BIM and construction activities. Other challenges are related to questions revolving around what fundamental CPS skills are and how educators can cultivate these skills in students to run BIM-related educational programmes and subjects more effectively.

This chapter aims to shed some light on these issues. An overview of the concepts and definitions associated with CPS skills is first presented to unpack the notion of CPS. Major theoretical frameworks and modes around CPS, as used within the educational context, are delineated for educators serving the construction industry. This is followed by a review of cases on applying collaborative problem solving in BIM teaching and learning. The chapter concludes by presenting ideas and guidelines for effective design and implementation of learning tasks within BIM-related programmes that might contribute to the development of students' CPS skills.

Collaborative problem solving

Concepts and definitions

CPS has been defined as "a joint activity where dyads or small groups [interact to] execute a number of steps in order to transform a current state into a desired goal state" (Hesse et al., 2015, p. 39). As the term suggests, there are two constituents underlying the concept of CPS: collaboration and problem solving. The collaboration part is referred to as individuals working together face-to-face or in digital networks towards a common goal (Hesse et al., 2015). Underpinning this definition is the interaction of people, their knowledge, and their resources. To reach a pre-determined goal, individuals need to participate in group activities, seek and negotiate information, and share their knowledge and opinions. These individuals are typically from the same discourse community and have in common similar concerns and intellectual approaches. Within a collaboration context, members of the community often learn from one another and co-construct knowledge that might contribute to more effective learning outcomes, compared with what could be achieved by individual work (Oraee et al., 2017). The notions of "discourse community" and "co-construction of knowledge" is further elaborated in the following section.

The problem-solving component, on the other hand, refers to the process of working out a solution to a complex context-specific issue using different strategies, from multiple perspectives, and with diverse modalities (Griffin, 2014). There is a hierarchical series of mental and behavioural processes that might operate in a problem-solving activity. One conceptualization proposed by Griffin (2014) is that the process begins with an exploration and understanding of the problem and its constituent elements. At this stage, an individual makes a mental representation of different elements of a particular problem and formulates a plan of the procedure and consecutive stages that might contribute to a solution. This is followed by discernment of patterns and relationships among the elements, and formulation of these patterns into rules. During this phase, an individual would be able to create and evaluate potential solutions.

Finally, there is the process of selection of the potential best solution and its implementation and verification. In lead up to finding a solution to the task, an individual critical thinking and decision-making capacity is a contributory factor. In most cases, however, the complexity of the problem might exceed any one individual's intellectual and cognitive capacity. To deal successfully with such cognitive-demanding problems (or tasks), multiple and diverse sets of knowledge and skills from different individuals might be needed (Care et al., 2016). This means that collaboration between community members and their complementary expertise is of paramount importance.

Theoretical foundations

Several theoretical standpoints are normally used in the educational context. These can be mapped into the learning practices of CPS skills, as discussed. In this chapter, two of the most relevant theories, namely *social constructivist* theory and *situated learning* theory, are explained.

Taken together, there are three interrelated assumptions underlying these theories:

(1) Individual cognition is socially mediated. This means that mental processes at an individual level have their origin in social processes.
(2) All knowledge is fundamentally situated in the context within which it was acquired.
(3) The third assumption, built upon the first two, is that learning occurs as a result of individuals' interactions in a social context in which they act and share ideas and experiences.

The core idea of social constructivism is that learning is focused on learners and it is a process constructed through learners' interactions within a dynamic social context. It is a theory of knowledge and learning which draws heavily upon the work of Vygotsky (1978). Vygotsky asserted that individuals' cognitive development is achieved initially through their functions in social or inter-psychological levels (a dynamic interaction between an individual and society), where the learners need to interact with others in a social setting. Once they have undertaken and experienced their social learning, learners will practice those mental functions at an individual or intra-psychological level (internal processes like self-reflection and critical thinking), leading to cognitive development. Such a cognitive development perspective, informed by the process of socialization and collaboration, is the crux of the social constructivist approach. The social constructivist approach, therefore, focuses on the epistemological stance; this stance provides a base for learners to learn concepts or construct meaning about ideas through their collaboration and interactions in social learning contexts and by generating interpretations of their experiences. From this perspective, learning is viewed as a shared process between learner, instructors, and peers, rather than a purely individual-based process in which knowledge is transmitted from instructors to students.

Derived from social constructivism, situated learning theory advocates that the setting/context within which the experience occurs shapes the main part of learning processes. According to this theory, knowledge is "situated" and is linked to the setting/context in which it is experienced. While both physical and social contexts can be considered here, the focus of situated learning theory is more on social context. That is, individuals learn knowledge, as well as related skills, in the social context where the experience occurs. According to Lave and Wenger (1991), social interaction within an authentic context is an essential element of the learning process, because it helps learners absorb the modes of action and meaning embedded within that specific context. For instance, through immersion in the law context and eliciting and negotiating information relevant in this context with other lawyers, a new lawyer will gradually realize how to navigate legal issues, what discourse to use in the court of law, and how to initiate and manage cases. It is argued that a collaboration process in context facilitates learning of such knowledge and skills as it provides the interpretative scaffolding required for sense-making of information (Lave and Wenger, 1991).

In general, a closer look at the literature reveals that the main tenets of situated learning theory can be summarized as follows:

- Learning evolves predominantly from experience in an authentic context of practice.
- Learning is an interactive activity involving massive collaboration and mentoring within social communities.
- Particular attention must be given to fostering group learning, rather than total dependence on individual activity.

Accordingly, it can be inferred that individual, activity, and social context – in which the activity is experienced – are interrelated components of the learning process.

Related skills development

Within the context of CPS, both social constructivism and situated learning theories point to the fact that developing CPS skills is not solely an inner-directed activity; it must be treated as a social activity that constitutes a mode of communication in an academic or discourse community. This means that CPS involves a type of social belonging within one's community. A fundamental aspect of developing CPS skills in this view is the concept of discourse community. This is the venue in which individuals interact with other members of a community to make meaning. Within higher education, a disciplinary community is comprised of individuals who share the same understanding of the concepts, norms, and conventions of a specific discipline such as peers, lecturers, or students in upper levels of study. Through collaboration, community members can collectively create a shared understanding that individual partners do not possess initially. In addition, they can gain an insight into group dynamics and processes, to develop their interpersonal and teamwork skills, such as communication, leadership, planning, and time management (Kostopoulos et al., 2013). CPS is therefore seen as a process through which members of a discourse community co-construct knowledge.

The concept of co-construction is in turn underpinned by two interrelated activities: (1) working out a solution to a complex context-specific problem collaboratively; and (2) creating and managing a joint problem space. Both these activities require constant negotiations and recreations of meaning. That is, through using the discourse specific to their community, individuals need to communicate information about the problem, make a mental representation of the constituent elements of the problem, and come up with a reasonable plan for the steps that might contribute to a solution. The interactions of individuals within a dynamic context to co-construct knowledge will help to activate students' potential level of development, or zone of proximal development (ZPD) (Vygotsky, 1978). ZPD is defined as an area of learning in which the learner can perform a range of tasks with the help and guidance of others. This is evidenced in problem solving tasks accomplished with the help of instructors or through collaboration with peers. It is the space between what the learner can do without help as an individual (i.e. actual development level), and what he or she can do in collaboration with members of the particular discourse community (i.e. potential development level). As argued by Lee and Smagorinsky (2000), learning in the zone of proximal development works when learners' intra-psychological level of understanding is developed through negotiation and dialogue. Through ongoing engagement in the process of generation and negotiation of meanings, the learners' ZPD regarding how to approach a problem-solving task will be gradually developed. This will gradually nurture their capability to meet the demands of certain tasks and apply the knowledge they have acquired in facing new situations without others' help; hence their individual learning is encouraged.

Though the concept of collaborative problem solving has received attention within the domain of education, BIM instructors and researchers of BIM education have not approached the problem from a theoretically informed approach (Hosseini et al., 2018a). That said, and in view of the crucial role of collaboration in handling BIM-related tasks, many educators have tried various methods for applying CPS in their programmes. A summary of major attempts on this topic, as reflected in the literature, is presented below.

Collaborative problem solving in BIM education

As discussed, various frameworks and approaches have been offered by several tertiary institutions, considering the need for skills in collaboration and problem-solving domains of CPS. Table 17.1 shows the list of existing studies on teaching collaboration in BIM education. As can be seen from the table, the United States figures strongly as the continuing major centre on education theory. By contrast, relatively few other countries have explored the challenge of BIM education, relying rather on the expertise and theoretical insights generated by current world-leading institutions in this domain.

The expectations by the industry of the BIM skill levels of new graduates have been steadily on the rise. At this point, no architectural or engineering firm recruiting graduates considers BIM capabilities an optional extra. Indeed, 95% of the industry requires new hires to be BIM literate (Ghosh et al., 2015). This means also that expectations surrounding the delivery of BIM education are high. While many institutions have responded with BIM curricula, the approaches vary and results are mixed. Of the institutions active within the domain of BIM education, most emphasize 3D coordination, but also scheduling and, to a lesser degree, estimating. The themes revolve around constructability, 4D scheduling, estimating, and, to a lesser degree, design visualization, cost control, and sustainability (Becerik-Gerber et al., 2012). There are few models available that effectively demonstrate how collaboration can be achieved with BIM. Moreover, BIM tools do not easily extend to associated specialities, such as heavy engineering, tunnel construction, and infrastructure (Solnosky et al., 2015). These challenges have rendered collaborative problem solving in BIM education an overlooked area, as discussed next.

Various approaches

Perhaps the core issue, however, at least with regards to BIM pedagogy, has been conflicts in ideology regarding how BIM should be taught: as a graphical application, or

Table 17.1 Current studies on teaching collaboration in tertiary BIM education

ID	Reference	University	Country/Region
1	Ghosh et al. (2015)	Arizona State University	USA
2	Becerik-Gerber et al. (2012)	Virginia Polytechnic Institute University of Southern California	USA
3	Dossick et al. (2015)	Indian Institute of Technology Madras National Cheng Kung University	India Taiwan
		National Taiwan University	Taiwan
		University of Twente	Netherlands
		University of Washington	USA
		Washington State University	USA
		Yonsei University	South Korea
4	Gnaur et al. (2015)	Aalborg University	Denmark
5	Zhao et al. (2015)	Virginia Polytechnic Institute and State University	USA
6	Pikas et al. (2013)	Technion-Israel Institute of Technology	Israel
7	Leite (2016)	The University of Texas	USA
8	Bozoglu (2016)	Illinois Institute of Technology	USA
9	Zhang et al. (2017)	Chang'an University	China
10	Zhang et al. (2018)	Chang'an University	China
11	Wang and Leite (2014)	The University of Texas	USA
12	Mathews (2013)	Dublin Institute of Technology	Ireland
13	Solnosky et al. (2015)	The Pennsylvania State University	USA
14	Solnosky et al. (2014)	The Pennsylvania State University	USA
15	McCuen and Pober (2016)	University of Oklahoma	USA
16	Wu and Luo (2016)	California State university	USA
17	Ahn et al. (2013)	Western Carolina University	USA
18	Adamu and Thorpe (2016)	Loughborough	UK
19	Jin et al. (2020)	Konyang University	South Korea

conceptually (Solnosky et al., 2014). Despite these well-known challenges, there is no shortage of piloted, trialled, and recommended BIM curricula in the education market-place. Research by Pikas et al. (2013) identifies at least 39 distinct competencies required of BIM managers, and therefore proposes these as the subject partitions required of a comprehensive BIM training programme. Others make similar observations; the point being that there is much material to cover (Solnosky et al., 2015). Process-oriented approaches, such as recommended by (Wang and Leite, 2014), aim to digest these large topical content inputs, and tend to emphasize their educational outcomes as delivering new career roles, expert in BIM: the BIM coordinator, the BIM manager, and the BIM engineer (Adamu and Thorpe, 2016).

BIM is ostensibly just a software package, yet it impacts so many aspects of a construction project, and at so many levels. Which of the many considerations that BIM touches on should be brought to the fore, and with what emphasis, is an open question. A review of any number of educational institutions offering BIM courses reveals important nuances and differences, while making clear it is also about content delivery. A sampling follows: (1) "Special topics in BIM", Texas A&M University – covers BIM principles and application in building life cycle, from design through to facilities management; (2) "BIM", University of Arkansas – covers BIM functions in residential and commercial construction, with an emphasis on virtual modelling for quantity take-offs; (3) "CAD and BIM for construction managers", Oklahoma State University – covers 3D design, scheduling, estimating and continuous improvement; (4) "BIM for commercial construction", Purdue University – covers geometry, spatial relationships, quantities, and building components; (5) "Revit fundamentals", University of Oregon – covers parametric building modelling for rendering and working drawings; (6) "Advanced project management concepts", University of Washington – covers 3D visualization for project planning, clash detection, and fabrication automation (Ahn et al., 2013, Zhang et al., 2017).

A variant pedagogical approach is "team-based learning" (TBL). The insight here is that BIM education is less about specific knowledge content transfer and more about facilitating an individual's efficacy when working in teams. TBL is a collaborative learning philosophy which brings students together in a symbiotic learning experience – where students learn through engagement with each other. In so doing, TBL aims to replicate and foster the working dynamics encountered in real-world BIM interactions. TBL was developed as an educational strategy by Larry Michaelson in the 1970s while he was teaching at the University of Oklahoma. The TBL approach involves a sequential cycle of four events. First, students prepare for lessons individually by reading and digesting preparatory material. Second, assessment tasks are assigned to the class, working in teams. Third, course concepts are applied to simulated activities, again carried out in teams. Finally, there may be optional end-of-course assessments (Michaelsen et al., 2011).

TBL was originally conceptualized for medical education, since patient diagnosis and treatment prescription were essentially problem-solving activities that benefited from interactive discussion and deliberations. However, problem-solving and creative thinking are at the foundation of engineering work. Consequently, TBL has been identified as suitable for engineering education, utilizing out-of-class preparation in combination with team project work. Clearly, this is especially suited to BIM learning. Standard or traditional approaches to BIM education stall at the overwhelming menu of material content that needs to be absorbed, whereas group work layered on project problems more closely simulates real-world dynamics, including the dilemmas of overcoming abstract knowledge deficits related to specific tasks. In other words, traditional education imparts abstract knowledge in anticipation of tackling specific real-world issues, but this falters where the required knowledge base is extensive. However, TBL uses specific task challenges as a medium to stimulate the assimilation of abstract knowledge. Groups of individuals, working on a shared project, can parcel out the task of seeking out missing knowledge. This very process simulates the BIM experience in practice, and learning to work through knowledge acquisition strategies becomes the desired outcome of the BIM education experience (Zhang et al., 2018).

A variant pedagogical approach to BIM education is "role-based learning" (RBL). Students again work in teams, but here the team roles are prescribed. Students take on the roles of the various professionals that would have input to a real-world BIM exercise. Roles include that of architect, engineer (mechanical, structural, electrical, hydraulic, etc.), scheduler, estimator, quantity surveyor, and project manager. Role-play learning can be expected to enhance student motivation, not merely because of its game-like qualities, but also because it offers some tangibility of the likely experiences that will be encountered when students eventually take up their professions. BIM conducted through RPL again offers students practical insight into what is required to deliver a successful BIM project (Becerik-Gerber et al., 2012).

The way forward

To this point in the discussion, we have seen that BIM is required of graduates by industry, yet academia and indeed the industry itself is slow to deliver on expectations. Difficulties exist in adjusting the prevailing educational status quo that resists change, or is otherwise unable to adequately adapt. Where innovation in education regarding BIM has taken place – and there are numerous examples – the issue becomes one of what pedagogical paradigm best suites. Standard knowledge-based curricula are observed to fall short, given that the extensive content is not easily grafted into current teaching platforms, and given that BIM is a collaborative skill more than mere abstract knowledge. Recognizing this, "team-based" and "role-based" learning have been proposed as a way forward. Intuitively, these approaches appeal, yet much literature that investigates the outcomes of the many trialled BIM educational models report mixed results; sometimes students and other stakeholders say that the programmes work, and sometimes they don't, and, mostly they achieve only a degree of success (Becerik-Gerber et al., 2012, Bozoglu, 2016, Zhang et al., 2018). The remaining question is, where does the problem lie? The evidence suggests that structural problems with the educational system and its interface with industry are to blame. The evidence also suggests that collaborative learning (TBL and RBL) is the right way to go (Ahn et al., 2013, Zhang et al., 2017). Solving the structural issues is left to another chapter. In these few closing paragraphs, the merits of collaborative learning as applied to BIM education are put forward as indeed the right way to deliver BIM education.

Pedagogically, interactive learning, collaborative learning, TBL, and RBL all fall under the banner of "social constructivist learning methodologies". Learning is here understood as "interpretation", or the process of constructing understanding, rather than the impersonal assimilation of objective (context-free) knowledge. In other words, learning does not come from an inert injection of facts and data, but through a personal process of interacting with the object of learning through experience. The key point here is that a group of individuals will all experience the same event somewhat differently (the event being imprinted on uniquely individual earlier formative experiences), and that learning is therefore a wholly unique and personal journey. The process of exposure, questioning, engagement, and answering facilitates knowledge transfer – or rather, learning – and this process is all the more enriched when the journey is taken with others; the shared learning experience of others further enriching one's own learning (Jin ct al., 2020).

A conceptual framework

At the core of the social constructivist paradigm are the ways in which continuous interactive learning can be conducted. There are five general methods: 1) flipped learning, 2) role-play learning, 3) scaffolding learning, 4) project-based learning, and 5) problem-based learning (Jin et al., 2020) (See Figure 17.1). Flipped learning is where the hierarchical relationship between students and teacher is dismantled. Learning is interactive, occurring through exchange between all, and the teacher acts as facilitator, on an equal footing in dynamic multi-participant discussions. Role-play learning is where students take on a social role (professional role) and perform with respect to a situational problem (the task at hand), in the capacity of that role, and with respect to other "actors" who may or may not have shared expectations as to the outcome. The value of role-play is that it vividly captures as closely as possible the dynamics and realities of real-life situations to be faced outside the classroom. Scaffolding is where an expert (teacher) aids the learner (student) with just as

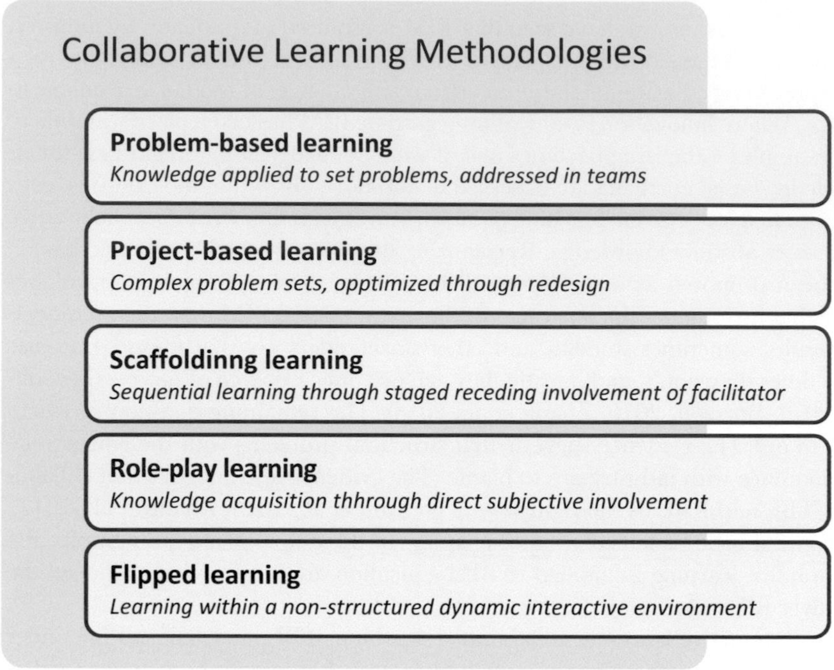

Collaborative Learning Methodologies

Problem-based learning
Knowledge applied to set problems, addressed in teams

Project-based learning
Complex problem sets, opptimized through redesign

Scaffoldinng learning
Sequential learning through staged receding involvement of facilitator

Role-play learning
Knowledge acquisition thhrough direct subjective involvement

Flipped learning
Learning within a non-strructured dynamic interactive environment

Figure 17.1 Five learning methodologies within the social constructivist paradigm.

much support as is needed for the learner to achieve the sought-after outcome. Over time, as the learner progresses, the expert withholds support until such time as support is no longer necessary. Consider piloting a plane; at first the veteran pilot flies the plane, incrementally discharging piloting functions to the novice pilot until the novice can safely take-off, fly, and land unaided. Project-based learning involves engaging with a complex undertaking, where no one attribute can be considered alone but must be resolved in totality with the whole project. This usually takes time, and requires circular reiterations of potential solutions until an optimum (or near optimum) outcome is realized. Problem-based learning is where the learner is confronted with a question, or problem, or dilemma, and must settle on an answer through innovation, imagination, and creative (lateral) thinking (Abdirad and Dossick, 2016, Jin et al., 2020).

These five methods can be considered a spectrum of approaches in which learning proceeds from a largely content-driven, semi-traditional format to a wholly experiential learning journey. In the early stages of learning, foundational content is a priority, and only with that content appreciated does it become possible to ingrain that content through experiential application. In other words, students new to BIM should first learn conceptually; texts, videos, lectures – absorbing the technical aspects of BIM and its uses. Only thereafter does it become meaningful to transcend conceptual knowledge in pursuit of practical application. And here, application also means working with others. In this way, BIM education not only promises mastery of the subject material but also empowers BIM practitioners with the tools to collaborate effectively in delivering construction projects (Abdirad and Dossick, 2016, Adamu and Thorpe, 2016).

Conclusion

This chapter contributes to the field of BIM education and teaching collaboration in BIM. The proposed "learning methodology framework" in this chapter can be considered as an educational framework consisting of five approaches: problem-based learning, project-based learning, scaffolding learning, role-play learning, and flipped learning. Indeed, the learning methodology framework can be included in both stand-alone and integrated BIM subjects curricula, which will provide significant benefits to the quality of teaching and learning. In other words, the framework increases both the collaboration motivation and collaboration skill level of the disciples studying BIM subjects at tertiary institutions, and eventually, this will address the industry's expectations of new graduates' BIM knowledge and collaboration skill levels.

References

Abdirad, H. and Dossick, C. S. (2016), BIM curriculum design in architecture, engineering, and construction education: A systematic review. *Journal of Information Technology in Construction*, Vol. 21 No. 17. pp. 250–271.

Adamu, Z. A. and Thorpe, T. (2016), How universities are teaching BIM: A review and case study from the UK. *Journal of Information Technology in Construction*, Vol. 21. pp. 119–139.

Ahn, Y. H. Cho, C.-S. and Lee, N. (2013), Building information modeling: Systematic course development for undergraduate construction students. *Journal of Professional Issues in Engineering Education and Practice*, Vol. 139 No. 4. pp. 290–300.

Azizan, M. T. Mellon, N. Ramli, R. M. and Yusup, S. (2018), Improving teamwork skills and enhancing deep learning via development of board game using cooperative learning method in reaction engineering course. *Education for Chemical Engineers*, Vol. 22. pp. 1–13.

Becerik-Gerber, B. Ku, K. H. and Jazizadeh, F. (2012), BIM-enabled virtual and collaborative construction engineering and management. *Journal of Professional Issues in Engineering Education and Practice*, Vol. 138 No. 3. pp. 234–245.

Bozoglu, J. (2016), Collaboration and coordination learning modules for BIM education. *Journal of Information Technology in Construction*, Vol. 21. pp. 152–163.

Care, E. Scoular, C. and Griffin, P. (2016), Assessment of collaborative problem solving in education environments. *Applied Measurement in Education*, Vol. 29 No. 4. pp. 250–264.

Cunningham, C. M. and Kelly, G. J. (2017), Epistemic practices of engineering for education. *Science Education*, Vol. 101 No. 3. pp. 486–505.

Dossick, C. S. and Neff, G. (2011), Messy talk and clean technology: Communication, problem-solving and collaboration using building information modelling. *Engineering Project Organization Journal*, Vol. 1 No. 2. pp. 83–93.

Dossick, C. S. Homayouni, H. and Lee, G. (2015), Learning in global teams: BIM planning and coordination. *International Journal of Automation and Smart Technology*, Vol. 5 No. 3. pp. 119–135.

Durdyev, S. Hosseini, M. R. Martek, I. Ismail, S. and Arashpour, M. (2019), Barriers to the use of integrated project delivery (IPD): A quantified model for Malaysia. *Engineering, Construction and Architectural Management*, Vol. 27 No. 1. pp. 186–204.

Elghaish, F. Abrishami, S. Hosseini, M. R. and Abu-Samra, S. (2020), Revolutionising cost structure for integrated project delivery: A BIM-based solution. *Engineering, Construction and Architectural Management*.

Emmitt, S. 2014. *Design Management for Architects*, Wiley, New York.

Ghosh, A. Parrish, K. and Chasey, A. D. (2015), Implementing a vertically integrated BIM curriculum in an undergraduate construction management program. *International Journal of Construction Education and Research*, Vol. 11 No. 2. pp. 121–139.

Gnaur, D. Svidt, K. and Thygesen, M. (2015), Developing students' collaborative skills in interdisciplinary learning environments. *International Journal of Engineering Education*, Vol. 31 No. 1 (B). pp. 257–266.

Griffin, P. (2014), Performance assessment of higher order thinking. *Journal of applied measurement*, Vol. 15 No. 1. pp. 53–68.

Hesse, F. Care, E. Buder, J. Sassenberg, K. and Griffin, P. (2015), A framework for teachable collaborative problem solving skills. *In*: Griffin, P. & Care, E. (eds.) *Assessment and Teaching of 21st Century Skills: Methods and Approach*, Springer, Dordrecht, Netherlands.

Hosseini, M. R. Chileshe, N. Zuo, J. and Baroudi, B. (2015), Virtuality in hybrid construction project teams: Causes and effects. *In*: The 4th International Scientific Conference on Project Management in the Baltic Countries, April 16–17, 2015, Riga, University of Latvia.

Hosseini, M. R. Maghrebi, M. Akbarnezhad, A. Martek, I. and Arashpour, M. (2018a), Analysis of citation networks in building information modeling research. *Journal of Construction Engineering and Management*, Vol. 144 No. 8. pp. 04018064.

Hosseini, M. R. Martek, I. Chileshe, N. Zavadskas, E. K. and Arashpour, M. (2018b), Assessing the influence of virtuality on the effectiveness of engineering project networks: "Big five theory" perspective. *Journal of Construction Engineering and Management*, Vol. 144 No. 7. pp. 04018059.

Jin, J. Hwang, K. E. and Kim, I. (2020), A study on the constructivism learning method for BIM/IPD collaboration education. *Applied Sciences*, Vol. 10 No. 15. pp. 5169.

Kostopoulos, K. C. Spanos, Y. E. and Prastacos, G. P. (2013), Structure and function of team learning emergence: A multilevel empirical validation. *Journal of Management*, Vol. 39 No. 6. pp. 1430–1461.

Lave, J. and Wenger, E. 1991. *Situated learning: Legitimate Peripheral Participation*, Cambridge University Press.

Lee, C. D. and Smagorinsky, P. 2000. *Vygotskian Perspectives on Literacy Research: Constructing Meaning through Collaborative Inquiry. Learning in Doing: Social, Cognitive, and Computational Perspectives*, ERIC.

Leite, F. (2016), Project-based learning in a building information modeling for construction management course. *Journal of Information Technology in Construction*, Vol. 21. pp. 164–176.

Mathews, M. (2013), BIM collaboration in student architectural technologist learning. *Journal of Engineering, Design and Technology*, Vol. 11 No. 2. pp. 190–206.

McCuen, T. and Pober, E. (2016), Process and structure: Performance impacts on collaborative interdisciplinary team experiences. *Journal of Information Technology in Construction*, Vol. 21. pp. 177–187.

Michaelsen, L. K. Sweet, M. and Parmelee, D. X. 2011. *Team-Based Learning: Small Group Learning's Next Big Step: New Directions for Teaching and Learning*, 116, Wiley.

Mignone, G. Hosseini, M. R. Chileshe, N. and Arashpour, M. (2016), Enhancing collaboration in BIM-based construction networks through organisational discontinuity theory: A case study of the new Royal Adelaide Hospital. *Architectural Engineering and Design Management*, Vol. 12 No. 5. pp. 333–352.

Oraee, M. Hosseini, M. R. Edwards, D. J. Li, H. Papadonikolaki, E. and Cao, D. (2019), Collaboration barriers in BIM-based construction networks: A conceptual model. *International Journal of Project Management*, Vol. 37 No. 6. pp. 839–854.

Oraee, M. Hosseini, M. R. Papadonikolaki, E. Palliyaguru, R. and Arashpour, M. (2017), Collaboration in BIM-based construction networks: A bibliometric-qualitative literature review. *International Journal of Project Management*, Vol. 35 No. 7. pp. 1288–1301.

Pikas, E. Sacks, R. and Hazzan, O. (2013), Building information modeling education for construction engineering and management. II: Procedures and implementation case study. *Journal of Construction Engineering and Management*, Vol. 139 No. 11. pp. 05013002. http://dx.doi.org/10.1061/(ASCE)CO.1943-7862.0000765.

Solnosky, R. Parfitt, M. K. and Holland, R. (2015), Delivery methods for a multi-disciplinary architectural engineering capstone design course. *Architectural Engineering and Design Management*, Vol. 11 No. 4. pp. 305–324.

Solnosky, R. Parfitt, M. K. and Holland, R. J. (2014), IPD and BIM-focused capstone course based on AEC industry needs and involvement. *Journal of Professional Issues in Engineering Education and Practice*, Vol. 140 No. 4. p. A4013001.

Vygotsky, L. S. 1978. Socio-cultural theory. *In*: *Mind and Society*. Harvard University Press, Cambridge, MA.

Wang, L. and Leite, F. (2014), Process-oriented approach of teaching building information modeling in construction management. *Journal of Professional Issues in Engineering Education and Practice*, Vol. 140 No. 4. p. 04014004.

Wu, W. and Luo, Y. (2016), Pedagogy and assessment of student learning in BIM and sustainable design and construction. *Journal of Information Technology in Construction*, Vol. 21. pp. 218–232.

Zhang, J. Wu, W. and Li, H. (2018), Enhancing building information modeling competency among civil engineering and management students with team-based learning. *Journal of Professional Issues in Engineering Education and Practice*, Vol. 144 No. 2. p. 05018001.

Zhang, J. Xie, H. and Li, H. (2017), Competency-based knowledge integration of BIM capstone in construction engineering and management education. *International Journal of Engineering Education*, Vol. 33 No. 6. pp. 2020–2032.

Zhao, D. McCoy, A. P.Bulbul, T.Fiori, C. and Nikkhoo, P. (2015), Building collaborative construction skills through BIM-integrated learning environment. *International Journal of Construction Education and Research*, Vol. 11 No. 2. pp. 97–120.

Zou, T. X. P. and Mickleborough, N. C. (2015), Promoting collaborative problem-solving skills in a course on engineering grand challenges. *Innovations in Education and Teaching International*, Vol. 52 No. 2. pp. 148–159.

18 BIM education assessment

Guidelines for making it authentic

Abbas Mehrabi Boshrabadi, Mehran Oraee, Igor Martek, and M. Reza Hosseini

Introduction

BIM competency comprises a wide range of skills in using software, collaborative problem solving, project management, and knowledge of construction methods and principles. BIM learning and education correspondingly require raising awareness of various topics along with skill development in software use and construction and project management learning practices. That is, while BIM is, in its essence, a technological process that relies heavily on various software packages, BIM education cannot be restricted to software skill development. BIM is designed to promote and support "integrated project delivery" and is touted as a panacea for bringing together many – typically – dispersed stakeholders. The required teaching approach hence should draw upon "team-based learning" (TBL) principles. With BIM, client, architects, contractors, and consultants must come together to work as a team (Oraee et al., 2019). Consequently, collaboration must dominate the learning outcome objectives, and as such, technical skills and the capability of collaboration with other team members are basic requirements of handling BIM-related tasks. Moreover, BIM-enabled projects are mostly complicated, large-sized projects where the best way of using BIM and team-working processes are defined based on the bespoke requirements of each project. Many decisions in using BIM on projects are context-specific and must be made in considering the unique nature of interactions and reciprocal impacts of perceptual, attitudinal, and structural components, compounded by alignments or misalignments among actors, tasks, and systems (Merschbrock et al., 2018). Projects that show comparable technical components and appear to be similar in terms of size and complexity might settle on divergent collaborative solutions. With this in mind, one essential component of learning BIM, apart from theoretical aspects, is exposure to real-life situations in BIM-enabled projects which assist students in appreciating the complex nature of BIM projects. They need to make decisions which are responsive to changing conditions and variables while at the same time being competent in the specific skills required for each project. Students should not engage with BIM-related aspects of a project in a vacuum. They should be exposed to all varying components and elements of a project and deal with different types of key players and participants. The skill set is not, for example, to use BIM to change the roof-drainage plumbing configuration, but to engage in a project – through BIM – to present, argue, negotiate, and come to a consensus decision with all stakeholders on the project's roof-drainage strategy and resultant action plan. But just as the teaching of software application skill sets is a process that must be tailored to the ability of students – from no knowledge all the way through to mastery – so too developing skills in dealing with unique features of projects for bespoke decision-making must be approached through a staged and gradual approach. The approach of creating meaningful tasks for students with relevance to the real-life context has a long history in many fields and disciplines like medicine. It was popularized by Larry Michaelsen of the University of Oklahoma as a means of inculcating creative thinking and problem-solving skills. The idea was that diagnosing a patient's illness, and subsequent treatment, could be a complex task which requires input from specialists in situations resembling real-life cases, who only collectively can bring together all the knowledge needed for a correct assessment (Michaelsen et al., 2011, Savery, 2006). These skills, similar to other aspects, must be systematically evaluated to ensure students fulfil the requirements of expected levels of proficiency (Jin et al., 2020). This highlights the crucial role of assessment, as discussed next.

The role of assessment

Assessment and feedback are widely regarded as significant components that influence students' learning and their experience of higher education (Boud and Dochy, 2010). Professional development in assessment design and practices is seen as pivotal to not only students' engagement in learning but also the development of higher order skills related to their future employability (Gill, 2015, Heeneman et al., 2015). Based on claims of the centrality of assessment and feedback in student learning and employability, there has been a shift of focus away from "assessment of learning" to "assessment for learning" (Boud, 2000, Sambell et al., 2013, Carless, 2015). Several influential factors have contributed to this paradigm shift. Chief among them is the need for developing students' long-term learning needs and employability skills.

Initially, a drive towards assessment for learning was put on the agenda as a response to calls for targeting the assessment methods at the "constructive alignment" of assessment, learning opportunities, and learning outcomes (Biggs and Tang, 2011b). In parallel with this, there has been an ongoing demand from higher education institutions to showcase their value to the broader community, in particular employers, by ensuring that their education provides competent and job-ready graduates. Taken together, these issues introduced the need to assess the type of learning and employment skills and abilities demanded from graduates to survive the challenges of today's workforce (Baartman and Kirschner, 2006). In this regard, Boud and colleagues proposed the establishment of an assessment environment in higher education where students' integration and participation in assessment practices is increased (Boud and Falchikov, 2007, Boud and Soler, 2016). The main objectives of such an assessment environment are (1) to shift students from surface approaches to learning, associated with conventional assessment methods, towards deeper levels of processing and learning how to learn; and (2) to ensure developing skills underpinning principles of professional practice in the job market after graduation. A critical aspect to be considered in designing such an assessment and learning environment is the authenticity of the assessment. This means that assessment activities should simulate real-work experience through assessing students' ability to apply knowledge or perform tasks under conditions that are somewhat similar to those found in real settings (Vu and Dall'Alba, 2014), a description of which follows.

Authentic assessment

Designing authentic tasks has been identified as one of the main characteristics of an effective assessment design which fosters lifelong learning and future employability (Bloxham, 2009, Sambell et al., 2013). As in other disciplines, the authenticity of tasks is crucial in built-environment educational programmes as the problems in real-world settings are

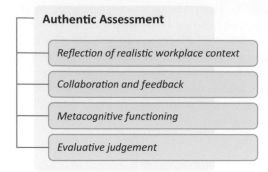

Figure 18.1 Authentic assessment framework.

ill-structured where the required information is not known in advance and the solution path is not obvious at the outset (Jonassen et al., 2006). It has been argued that designing authentic tasks can help students deal successfully with real-world problem-solving issues by improving their reflexivity, metacognition, and collaboration skills (Ashford-Rowe et al., 2014, Villarroel et al., 2018).

In general, authentic assessment aims to align teaching and learning outcomes with standards of performance that students face in the workplace context (Wiggins, 1990). The term emerged in the literature as a response to the ongoing criticism of higher education in terms of its failure to equip students with knowledge and skills needed for employment (Andrews and Higson, 2014). While various categorizations of these skills have been proposed, a quick review of the literature indicates at least four basic skills underlying these classifications: critical thinking, problem solving, collaboration, and communication skills. Inspired by conceptualizations of these skills and drawing on theoretical and practical issues discussed in the literature on assessment authenticity, a framework comprising four interrelated elements of authentic assessment is proposed (Figure 18.1).

This framework (Figure 18.1) can be used as a blueprint for designing and implementing assessment activities that can lead to an authentic learning environment in BIM-related subjects and programmes. The necessary elements to be included are illustrated in Figure 18.1 and described in detail as follows.

Elements of authentic assessment design

Reflection of a realistic workplace context

The first element that characterizes an authentic assessment activity is its resemblance to what students are required to perform within the physical and social contexts of their profession after graduation (Bosco and Ferns, 2014, Villarroel et al., 2018). Students need to be engaged in problem-solving or divergent activities that are replicas or analogies of the sorts of problems they encounter in the real-world context of their future jobs (Adams et al., 2013). From this perspective, the notions of "real-world context" and "similar task" are both representatives of an authentic assessment. As a real-world context might be in some cases difficult to create, it can include a simulated environment within which an assessment occurs, or a contextual anchor which reflects the conventions of the environment (Ashford-Rowe et al.,

2014). Such an assessment context can set the ground for transferring individuals' skills and experiences to new situations and tasks in professional life.

The other factor that contributes to the realistic element of authentic assessment is implementing tasks that are similar to the professional context in various assessment situations (Bosco and Ferns, 2014). "Similar tasks" here refers to the most common activities and functions demanded of a successful employee within the workplace setting. Within the engineering context, for example, students need to be equipped with interpersonal and teamwork skills such as communication, leadership, planning, and time management to be able to deal successfully with complex and ill-structured problems (Jonassen et al., 2006, Cunningham and Kelly, 2017). Therefore, designing tasks that provide students with extended opportunities to practice these skills while completing the task is important.

A critical factor to be considered in designing realistic, or near-realistic, activities is to place the focus on the purpose of the activity. The idea is that the activity should answer the following questions: Why are students practising this activity? How can the knowledge learned through the activities help students with their future professional performance within the discipline? Reflecting on such questions might help assessment designers to consider the expectations of other stakeholders such as clients and employers of the graduates. This would help to align assessments with the demands of the contemporary job market and to prepare job-ready graduates. This is echoed by assessment scholars, who contend that a quality assessment design relies on managing the complexity of taking the requirements of the third party (the work organization) into account (Clarke et al., 2010, Ajjawi et al., 2020).

Collaboration and feedback

Assessment authenticity involves formulating a collaborative learning environment in which students can share and exchange ideas. The collaboration dimension of assessment activities is crucial in terms of individual successful performance in contemporary workplace contexts. This is mainly because current workplace settings require individuals to be equipped with certain abilities, including coordinating schedules, facilitating communication, sharing information, and exchanging opinions with diverse groups of people, as there might be multiple team members participating in one project (Kostopoulos et al., 2013). Engineers, for example, often engage in teamwork and communicate with different personnel, including engineers who may be experts in other disciplines, technical professionals (e.g., draftsperson, survey designers), or administrators. These team members might come from various disciplines and might have various learning experiences, while being located in various locations around the world (Hosseini et al., 2018). Essential for such communication and teamwork abilities are students' identification of mutual knowledge, their understanding of the perspectives of others, and their use of effective communication means such as asking questions, responding to requests, sharing knowledge, clarifying misunderstandings, and the like. To develop such skills, therefore, assessment tasks need to be designed in ways that engage students in an interactive environment in which they can communicate and collaborate with others who might offer complementary skills, ideas, expertise, and resources.

This view of assessment design aligns with the basic tenets of the social constructivist theory of learning. Under this paradigm, learning is not best served by the passive reception of instruction. Rather, learning occurs through interaction with others. Knowledge does not come from outside; it is generated internally through experience. Contemporary conceptions of social constructivist theory have been primarily informed by the work of Vygotsky (1978) and his social theory of mind (Smagorinsky, 1995). Vygotskian social theory of mind embeds thought, cognition, and mental processes in a social context and explains how the individual's interaction with the social context in which he or she acts and shares ideas/experiences acts as a crucial element in the learning process (Smagorinsky, 1995). According to Smagorinsky (1995), Vygotsky's theory has two main components:

1. The learner develops skills and abilities as changes in social context impact cognition.
2. Cognition is socially mediated; mental processes in the individual have their origin in social processes.

This means that learning certain skills and abilities cannot be seen solely as an inner-directed activity, but also as a social activity that constitutes a mode of collaboration in an academic or discourse community. Within engineering education, a disciplinary community is comprised of individuals who share the same understanding of the concepts, norms, and conventions of the discipline, such as peers, lecturers, or students in upper levels of study. Assessment activities must therefore encourage students to collaborate and converse with members of their disciplinary community to exchange and construct the meaning. An example of such activities is dialogic and group discussions within a collaborative problem-solving context in which students practice negotiating ideas and relevant information about different aspects of the project drawing on diverse tools and formats. The ideas and thoughts can be elicited and articulated formally or informally among members of the group across various settings. Relevant information such as how to organize activities or define responsibilities can also be exchanged during the process of collaboration.

One of the great benefits of collaborative activities is the engagement of students in a dialogic feedback process. The feedback that students produce and receive can be in most cases a crucial determinant of their learning success in a collaborative environment (Sridharan and Boud, 2019). Previous studies have revealed the effect of dialogic feedback in engineering problem-solving contexts, with benefits mounting towards the producers of feedback (Mostert and Snowball, 2013, Donia et al., 2018). It is reported that in an attempt to provide feedback, students get involved in a range of higher order mental activities like analyzing the problem, generating original opinions and expressing their standpoints, proposing a solution, and justifying the solution (Mostert and Snowball, 2013). Engagement in and practising these activities have the potential to help students develop their communication and teamwork skills.

The influence of receiving feedback has also been reported in some studies. Attention has been focused on the feedback acting as a scaffold for complex collaborative problem-solving tasks. To solve complex ill-structured problems in engineering settings, students need to be scaffolded through external sources of information including peer ideas (Hall et al., 2018). An example of peer feedback scaffolding in collaborative tasks is when a fellow student shares an idea on how to break down a complex problem into smaller tasks or steps and provides feedback on how to address these steps prior to completing the overall task. This initial guided practice can act as a stepping stone for opening up an interactive space where other students can explore and justify their thoughts, negotiate and exchange their ideas, and come up with alternative points of view until the desired goal is achieved.

Metacognitive functioning

An authentic assessment design proposes learning activities that encourage students' use of metacognitive knowledge (Ashford-Rowe et al., 2014). Research on workplace problem solving has shown that it is a complicated intellectual task and involves a wide range of meta-cognitive skills. In any problem-solving situation, a variety of skills and strategies are applied, including setting goals; evaluating progress towards the goals; processing a huge, sometimes conflicting, amount of information; and employing appropriate strategies in different situations. In such complex problem-solving situations, metacognition has been identified as an important success factor. It has been suggested that developing students' metacognitive knowledge would help them monitor and control their information processing in solving actual problems, and to adjust their behaviours under different circumstances (Martinez, 2006, Tempelaar, 2006). The question that arises here is what metacognitive knowledge is and how it can be conceptualized within an authentic assessment context.

The term *metacognitive knowledge* is widely used for referring to thinking about thinking, i.e. reflecting on or monitoring one's own cognition (Flavell, 1979). It has been defined operationally and conceptually in a variety of ways, with different categories emerging from these conceptualizations. However, an integral part of metacognitive knowledge, which has often been taken for granted in the literature on problem-solving, is strategy knowledge. As suggested by Schraw and Dennison (1994), strategy knowledge has two domains: declarative knowledge and procedural knowledge. Declarative knowledge refers to awareness of strategies, while procedural knowledge refers to awareness of how, when, and why to apply strategies in a specific situation. Taking on this perspective, metacognitive knowledge involves students' ability to identify quality learning strategies and to determine the relative benefits of one strategy over another to plan, monitor, and evaluate their problem-solving process more effectively.

In applying this conceptualization of metacognitive knowledge to an authentic assessment context, we might expect the activities to help students formulate the link between new ideas and previous knowledge to analyse problem-solving tasks, and make decisions about the most effective strategies required for the completion of a specific outcome when no obvious method of solution is available. Such strategies might include managing and regulating resources such as peer learning, online learning, help seeking, and learning environment (i.e., an environment where it is possible to negotiate and share ideas with others including online platforms). It must be noted that developing knowledge of how, when, and why to use these strategies effectively requires sustained effort on the part of students, that is, an active engagement of students in solving actual problems and providing them with extended opportunities to exchange their ideas and elicit information and resources held by their peers.

Evaluative judgement

Authentic assessment should develop students' independent decision-making ability, or what Tai et al. (2017) refer to as *evaluative judgement capability*. It has been suggested that for assessment to have sustainable learning effects, activities should focus on both the immediate content and on implications for meeting students' future learning needs in the job market (Boud and Soler, 2016). They should focus on developing learning-how-to-assess skills and nurturing flexible learners who can move beyond the boundaries of their current module objectives and make proper decisions in challenging situations in the workplace. Students need to become effective independent decision-makers, namely, to be able to independently approach problem-solving situations in a real-world context and make evaluative judgements. In any workplace setting, evaluative judgement might include a student's capacity to use evidence to analyse various aspects of conflict situations and circumstances, negotiate ideas with others, evaluate potential solution paths, and draw sound conclusions.

Within the engineering context, a range of abilities can be used to define and conceptualize the notion of evaluative judgement. An individual ability to resolve conflicts can be a common example across various engineering workplace settings (Mehrabi Boshrabadi & Hosseini, 2020). Conflicts and misunderstandings are inevitable in team works as a diverse group of people with multiple opinions and ideas are involved. Therefore, individuals need to be equipped with the ability to deal with such tough situations and make proper decisions on how and which strategies to use to confront the conflicts. They need to take the benefits of all parties into account, or to incorporate the viewpoints of all team members, while proposing a given solution. As another example, students' ability in team organization can be central to the notion of evaluative judgement. To advance the project goals and address the problem successfully, individuals need to interact with each other to organize activities and define responsibilities, monitor their own and others' performance (Hesse et al., 2015), and make critical decisions on how to manage time and resources constraints.

In order to develop students' capacity to make evaluative judgements, an important consideration is to provide them with extended opportunities from the early stages of their undergraduate studies to practice authentic activities within a meaningful environment and across multiple contexts (Sadler, 2009). This requires integrative assessment tasks that actively engage students in a range of problem-solving activities, with the focus on developing both metacognitive and collaboration skills. Students need to be afforded opportunities to work and communicate with others, exchange ideas, elicit information, and provide feedback. Such opportunities can be provided in virtual environments that support discussion-based interactions through synchronous or asynchronous communication. A simple example here may help clarify the point. An authentic activity might be engaging two teams of students to work collaboratively to arrive at a decision on how to generate electricity from clean and renewable resources. To reach an outcome, students are required to negotiate the possibility and effect of energy resources such as wind, nuclear power, and solar energy. Teacher guidance might be required at this stage to support students in the scaffolding process of clarifying the expectations of the tasks. Each group will then work out an initial solution, discuss the solution with the other group, seek comments and feedback, and justify their ideas. Following this negotiation, both groups will then exchange their ideas and collaboratively generate a final effective solution.

Providing such interactional opportunities that are to some extent like the kinds of activities students might regularly face in their workplace is a critical element that must be considered in designing assessment tasks in engineering setting. This would contribute to assessment authenticity in two ways: first, it helps students gain insights into what quality performance means in a real-world environment; second, over time it develops students' capacity to make effective decisions and regulate their performance in any complex problem-solving situation.

The advantages of authentic assessment, as discussed, are diverse and well documented, including enhancing students' metacognitive, problem-solving, and collaboration skills. Several challenges, however, have been identified with respect to the design and implementation of authentic assessments. These challenges are discussed below.

Challenges of designing authentic assessments

First, there is a tendency among teaching staff to resist changing their assessment methods (Deneen and Boud, 2014, Rodríguez-Gómez et al., 2016). Integrating authentic assessment approaches needs a willingness to embrace replanning of conventional approaches; however, staff may not perceive a need to rethink approaches with which they are familiar and comfortable. Quesada-Serra et al. (2016) echoed this view, reporting that while staff were quite comfortable implementing more traditional assessment genres and practices, they felt less comfortable with those requiring students' involvement and, therefore, rarely included them in their evaluation methods. Staff may resist changes to assessment philosophy and practices for various reasons. The key reason is that change of an assessment approach requires a great deal of time, energy, and cognitive flexibility to monitor, challenge, and guide learners in complex problem-solving tasks (Brush and Saye, 2008). With current modularized course structures, which have led to large class sizes and a heavy workload for educators, this has proven problematic.

Another challenge of authentic assessment is designing and implementing activities for large groups of students (Palmer, 2004). Reports from other studies indicate that the use of detailed, individual authentic assessment in large classes is restricted. As such, educators tend to use traditional testing approaches (e.g., multiple-choice tests) suitable for large-scale assessment, with ease of programming, data gathering, and scoring individual performances (Biggs and Tang, 2011a). However, questions might be raised regarding the potential of such traditional approaches to developing students' longer-term learning needs – the skills in the collaboration process demanded from students after graduation in real-world projects. Indeed, as discussed earlier, an important factor determining student success in the real-world job market is their mastery of collaboration skills such as seeking and eliciting information, exchanging and negotiating ideas, evaluating conflict situations, and making critical decisions. The lack of actions and limitations in interactions in traditional approaches, however, would make it difficult to properly address a number of these complex skills.

To address the aforementioned challenges, the use of effective pedagogical strategies such as peer assessment and self-assessment has been proposed (Sadler, 2009, Boud et al., 2015). In addition to reducing the amount of time educators have to devote to assessing individual students, the proper implementation of these strategies can be beneficial to students in many ways. Peer assessment, for example, can improve students' reflexivity and collaboration skills as it encourages a reflective knowledge-building approach so that students ask questions, elicit knowledge, and figure out solutions and improvements to complex tasks (Donia et al., 2018). This is in line with more recent theories of cognition that focus on information processing and learning as an act that is distributed among people, their history, conversations, and social interactions (Jonassen and Henning, 1999). Such conversations and interactions may not happen within the formal context of academia, and/or may be solely intended for meeting the requirements of the final assessment task. Students can draw upon a range of interpersonal activities at any time during the course of study both within and beyond the academic context to negotiate their options and viewpoints and bring their performance closer to the requirements of a problem-solving task. An example of a peer assessment activity conducted with large groups is given by He et al. (2017), who tried to simulate peer interaction through designing chat-based tasks where students could discuss a range of solution paths and various strategies during the problem-solving task.

Conclusion

This chapter identified a framework for authentic assessment in BIM-related educational programmes. The framework comprises four elements: (1) reflection of a realistic workplace context, (2) collaboration and feedback, (3) metacognitive functioning, and (4) evaluative judgement. This framework and its elements are generalizable and can be applied to any BIM assessment within construction education. The framework identified here is focused on ensuring that students see BIM and its collaboration process as positively contributing to the quality of their output. As this framework applies in BIM education assessment, the process must be managed so that students do not frame the act of working collaboratively as inherently negative or an inconvenience. Indeed, the proposed framework and its elements are aimed towards achieving this goal.

References

Adams, R. J. Lietz, P. and Berezner, A. (2013), On the use of rotated context questionnaires in conjunction with multilevel item response models. *Large-Scale Assessments in Education*, Vol. 1 No. 1. pp. 5.

Ajjawi, R., Tai, J., Huu Nghia, T. L. Boud, D., Johnson, L. and Patrick, C.-J. (2020), Aligning assessment with the needs of work-integrated learning: The challenges of authentic assessment in a complex context. *Assessment & Evaluation in Higher Education*, Vol. 45 No. 2. pp. 304–316.

Andrews, J. and Higson, H. (2014), Is Bologna working? Employer and graduate reflections of the quality, value and relevance of business and management education in four European union countries. *Higher Education Quarterly*, Vol. 68 No. 3. pp. 267–287.

Ashford-Rowe, K. Herrington, J. and Brown, C. (2014), Establishing the critical elements that determine authentic assessment. *Assessment & Evaluation in Higher Education*, Vol. 39 No. 2. pp. 205–222.

Baartman, L. and Kirschner, P. (2006), Innovative assessment in higher education: Cordelia Bryan & Karen Clegg. *British Journal of Educational Technology*, Vol. 37 No. 6. pp. 977–978.

Biggs, J. and Tang, C. (2011a), Train-the-trainers: Implementing outcomes-based teaching and learning in Malaysian higher education. *Malaysian Journal of Learning and Instruction*, Vol. 8. pp. 1–19.

Biggs, J. B. and Tang, C. S.-k. (2011b). *Teaching for Quality Learning at University: What the Student Does*, 4th ed. McGraw-Hill/Society for Research into Higher Education/Open University Press, Maidenhead, Berkshire, England.

Bloxham, S. (2009), Marking and moderation in the UK: False assumptions and wasted resources. *Assessment & Evaluation in Higher Education*, Vol. 34 No. 2. pp. 209–220.

Bosco, A. M. and Ferns, S. (2014), Embedding of authentic assessment in work-integrated learning curriculum. *Asia-Pacific Journal of Cooperative Education*, Vol. 15 No. 4. pp. 281–290.

Boud, D. (2000), Sustainable assessment: Rethinking assessment for the learning society. *Studies in Continuing Education*, Vol. 22 No. 2. pp. 151–167.

Boud, D. and Associates (2010), *Assessment 2020: Seven Propositions for Assessment Reform in Higher Education*. Sydney: Australian Learning and Teaching Council.

Boud, D. and Falchikov, N. (2007). *Rethinking Assessment in Higher Education, Learning for the Longer Term*. Oxford, UK: Routledge.

Boud, D. and Soler, R. (2016), Sustainable assessment revisited. *Assessment & Evaluation in Higher Education*, Vol. 41 No. 3. pp. 400–413.

Boud, D. Lawson, R. and Thompson, D. G. (2015), The calibration of student judgement through self-assessment: disruptive effects of assessment patterns. *Higher Education Research and Development*, Vol. 34 No. 1. pp. 45–59.

Brush, T. and Saye, J. (2008), The effects of multimedia-supported problem-based inquiry on student engagement, empathy, and assumptions about history. *Interdisciplinary Journal of Problem-Based Learning*, Vol. 2 No. 1. pp. 21–56.

Carless, D. (2015), Exploring learning-oriented assessment processes. *Higher Education*, Vol. 69 No. 6. pp. 963–976.

Clarke, D. Litchfield, C. andDrinkwater, E. (2010). Supporting exercise science students to respond to the challenges of an authentic work-integrated learning (WIL) assessment. *Asia-Pacific Journal of Cooperative Education*, Vol. 12 No. 3. pp. 153–157.

Cunningham, C. M. and Kelly, G. J. (2017), Epistemic practices of engineering for education. *Science Education*, Vol. 101 No. 3. pp. 486–505.

Deneen, C. and Boud, D. (2014), Patterns of resistance in managing assessment change. *Assessment and Evaluation in Higher Education*, Vol. 39 No. 5. pp. 577–591.

Donia, M. B. L. O'Neill, T. A. and Brutus, S. (2018), The longitudinal effects of peer feedback in the development and transfer of student teamwork skills. *Learning and Individual Differences*, Vol. 61. pp. 87–98.

Flavell, J. H. (1979), Metacognition and cognitive monitoring: A new area of cognitive–developmental inquiry. *American Psychologist*, Vol. 34 No. 10. pp. 906–911.

Gill, B. (2015), Talking about the elephant in the room: Improving fundamental assessment practices. *Student Success*, Vol. 6 No. 2. pp. 53.

Hall, K. L. Vogel, A. L. Huang, G. C. Serrano, K. J. Rice, E. L. Tsakraklides, S. P. and Fiore, S. M. (2018), The science of team science: A review of the empirical evidence and research gaps on collaboration in science. *American Psychologist*, Vol. 73 No. 4. pp. 532–548.

He, Q. von Davier, M. Greiff, S. Steinhauer, E. W. and Borysewicz, P. B. (2017). Collaborative problem solving measures in the programme for international student assessment (PISA). *In*: Von Davier, A. A., Zhu, M. & Kyllonen, P. C. (eds.) *Innovative Assessment of Collaboration*. Springer International Publishing, Cham.

Heeneman, S. Oudkerk Pool, A. Schuwirth, L. W. T. van der Vleuten, C. P. M. and Driessen, E. W. (2015), The impact of programmatic assessment on student learning: theory versus practice. *Medical Education*, Vol. 49 No. 5. pp. 487–498.

Hesse, F. Care, E. Buder, J. Sassenberg, K. and Griffin, P. (2015). A framework for teachable collaborative problem solving skills. *In*: Griffin, P. & Care, E. (eds.) *Assessment and Teaching of 21st Century Skills: Methods and Approach*. Springer, Dordrecht Netherlands.

Hosseini, M. R. Martek, I. Chileshe, N. Zavadskas, E. K. and Arashpour, M. (2018). Assessing the influence of virtuality on the effectiveness of engineering project networks: "Big Five theory" perspective. *Journal of Construction Engineering and Management*, Vol. 144 No. 7. pp. 04018059.

Jin, J. Hwang, K. E. and Kim, I. (2020), A study on the constructivism learning method for BIM/IPD collaboration education. *Applied Sciences*, Vol. 10 No. 15. pp. 51–69.

Jonassen, D. and Henning, P. (1999), Mental models: Knowledge in the head and knowledge in the world. *Educational Technology*, Vol. 39 No. 3. pp. 37.

Jonassen, D. Strobel, J. and Chwee Beng, L. (2006), Everyday problem solving in engineering: Lessons for engineering educators. *Journal of Engineering Education*, Vol. 95 No. 2. pp. 139–151.

Kostopoulos, K. C. Spanos, Y. E. and Prastacos, G. P. (2013), Structure and function of team learning emergence: A multilevel empirical validation. *Journal of Management*, Vol. 39 No. 6. pp. 1430–1461.

Martinez, M. E. (2006), What is metacognition? *Phi Delta Kappan*, Vol. 87 No. 9. pp. 696–699.

Mehrabi Boshrabadi, A. and Hosseini, M. R. (2020), Designing collaborative problem solving assessment tasks in engineering: An evaluative judgement perspective. *Assessment & Evaluation in Higher Education*, pp. 1–15.

Merschbrock, C. Hosseini, M. R. Martek, I. Arashpour, M. and Mignone, G. (2018), Collaborative role of sociotechnical components in BIM-based construction networks in two hospitals. *Journal of Management in Engineering*, Vol. 34 No. 4. pp. 05018006.

Michaelsen, L. K. Sweet, M. and Parmelee, D. X. (2011). *Team-Based Learning: Small Group Learning's Next Big Step: New Directions for Teaching and Learning*, Vol. 116. Hoboken, NJ: Wiley.

Mostert, M. and Snowball, J. D. (2013), Where angels fear to tread: Online peer-assessment in a large first-year class. *Assessment & Evaluation in Higher Education*, Vol. 38 No. 6. pp. 674–686.

Oraee, M. Hosseini, M. R. Edwards, D. J. Li, H. Papadonikolaki, E. and Cao, D. (2019), Collaboration barriers in BIM-based construction networks: A conceptual model. *International Journal of Project Management*, Vol. 37 No. 6. pp. 839–854.

Palmer, S. (2004), Authenticity in assessment: Reflecting undergraduate study and professional practice. *European Journal of Engineering Education*, Vol. 29 No. 2. pp. 193–202.

Quesada-Serra, V. Rodríguez-Gómez, G. and Ibarra-Sáiz, M. S. (2016), What are we missing? Spanish lecturers' perceptions of their assessment practices. *Innovations in Education and Teaching International*, Vol. 53 No. 1. pp. 48–59.

Rodríguez-Gómez, G. Quesada-Serra, V. and Ibarra-Sáiz, M. S. (2016), Learning-oriented e-assessment: the effects of a training and guidance programme on lecturers' perceptions. *Assessment & Evaluation in Higher Education*, Vol. 41 No. 1. pp. 35–52.

Sadler, D. R. (2009), Grade integrity and the representation of academic achievement. *Studies in Higher Education*, Vol. 34 No. 7. pp. 807–826.

Sambell, K. McDowell, L. and Montgomery, C. (2013). *Assessment for Learning in Higher Education*. Abingdon, UK: Routledge, Taylor & Francis.

Savery, J. R. (2006), Overview of problem-based learning: Definitions and distinctions *Interdisciplinary Journal of Problem-Based Learning*, Vol. 1 No. 1. pp. 9–20.

Schraw, G. and Dennison, R. S. (1994), Assessing metacognitive awareness. *Contemporary Educational Psychology*, Vol. 19 No. 4. pp. 460–475.

Smagorinsky, P. (1995), The social construction of data: Methodological problems of investigating learning in the zone of proximal development. *Review of Educational Research*, Vol. 65 No. 3. pp. 191.

Sridharan, B. and Boud, D. 2019. *The Effects of Peer Judgements on Teamwork and Self-Assessment Ability in Collaborative Group Work*. Oxfordshire, UK: Taylor & Francis.

Tai, J. Ajjawi, R. Boud, D. Dawson, P. and Panadero, E. (2017), Developing evaluative judgement: Enabling students to make decisions about the quality of work. *Higher Education*, Vol. 76 No. 3. pp. 467–481.

Tempelaar, D. T. (2006), The role of metacognition in business education. *Industry and Higher Education*, Vol. 20 No. 5. pp. 291–297.

Villarroel, V. Bloxham, S. Bruna, D. Bruna, C. and Herrera-Seda, C. (2018), Authentic assessment: Creating a blueprint for course design. *Assessment & Evaluation in Higher Education*, Vol. 43 No. 5. pp. 840–854.

Vu, T. T. and Dall'Alba, G. (2014), Authentic assessment for student learning: An ontological conceptualisation. *Educational Philosophy and Theory*, Vol. 46 No. 7. pp. 778–791.

Vygotsky, L. (1978), Interaction between learning and development. *Readings on the Development of Children*, Vol. 23 No. 3. pp. 34–41.

Wiggins, G. (1990), The case for authentic assessment. *Practical Assessment, Research and Evaluation*, Vol. 2 No. 1. pp. 2.

19 Using gamification and competitions to enhance BIM learning experience

*Ajibade A. Aibinu, Teo Ai Lin Evelyn,
Juan S. Rojas-Quintero, M. Reza Hosseini,
Chiranjib Dey, Reza Taban, and Tayyab Ahmad*

Introduction

Building information modelling (BIM) methodology and processes enabled by technological innovations represent the most advanced common new way of work in the architecture, engineering, and construction (AEC) industry; BIM is currently at the forefront of digital transformation, transforming how information is created, used, shared, and reused across planning, design, construction, operations, and end-life of built assets (Aibinu and Papadonikolaki, 2020, Hosseini et al., 2021). The selling point of BIM is that it facilitates simultaneous work by multiple disciplines, minimizes project coordination difficulties, improves information management problems, and enables smoother sharing of information across the supply chain. Theoretically, BIM features and associated smart technologies reduce complexity by providing better visualization and digital information flow (Eastman et al., 2018). The use of BIM can lead to faster, cheaper, and better buildings that are environmentally sustainable. BIM-enabled project delivery can facilitate the integration of the project team and of the building's design, construction, and operations (Aibinu and Papadonikolaki, 2020). BIM can bring cost and time savings on projects given its potential for (a) higher quality of design documentation and well-coordinated project team; (b) the ability to facilitate the exchange of information among various design disciplines, thereby reducing design errors and omissions with significant reduction in design time; and (c) the ability to collate multiple components of designs – with BIM tools – to identify clashes early during the design stage (Elghaish et al., 2019).

BIM features can also make project teams more efficient in their individual work. That is, BIM can enable project teams and owners to visualize initial concepts and designs, and various options and alternatives. Visualization can support design decisions regarding the choice of alternative materials, technology, and building systems (Hamidavi et al., 2020). An advanced level of BIM implementation can assist owners and the design team to understand the impacts and location of a proposed facility in relation to existing facilities and road network in a precinct, town, or city. The impact of the newly proposed facility on neighbourhood infrastructure can be interrogated to potentially improve the sustainability of cities and the built environment. Facility managers can use the information during operations. For example, upon project completion, the builder's coordinated construction models can become as-built models, including electronic specifications, catalogues, maintenance manuals, testing, and commissioning documents. These are passed on to clients and facility managers for asset operations, maintenance, and later refurbishment when required (Pärn et al., 2017).

Being a new way and the future approach for procuring built assets, students need to appreciate the benefits of BIM, its possibilities, potential, and impacts too. For students to be BIM ready, they must gain knowledge of all the above areas, develop skills, and acquire abilities in applying BIM in various dimensions related to each of these concepts. These include understanding new workflows, technologies, and collaboration skills, to mention a few (Abdirad and Dossick, 2016). In traditional teaching, participants are passive learners; these methods do not facilitate engagement and are not suitable for BIM-related learning and teaching (Casasayas et al., 2021). That said, research into developing alternative and effective methods of providing BIM teaching and learning to students and trainees is limited; very few case studies are available in the literature that provide experience and lessons learned of using alternative methods based on tested and proven methods (Wu et al., 2018). This chapter aims at addressing this gap, which leaves educators with few options other than pursuing traditional methods of teaching. We propose that gamification and competition represent a proven approach that can positively change students' attitudes towards and experience with BIM learning, and accordingly improve motivation and engagement in a setting that simulates real-life projects. The remainder of the chapter is structured as follows. First, the principles and fundamental aspects of gamification and competition in learning environments are briefly discussed. The second part of the chapter provides an account of a gamification and competition event designed for BIM learning, and examines the challenges, drawbacks, and lessons learned in four years of co-organising this event by the University of Melbourne, Australia and National University of Singapore.

Key concepts and context

Key learning methodologies

Most activities in the AEC context require collaborative efforts by several parties who are driven by a common goal but work independently across organizations, time zones, and space boundaries (Allen et al., 2005). As such, developing skills in "working and learning with others from different disciplines and backgrounds" has been included among learning outcomes (LO) for almost all subjects in the AEC-related curriculum in various universities. According to Davey et al. (2016, p. 89), "the prevalence of teamwork in industry makes it incumbent upon universities to better prepare students for real life projects". This LO has been translated to group projects to drive teamwork learning among students in subjects across the curriculum according to the constructive alignment principle (Biggs and Tang, 2011). In doing so, several learning methodologies can inform the design of these teaching and learning initiative. Of these, the major ones are listed below.

- Collaborative-based learning (CBL);
- Problem-based learning (PBL);
- Project-based learning (PjBL);
- Competition-based learning (CnBL).

In collaborative-based learning (CBL), learners work in small groups and are given activities that maximize collaboration (Burguillo, 2010). Collaboration promotes the exchange of information and knowledge and motivates individual learners. So too, using CBL ensures a common reinforcement. Problem-based learning (PBL) focuses on collaborative problem solving, which enables learners to reflect on their experiences. PBL uses open-ended problems requiring learners to work in small collaborative teams and be responsible for organizing their team. They also have to manage the learning process with a facilitator (Mehrabi Boshrabadi and Hosseini, 2020).

Project-based learning (PjBL) involves complex tasks based on challenging questions. Learners use problem solving, decision making, investigative skills, and reflection to address the problem with support from their instructors. The project results in deep learning about issues directly relating to the topic of learning and education. In PBL, instructors specify the task to be performed at a basic granularity level, whereas in PjBL instructors set a relatively complex task to get the students to design and self-organize the division of their overall task into sub-tasks and activities. These approaches have been used extensively in various fields like medicine and engineering, among others (Biggs and Tang, 2011, Mehrabi Boshrabadi and Hosseini, 2020).

In the case of competition-based learning (CnBL), learning is achieved through a competition; learners' scores are not directly related to the learning result. In contrast, competitive-based learning or competitive programming implies that learning depends on the result of the competition itself (Johnson et al., 1985). According to Burguillo (2010) and Sailer et al. (2017), CnBL can be combined with other learning methodologies as CBL, PBL, or PjBL, to motivate students and help to improve their performance. Cagiltay et al. (2015) found that when a competitive environment is created in a serious game, the motivation and post-test scores of learners improved significantly. Ediger (2001) addressed healthy competition in the classroom and set forth five guidelines for stressing competitive events. According to Ediger (2001), those competing should be somewhat equivalent in talents, skills, and abilities; have positive attitudes towards each other; desire to participate and learn; have definite goals to achieve in the competitive event; and realize that not all individuals can be winners. Against this backdrop, the literature provides encouraging support for the positive impacts of CnBL in higher education used in the form of gamification and game-based learning. The successful implementation of these approaches gives reason to be enthusiastic about their application in higher education across various country/student cultures, subjects, and formats (Subhash and Cudney, 2018). The context of AEC is no exception.

The facts above provide theoretical justification for and speak to the effectiveness of using gamification and competitions for BIM learning and teaching. That said, in the context of the case study presented in this chapter, we adopted a combination of learning methodologies, including competition-based learning (CnBL), project-based learning (PjBL), and collaborative-based learning (CBL). However, teaching was organized around competition, where participants engaged with tasks to understand BIM concepts and workflows via learning by doing in a competitive environment.

Competition and learning

Competition has historically been used to inspire learning. Youngsters impulsively seek competition with their companions. This inherent urge to compare themselves with their peers leads them to intellectual and physical contests such as sports, where they participate for their interest. Initially, formal competitions were limited to sports activities. The function of competition in other fields is more of a recent phenomenon. It is worth noting that even now, informal competition plays a key role.

The idea of competition started in the scientific world from ancient activities like sports, arts, and military actions. According to Verhoeff (1997), academies were the key scientific institutions in the eighteenth century, to be later succeeded by universities. Academies used to organize competitions to find solutions for important scientific and mathematical problems.

Mathematicians such as Daniel Bernoulli, Lagrange, and Jean d' Alembert have won several prize competitions.

It is debatable whether competition influences the learning process positively or negatively. A significant number of studies have revealed that competition positively influences learning outcomes (Verhoeff, 1997, Lawrence, 2004, Fasli and Michalakopoulos, 2005, Fulu, 2007, Cantador and Conde, 2010, Janssen et al., 2015). According to Verhoeff (1997) and Lawrence (2004), a well-designed competition enhances students' motivation and learning by stimulating participants to give their best. A competitive element drives students to put in more effort; even weaker students persist with participating in the activity (Fasli and Michalakopoulos, 2005). Participants enjoy the competitive aspect of the activity, particularly the opportunity to compete against peers they consider their academic equals (Janssen et al., 2015). For participants, competitions can also result in recognition, self-esteem, motivation, active engagement, and responsible learning (Fulu, 2007). When competition is organized in teams, it engages each team member due to members' interest in their team's tangible final goals. The competition concept creates a dramatic increase in students' motivation and participation (Morgan and Porter, 2003). For the sake of their group, team members undertake responsibilities and tasks to achieve their cooperative goals (Thousand et al., 2002).

According to some studies, however, competitions' learning outcomes may not be the same for all participants involved. Sternberg and Baalsrud-Hauge (2015) found that a monetary prize further motivates already motivated students but does not affect less motivated students in the same way. Competition may also adversely affect the learning opportunities, as students tend to concentrate more on the performance rather than on the learning process (Lam et al., 2001). Bergin (1995) emphasized that students with lower ability learn better from a non-competitive environment than a competitive environment.

Gamification: Types and benefits

Competitive environment can be created to motivate learning via games in a learning exercise. *Gamification* can be defined as the techniques used in integrating game elements or game-like characteristics into education (Deterding et al., 2011). It is a process that optimizes learners' motivation along with learners' efficiency. Gamification creates an interactive and engaging learning process and makes learning less monotonous and more exciting. The use of gamified learning can result in shorter feedback cycles, clear learning pathways, enhanced collaboration, and recognition of talent among learners (Sailer et al., 2017). In higher education, gamification techniques have been used to increase learners' motivation and engagement in learning tasks to achieve better outcomes (Kuo and Chuang, 2016, Kim et al., 2018).

Gamified learning experiences include points, prizes, badges, leaderboards, scoreboards, challenges, levels, and feedbacks (Barata et al., 2015, Subhash and Cudney, 2018, Yildirim, 2017). Gamification techniques can influence learners' sense of competition, interaction, and motivation (Davis et al., 2018). In gamified learning, more goals are achieved by participants because of recognition and reward (Subhash and Cudney, 2018). In this regard, Reeves and Read (2009) identified 10 ingredients for a successful game design. These are self-representations, three-dimensional environments, narrative, feedback, reputations, ranks and levels, marketplaces and economies, competition under rules, teams, communication, and time pressure.

When considering the gamification of learning exercise, it is possible to add game elements to the structure of content or the content itself, resulting in *structural gamification* or *content gamification*. *Structural gamification* is characterized by game mechanics like rewards, namely, incentives to engage and encourage learners to continue the learning exercise, where a reward can motivate a person to complete one more task in a learning exercise. The content of the learning material does not become game-like, only the structure around the content. On the other hand, in *content gamification*, game elements and game thinking are applied to change content to make it more game-like, for example, when a course commences with a challenge instead of a list of objectives. In the BIM Immersion and Competition event, we used both structural and content gamification.

Gamification frameworks

Principles of game design theory are at the core of gamified learning. Historically, several frameworks have been developed to enable successful gamification designs (Subhash and Cudney, 2018). Mora et al. (2015) identified 10 game design items and clustered/organized them into five categories: economic, logic, measurement, psychology, and interaction.

According to Mora et al. (2015), most gamification frameworks have adopted the human-centric approach. Principles of social psychology underpins most frameworks too. Self-determination theory (SDT) by Ryan and Deci (2000) is a general approach in gamification. It assumes that people are motivated to grow and change by psychological needs such as competence, connection, and autonomy. Hence, successful gamification significantly relies on the psychological experience it creates. Zichermann and Cunningham (2011) also found that gamification is made up of 75% psychology and 25% technology.

The Octalysis gamification framework

Octalysis framework by Chou (2015) is one of the many gamification frameworks. It is a complete gamification framework about analyzing and building strategies around various systems that make a game fun. Chou (2015) perceives gamification as a design that emphasizes human motivation in the process. The idea is that games are fun because they appeal to specific core drives within people that motivate them towards certain activities. The Octalysis framework is based on an octagon shape with eight core drives represented by each side, as illustrated in Figure 19.1.

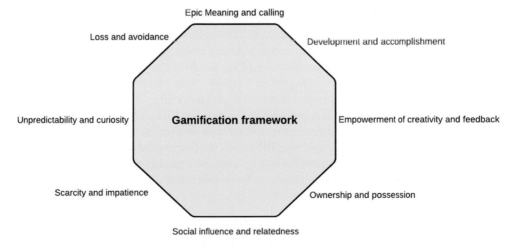

Figure 19.1 The Octalysis gamification framework (adapted from Chou [2015]).

These drives in Figure 19.1 are defined as follows:

I.   Epic meaning and calling: where a learner believes that he/she is doing something greater than himself, or that he/she was "chosen" to do something.

II. Development and accomplishment: learners have an internal drive to make progress, develop skills, and eventually overcome challenges.

III. Empowerment of creativity and feedback: learners are engaged in a creative process to figure things out and try different combinations repeatedly. They not only need ways to express their creativity, but also need to be able to see the results of their creativity, receive feedback, and respond in turn.

IV. Ownership and possession: learners are motivated because they feel like they own something. A sense of ownership leads learners to make what they own better and to own even more.

V.   Social influence and relatedness: this is the drive that incorporates all the social elements that drive people, including mentorship, acceptance, social responses, companionship, competition, and envy.

VI. Scarcity and impatience: this is the drive of wanting something because one cannot have it. Many games have appointment dynamics (come back two hours later to get your reward); the fact that people cannot get something right away motivates them to think about it continuously for some time.

VII. Unpredictability and curiosity: this is a harmless drive of wanting to find out what will happen next. If you do not know what will happen, your brain is engaged, and you often think about it.

VIII. Loss and avoidance: this is a core drive based upon the avoidance of something negative happening. On a small scale, it could be to avoid losing previous work. On a larger scale, it could be to avoid admitting that everything you did up to this point was useless because you are now quitting.

The case study

The competition and its objectives

A five-day event, *BIM Immersion and Competition*, was initiated in 2015, involving student participants and construction industry professionals. The event was a combination of three days of intensive BIM education and training and two days of competition. The intensive training comprised lectures, seminars, and hands-on use of various BIM tools to explore 3D, 4D, and 5D BIM and model coordination concepts in a studio environment. Student participants were exposed to theories and concepts related to digital modelling (3D) and BIM; the concept of model-based scheduling (4D); model-based estimation (5D); and model coordination workflow (model analysis). The initiative focused on both theoretical concepts and practices of BIM alike, with a bias towards applying BIM tools to reinforce learning via competition. The fundamental aim of the competition was to increase digital literacy across the supply chain in design, construction, and operations in the construction industry.

The main objectives of the event were to

1 Explore BIM as a tool and process;
2 Increase participants' understanding of 3D, 4D, and 5D BIM concepts and provide participants with the skills needed to apply these concepts in practice;

3 Promote the adoption of advanced digital technologies by educating construction students and making them BIM literate;

4 Raise awareness of the opportunities and benefits of digital technologies through education and simplify complex digital concepts and scenarios;

5 Demonstrate the impacts of digital processes on practice and provide a picture of the future of the industry and what is possible in adopting BIM.

Key considerations and assumptions

Competition was the central gamification element incorporated into the BIM event. It was assumed that the competition aspect would dramatically enhance student motivation, participation, learning, and performance. We used the Octalysis framework to design and implement the competition because of its emphasis on human motivation. We assumed that working in teams would improve the social climate among the participants and improve the learning process, while competing in teams would encourage collaboration because of shared success feelings when working in teams. Participants were rewarded with cash prizes, badges, certificates, and a trophy (Hosseini et al., 2021 following Alomari et al., 2019). Psychological considerations were also primary in designing and implementing the competition. The participants were required to participate in teaching and learning sessions comprising of all the teams. Before the competition period, the cohort engaged in joint exercises. During the competition, they worked in teams collocated in the same facility to promote interaction. The competition tasks focused on the development of BIM competency and skills which participants can immediately put to work when they graduate, making the task authentic and promoting commitment and determination to achieve the goals.

The details of competition delivery followed both structural and content gamification principles, as discussed. The core drives of the Octalysis framework were also incorporated in organizing the event to facilitate achieving the defined objectives. Table 19.1 shows the alignment between each major aspect of the event with elements of the Octalysis framework (see Figure 19.1).

Team configuration

A total of 24 students were selected for this event. Students worked in teams of four. Therefore, there were six teams, and each team comprised a mix of students from all the participating universities (see Figure 19.2). In an ideal competition, all participants should feel like they have a chance to win (Cantador and Conde, 2010). Accordingly, the organizers created six teams prior to the event's commencement to ensure that all the teams were well balanced in terms of knowledge and skills, to provide a healthy and fair competition.

To ensure healthy competition, fair opportunities for success, and effective learning, student participants were required to view the online videos on ArchiCAD™ provided by Graphisoft and videos on 5D BIM provided and sponsored by Exactal before their arrival for the event. They were also provided with related literature to ensure they had a basic understanding of 4D BIM software like Synchro™. Students were trained on the ArchiCAD™ application for handling building models in IFC format, the Solibri™ application for clash detection, the Synchro™ application for 4D BIM, and the CostX™ application for 5D BIM during the intense three-day training sessions – training sessions continued the entire day with short breaks (Figure 19.3). On the fourth and the fifth day of the competition, teams were provided with a building model in IFC format to prepare a 4D BIM model with Synchro™ and 5D BIM model with CostX™, as two parts of the competition. The task also involved

Table 19.1 The BIM competition elements implemented according to Octalysis framework (refer to Figure 19.1 and the notes below the figure)

Octalysis framework drive	Aspect of BIM competition
Epic meaning and calling	Students were asked to register their interest, and participants were selected via a raffle where the organizers draw names at random. Participants were aware of the innovative nature of BIM in theory but had little or no prior working knowledge of BIM processes and workflows, and each team comprised a mix of students from both undergraduate and postgraduate levels. The selection and arrangement gave participants a greater sense of worth. The lack of prior working knowledge of BIM practice provided a sense that participants were doing something that is topical, current, and important and something that they will immediately put to work when they graduate.
Development and accomplishment	Teams were required to complete new tasks they had never engaged with in the past, using digital technology tools. Mentoring was provided by industry practitioners who have working experience in BIM. The process and the deliverables replicated BIM workflows in real-life contexts. Completing the task was challenging, yet enabled teams to accomplish something similar to what they will do in the industry after graduation. The tasks/deliverables were therefore deemed to be authentic (see chapter 18).
Empowerment of creativity and feedback	The tasks were new and challenging; tasks required participants to navigate models and identify and resolve clashes. Each team needed to decide on the best workflow for completing their tasks. Teams received ongoing feedback as well via discussion with their mentors
Ownership and possession	Teams were responsible for planning and scheduling their time and tasks, including allocation of sub-tasks. They kept meeting diaries, which they submitted as an essential part of their deliverables.
Social influence and relatedness	There was peer pressure within teams to win. Teams were collocated in the same facility. During the training period, all participants engaged with teaching and learning activities in one space. However, teams worked in separate rooms during the competition period.
Scarcity and impatience	The final formal assessment occurred on the last day of the competition after oral presentations. Delaying the reward increased engagement throughout and created motivation to continuously work hard towards the final end of competition assessment and award.
Unpredictability and curiosity	The final formal assessment occurred on the last day of the competition after the oral presentations, thereby creating curiosity and unpredictability across the teams throughout.
Loss and avoidance	Teams avoiding losing.

model analysis using Solibri™ software. The assessment criteria were shared and explained to all student participants at the competition's commencement.

Learning activities

The event benefited from principles of competition-based learning, project-based learning, and collaborative-based learning, as discussed. On the first day of the event, participants

Figure 19.2 Participants and mentors in 2015.

Figure 19.3 Digital modelling training session.

were engaged in a team-building activity to promote interaction and communication. All participants were randomly divided into two groups, with 12 students in each group. In the first stage of this activity, a useful metaphor was introduced to participants to raise awareness of the value of collaboration amongst all stakeholders on projects as a fundamental requirement to achieve BIM success. To demonstrate this, each group was given some A4 papers with different body parts written on them, such as head, neck, chest, arms, stomach, legs, and feet. Everyone had to sketch the body part independently and assemble all the features on the display board. Because of the lack of coordination among participants, the body parts were drawn to different scales, resulting in some large and some small parts. The assembled human body looked distorted and out of proportion, as illustrated in Figure 19.4.

At the second stage, they were asked to coordinate with each other, draw the same body parts on A4 paper, and assemble it on the display board. This time, the human figure was much better proportioned and scaled compared to the previous one, proving that teamwork and coordination are crucial in integrating all the human body parts to the right shape and size (see Figure 19.4).

On the third day, participants were engaged in another team-building activity called the *Helium Stick* (Figure 19.5). The participants were divided randomly into two groups. The participants of each group were asked to stand facing each other with their index finger raised to the chest level, pointing outwards. Then a long thin pipe made of paper was placed on

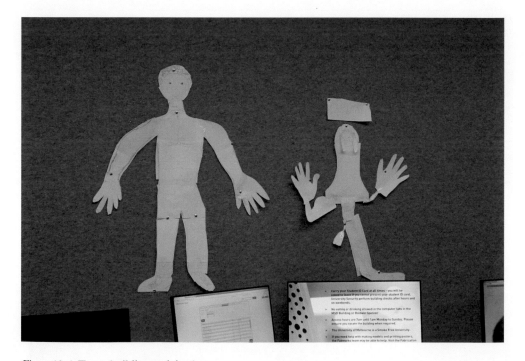

Figure 19.4 Team-building activity 1.
On the right in the above image is the assembled body drawn from parts of disproportionate scale. Lack of coordination resulted in parts looking distorted and out of proportion. On the left in the above image is the body assembled from parts drawn to same scale in a collaborative effort by participants. Coordination resulted in parts of the right shape and size.

Figure 19.5 Team-building activity 2.
The Helium Stick game where participants are working collaboratively, to take the stick down.

the index fingers such that the index fingers of all the participants were in contact with the pipe. This activity requires participants to work together and lower the pipe to the ground level without losing contact between their index finger and the pipe. The two teams had to simultaneously start this activity, and whoever did it first was declared the winner. This activity also highlighted the need for cooperation, communication, reflective practice, and leadership, raising awareness of and creating first-hand exposure to these critical ingredients of successful collaboration.

Both the sketching and Helium Stick exercises were designed to prepare participants for competition-based, project-based, and collaborative-based learning.

Assessment criteria

The first part of the competition focused on 4D BIM where teams were required to plan, develop, and implement a 4D BIM workflow for a given project. The deliverable also included a construction sequencing simulation video. Teams were expected to demonstrate, at least, the following aspects as parts of the assessment:

1. Import and resource assignment of CAD model (IFC models will be auto-assigned to resources) to Synchro™.
2. Resource refinement, including simplification of detailed resources.
3. Creation of CPM (critical path method) schedule.
4. Assignment of 3D resources to tasks using applicable Resource Appearance Profile.

5. With the Focus Time positioned before the first schedule task, the appearance of the 3D model should resemble a cleared site (or empty model).
6. With the Focus Time positioned after the last schedule task, the appearance of the 3D model should resemble a completed project.

Bonus points were awarded to those groups that demonstrated the following:

1. Logical order of construction.
2. Reasonable estimations of task durations.
3. Additional visual elements such as Resource Appearance Profile transparency and growth.
4. 3D path animation using mobile plant (i.e. excavator and crawler crane).
5. Animation exported to Audio Video Interleave file (AVI).
6. Commentary on how the 4D model could be used to promote communication and interfacing between different contractors.

Super-bonus points were awarded to any team that demonstrated the baseline vs. actual 4D schedule.

The second part focused on 5D BIM, where teams were required to plan, develop, and implement a 5D BIM workflow for the project case study. Teams were required to demonstrate the following:

1. Accuracy – spot errors in model (if any), allow for missing items on drawings (if any);
2. Cost plan formatting;
3. Presentation – appropriate description and generation of cost plan report;
4. Process/workflow explanation.

Team were also required to demonstrate how the model was analyzed and checked before implementing 4D and 5D workflow.

At the end of the competition period, each team prepared a 20-minute oral presentation describing their 5D workflow, their deliverables and the 4D workflow, and the schedule and simulation videos (Figure 19.6). Teams highlighted the challenges and lessons learned in using model-based processes and highlighted barriers to BIM adoption by the industry at large. They also discussed their views about the possible changes in cost management and scheduling practices that could result from digital technology. Teams were given scores based on the above parameters, and the final scores were the aggregate of all the sub-scores for various assessment parameters.

Rewards

The winning student team was awarded \$AUD2,400, free full version CostX™ software licences, free RICS membership for one year, a certificate, and a trophy (Figure 19.7). Teams winning second and third positions were also awarded cash prizes and certificates. All other participants were given a certificate of participation.

Evaluating the BIM competition outcome

A survey of the participants was conducted to elicit information about the experiences of all 24 participants. They were asked to complete a questionnaire at two stages, first after the

Figure 19.6 Teams presenting BIM task deliverables to the jury and other participants.

Figure 19.7 The winning team being awarded $AUD2,400, a certificate, and a trophy.

training session (post-training), and second immediately after the competition (post-competition), yet before the result was announced. The post-training evaluation ascertained if the application of gamification elements (such as competition) can lead to a more in-depth learning experience for students. The anonymous responses post-training also helped in assessing the event's coordination and organization. It also provided a means towards assessing the extent to which participants' expectations were met. The questionnaire post-competition determined the degree of collaboration and competition among teams, level of satisfaction, and challenges and motivation both at the individual and team level.

Participants' feedback

The survey findings have been summarized based on submitted student participants' feedback at the post-training and post-competition stages.

Post-training survey

The post-training questionnaire was intended to assess two main aspects: first, whether the event was well-organized and thought-provoking; and second, whether there was an improvement in the participants' understanding of the meaning, concept, and principles of 4D and 5D BIM. Table 19.2 shows a summary of responses to the key questions in the questionnaire.

It can be inferred from Table 19.2 that most participants found the training session to be informative, satisfactory, and thought-provoking. A significant number of student participants (91%) admitted that the event was well-organized and structured, providing an appropriate learning environment and experience. The training sessions and activities created a suitable environment for the competition. All the students agreed that the training session on BIM applications had improved their understanding of 4D and 5D BIM concepts and principles. According to 79% of students, BIM can play a vital role in the delivery process of construction projects. Many participants (71%) agreed that their understanding of the meaning, concept, and BIM philosophy was below 50% before the training, and rose above 75% after the training.

Table 19.2 Selected responses obtained from the post-training questionnaire

Question	Percentage of respondents		
	Strongly agree, agree	*Neither*	*Disagree, strongly disagree*
Overall, the event was intellectually stimulating.	**100%**	0%	0%
Overall, the event had been well coordinated and well structured.	**91%**	9%	0%
Focusing on my own growth and development, I have felt part of a network.	**83%**	17%	0%
The event improved my understanding of the concepts and principles of 4D and 5D BIM.	**100%**	0%	0%
I improved in my understanding of concepts and principles of 5D BIM.	**100%**	0%	0%
The event challenged my way of thinking about the construction project delivery process.	**79%**	13%	8%

As illustrated in Table 19.3, most of the participants (88%) were more than 60% confident of using BIM technology for 4D planning and many participants (75%) were more than 60% confident of using BIM technology for 5D planning (as indicated in Table 19.3). So too, 79% of the participants agreed that they gained more than 60% knowledge on the concept and theory of model coordination post-training. Many participants (87.5%) believed that learning from training sessions was the best aspect of this event. Half of the participants (50%) suggested that increasing the training duration could improve this event. Almost all the participants (96%) felt that the software applications they learned during the training would be beneficial from quantity surveying, cost engineering, and cost management viewpoints. They also agreed that they would be strongly interested in working with BIM features when they graduate.

Post-competition survey

Upon completion of the competition and submitting the deliverables, all the participants were given another questionnaire to fill. In this questionnaire, all the teams were required to respond depending on their position in the competition ranking. Table 19.4 and 19.5 summarize the findings related to the key questions. The majority of participants (86%) agreed that they worked collaboratively within their teams during the competition. In the case of the winning teams, 100% of participants agreed to this. Students felt more confident in competing as a team, rather than competing as individuals; 73% of them admitted that their performance improved when working in a team during the competition. The winning teams also provided a similar response.

Table 19.6 summarizes the responses submitted by the students about their motivation behind participation in the competition. It is evident that 71% of the students participated with the intent to learn, which infers that the prize is more symbolic for the majority of the students; it is not the main driving force for participating in the competition. Their key focus

Table 19.3 Skill development post-training

Question	Percentage of respondents
More than 60% increase in ability to use BIM technology for 4D planning	88%
More than 60% increase in ability to use BIM technology for 5D planning	75%
More than 60% increase in knowledge of the concept and theory of model coordination	79%

Table 19.4 Individual response obtained from the post-competition questionnaire

Question	Percentage of respondents		
	Strongly agree, agree	*Neither*	*Disagree, strongly disagree*
The team worked in a collaborative way.	86%	5%	9%
A competitive environment enhanced your performance as a team.	73%	23%	5%
A competitive environment enhanced your performance as an individual.	59%	27%	14%

Table 19.5 Team response obtained from the post-competition questionnaire

Question	Percentage of response of winning teams
The team worked in a collaborative way.	100%
A competitive environment enhanced your performance as a team.	81%
A competitive environment enhanced your performance as an individual.	72%

Table 19.6 Individual response statistics obtained from the post-competition questionnaire about students' motivation

Question	Answers	Percentage of respondents
Your main motivation as an individual during this competition was … ?	Winning prize	21%
	Learning	71%
	Networking	25%

was to enjoy the learning process rather than the goal. This therefore means that a healthy competition resulted from this event, and the negative impacts of competition highlighted by Lam et al. (2001) were avoided.

Another interesting observation from the questionnaire was that 73% of students admitted that they had put in more effort than asked for in the competition. This agrees with Fasli and Michalakopoulos (2005), demonstrating that the sense of competition drives the students to undertake more efforts and to achieve more goals.

The outcomes of the questionnaire responses were also analyzed to identify any contrasting trends between the responses of the winning and losing teams. No noticeable difference between the teams with high and low ranking was found, suggesting that the goal of executing a healthy competition was achieved where all student participants worked collaboratively within their team and experienced an optimized learning opportunity in the process.

Discussion

Several issues emerged from the analysis of the findings:

Benefit: The event was intellectually stimulating and increased the participants' knowledge and understanding of 4D and 5D BIM concepts, workflows, and processes. Participants perceived that BIM applications are beneficial for quantity surveyors/cost engineers/cost managers, and 96% of the participants said that they would use BIM when they work in practice, if they had the opportunity. Most of the participants stated that they had put in additional effort than was asked for to be a winner even though their motivation during this competition was to learn. This indicates that there is potential in the competition for added levels of complexity or for more tasks to be included.

Duration: In general, nearly half of the students felt that the event's duration was short and should be increased for a better learning experience. It appears that too much information was being imparted quickly; therefore, it was challenging for the students to digest and retain the knowledge and use it in an optimized manner during the competition period. We recommend an extended duration where participants are provided with learning material beforehand to learn some concepts and complete some tasks within the weeks or months before the official competition. This should increase the participants' opportunity to reflect on their learning.

Challenges: During the competition, there were differences in the level of competency amongst team members. The software applications used as a media of learning were new to most participants. Everyone had varying levels of understanding and capacity to learn the software, so 40% found it challenging. The students overcame this challenge by allocating particular sub-tasks to each team member, depending on areas of interest. At the individual level, only 33% of the students felt the pressure of time, and 14% felt anxiety during the competition. An extended competition period could minimize these feelings.

Social atmosphere, cooperation, and collaboration: The participants were from Australia, Singapore, and Colombia, and the common language of communication was English. The participants did not know each other before the event; therefore, it was an excellent opportunity to network and socialize. Most of the students admitted that the social atmosphere was good within and amongst the groups. It is common for AEC project teams to be diverse with no prior working experience. The diversity within each team helped promote and increase participants' ability to work in a team. It mimics how they will work in practice to solve design, construction, and operations problems using BIM. The success of BIM lies in collaboration, which is not possible without an excellent cooperative environment. An average of 86% of the students said there was good collaboration amongst team members. Collaboration was 100% amongst members of the winning teams. This shows that all students were very cooperative within their teams, despite being from diverse backgrounds. This corroborates findings in the literature that for AEC teams, cultural background has no adverse impact on the ability for collaboration and delivering results (Hosseini et al., 2018). The competition, the expected reward, and the activities encouraged commitment and cooperation within teams.

Competitive environment: The opinion about competition was positive amongst most of the participants. Most admitted that their performance was much better working as a team than working as individuals for the competition and that they were motivated to learn about 4D and 5D BIM concepts and workflows. These results corroborate Lam et al. (2001), suggesting that a healthy and well-organized competition encourages students and enhances their learning BIM skills.

Conclusion and future directions

The students' experience survey shows that a good teaching and learning environment underlaid with healthy competition intrinsically motivates students to learn more. It promotes better student engagement and enhances their learning experience of BIM. Most students believed that the competitive environment improved their performance both as individuals and team members alike. Pedagogical results indicate that the combination of hands-on BIM exercise and student-friendly competition is a significant motivator for increased student engagement with BIM education, contributing to students' learning and professional development.

This event made BIM an exciting learning experience for the students. The experience can prepare students so that they can immediately apply what they have learned to work when they graduate. In this chapter, we have described how a structured, transparent, and learner-cantered competition can motivate, engage, and enhance BIM learning. The competition's objectives are not directly related to the deliverables (winning); they in fact lie in the learning process. Competition has positive learning outcomes and fosters the view that there is more learning rather than merely focusing only on winning as the goal. Students also acquire the skills of cooperative and life-long learning through the competition. They feel more assertive in using the software application used during the training session and can immediately work with the same application during the competition. Since participants are from different educational systems and backgrounds, it would be interesting to analyse whether these differences and different educational systems affect the competitive elements.

Based on the feedback, we identified time as a limitation in our approach. Several students recommended that the event's duration be increased as there was too much to do in a short period. This feedback has been taken on board and implemented in the 2020/2021 Internet of Things in Construction Management Case Competition at the University of Melbourne. Instead of a one-week event, the competition is being implemented over four months, including a one-week intensive held virtually.

Acknowledgement

The authors would like to thank the University of Melbourne for providing the funds and resources to organize the competition. Many thanks to all the universities that have participated in the competition since its inception in 2015: the National University of Singapore, the University of Los Andes, Colombia; Deakin University; Chongqing University China; and Bandung Institute of Technology (ITB) Indonesia. We also want to thank our industry partners and sponsors, including Graphisoft, Central Innovation, Glodon International, Exactal (now RIB), Synchro, and Solibri for their financial and in-kind support.

References

Abdirad, H. and Dossick, C. S. (2016), BIM curriculum design in architecture, engineering, and construction education: A systematic review. *Journal of Information Technology in Construction*, Vol. 21. pp. 250–271.

Aibinu, A. A. and Papadonikolaki, E. (2020), Conceptualizing and operationalizing team task interdependences: BIM implementation assessment using effort distribution analytics. *Construction Management and Economics*, Vol. 38 No. 5. pp. 420–446. 10.1080/01446193.2019.1623409

Allen, R. K. Becerik, B. Pollalis, S. N. and Schwegler, B. R. (2005), Promise and barriers to technology enabled and open project team collaboration. *Journal of Professional Issues in Engineering Education and Practice*, Vol. 131 No. 4. pp. 301–311.

Alomari, I. Al-Samarraie, H. and Yousef, R. (2019), The role of gamification techniques in promoting student learning: A review and synthesis. *Journal of Information Technology Education*, Vol. 18. pp. 395–417.

Barata, G. Gama, S. Jorge, J. and Gonçalves, D. (2015), Gamification for smarter learning: tales from the trenches. *Smart Learning Environments*, Vol. 2 No. 1. pp. 10.

Bergin, D. A. (1995), Effects of a mastery versus competitive motivation situation on learning. *The Journal of Experimental Education*, Vol. 63 No. 4. pp. 303–314.

Biggs, J. B. and Tang, C. 2011. *Teaching for Quality Learning at University*, Open University Press.

Burguillo, J. C. (2010), Using game theory and competition-based learning to stimulate student motivation and performance. *Computers & Education*, Vol. 55 No. 2. pp. 566–575.

Cagiltay, N. E. Ozcelik, E. and Ozcelik, N. S. (2015), The effect of competition on learning in games. *Computers & Education*, Vol. 87. pp. 35–41. https://doi.org/10.1016/j.compedu.2015.04.001

Cantador, I. and Conde, J. M. (2010), Effects of competition in education: A case study in an e-learning environment. In Proceedings of the IADIS international conference e-learning 2010. Freiburg, Germany.

Casasayas, O. Hosseini, M. R. Edwards, D. J. Shuchi, S. and Chowdhury, M. (2021), Integrating BIM in higher education programs: Barriers and remedial solutions in Australia. *Journal of Architectural Engineering*, Vol. 27 No. 1. pp. 05020010. doi:10.1061/(ASCE)AE.1943-5568.0000444

Chou, Y.-K. (2015), *Actionable Gamification. Beyond Points, Badges, and Leaderboards*, Octalysis Group, Fremont, CA.

Davey, B. Bozan, K. Houghton, R. and Parker, K. R. (2016), Alternatives for pragmatic responses to group work problems. *Informing Science the International Journal of an Emerging Transdiscipline*, Vol. 19. pp. 89–102.

Davis, K. Sridharan, H. Koepke, L. Singh, S. and Boiko, R. (2018), Learning and engagement in a gamified course: Investigating the effects of student characteristics. *Journal of Computer Assisted Learning*, Vol. 34 No. 5. pp. 492–503.

Deterding, S. Khaled, R. Nacke, L. E. and Dixon, D. (2011), Gamification: Toward a definition. In CHI 2011 Gamification Workshop Proceedings. Vancouver BC, Canada.

Eastman, C. M. a. Lee, G. Sacks, R. and Teicholz, P. M. (2018), *BIM Handbook: A Guide to Building Information Modeling for Owners, Managers, Designers, Engineers and Contractors*, Wiley.

Ediger, M. (2001), Cooperative learning versus competition: Which is better? Arlington, VA. (ERIC Document Reproduction No. ED 461894).

Elghaish, F. Abrishami, S. Abu Samra, S. Gaterell, M. Hosseini, M. R. and Wise, R. (2019), Cash flow system development framework within integrated project delivery (IPD) using BIM tools. *International Journal of Construction Management*. pp. 1–16. DOI: 10.1080/15623599.2019.1573477.

Fasli, M. and Michalakopoulos, M. (2005), Supporting active learning through game-like exercises. In Fifth IEEE International Conference on Advanced Learning Technologies (ICALT'05). pp. 730. Kaohsiung, Taiwan.

Fulu, I. (2007), *Enhancing Learning through Competitions. School of InfoComm Technology*, Ngee Ann Polytechnic.

Hamidavi, T. Abrishami, S. and Hosseini, M. R. (2020), Towards intelligent structural design of buildings: A BIM-based solution. *Journal of Building Engineering*, Vol. 32. pp. 101685. https://doi.org /10.1016/j.jobe.2020.101685

Hosseini, M. R. Jupp, J. Papadonikolaki, E. Mumford, T. Joske, W. and Nikmehr, B. (2021), Position paper: Digital engineering and building information modelling in Australia. *Smart and Sustainable Built Environment*.

Hosseini, M. R. Martek, I. Chileshe, N. Zavadskas, E. K. and Arashpour, M. (2018), Assessing the influence of virtuality on the effectiveness of engineering project networks: "Big Five theory" perspective. *Journal of Construction Engineering and Management*, Vol. 144 No. 7. pp. 04018059.

Janssen, A. Shaw, T. Goodyear, P. Kerfoot, B. P. and Bryce, D. (2015), A little healthy competition: Using mixed methods to pilot a team-based digital game for boosting medical student engagement with anatomy and histology content. *BMC Medical Education*, Vol. 15 No. 1. pp. 1–10. 10.1186/ s12909-015-0455-6

Johnson, R. T. Johnson, D. W. and Stanne, M. B. (1985), Effects of cooperative, competitive, and individualistic goal structures on computer-assisted instruction. *Journal of Educational Psychology*, Vol. 77 No. 6. pp. 668–677. 10.1037/0022-0663.77.6.668

Kim, S. Song, K. Lockee, B. and Burton, J. (2018), What is gamification in learning and education? In *Gamification in Learning and Education*, Springer.

Kuo, M.-S. and Chuang, T.-Y. (2016), How gamification motivates visits and engagement for online academic dissemination – An empirical study. *Computers in Human Behavior*, Vol. 55. pp. 16–27.

Lam, S.-f. Yim, P.-s. Law, J. S. and Cheung, R. W. (2001), *The Effects of Classroom Competition on Achievement Motivation*, ERIC.

Lawrence, R. (2004), Teaching data structures using competitive games. *IEEE transactions on education*, Vol. 47 No. 4. pp. 459–466.

Mehrabi Boshrabadi, A. and Hosseini, M. R. (2020), Designing collaborative problem solving assessment tasks in engineering: an evaluative judgement perspective. *Assessment & Evaluation in Higher Education.* pp. 1–15. DOI: 10.1080/02602938.2020.1836122.

Mora, A. Riera, D. Gonzalez, C. and Arnedo-Moreno, J. (2015), A literature review of gamification design frameworks. In 2015 7th International Conference on Games and Virtual Worlds for Serious Applications (VS-Games). IEEE, 1–8. Skövde, Sweden.

Morgan, J. and Porter, J. (2003), Education through competition: Mobile platform technology. In Proceedings of the 2003 American Society for Engineering Education Annual Conference and Exposition. Nashville, TN.

Pärn, E. Edwards, D. and Sing, M. (2017), The building information modelling trajectory in facilities management: A review. *Automation in Construction*, Vol. 75. pp. 45–55.

Reeves, B. and Read, J. L. 2009. *Total Engagement: How Games and Virtual Worlds Are Changing the Way People Work and Businesses Compete*, Harvard Business Press.

Ryan, R. M. and Deci, E. L. (2000), Self-determination theory and the facilitation of intrinsic motivation, social development, and well-being. *American Psychologist*, Vol. 55 No. 1. pp. 68.

Sailer, M. Hense, J. U. Mayr, S. K. and Mandl, H. (2017), How gamification motivates: An experimental study of the effects of specific game design elements on psychological need satisfaction. *Computers in Human Behavior*, Vol. 69. pp. 371–380. https://doi.org/10.1016/j.chb.2016.12.033

Sternberg, H. and Baalsrud-Hauge, J. (2015), Does competition with monetary prize improve student learning? – An exploratory study on extrinsic motivation. *LU: s femte högskolepedagogiska utvecklingskonferens*, Lund University.

Subhash, S. and Cudney, E. A. (2018), Gamified learning in higher education: A systematic review of the literature. *Computers in Human Behavior*, Vol. 87. pp. 192–206. https://doi.org/10.1016/j.chb.2018.05.028

Thousand, J. S. Villa, R. A. and Nevin, A. I. 2002. *Creativity and Collaborative Learning: The Practical Guide to Empowering Students, Teachers, and Families*, ERIC.

Verhoeff, T. (1997), The role of competitions in education. In *Future World: Educating for the 21st Century*. The University of CapeTown.

Wu, W. Mayo, G. McCuen, T. L. Issa, R. R. and Smith, D. K. (2018), Building information modeling body of knowledge. I: Background, framework, and initial development. *Journal of Construction Engineering and Management*, Vol. 144 No. 8. pp. 04018065.

Yildirim, I. (2017), The effects of gamification-based teaching practices on student achievement and students' attitudes toward lessons. *The Internet and Higher Education*, Vol. 33. pp. 86–92.

Zichermann, G. and Cunningham, C. 2011. *Gamification by Design: Implementing Game Mechanics in Web and Mobile Apps*, O'Reilly Media, Inc.

20 An Australian consolidated framework for BIM teaching and learning

Sas Mihindu and Farzad Khosrowshahi

Introduction

Higher and further education institutions are facing many challenges to their efforts to intro-duce building information modelling (BIM) in the architecture, engineering, construction, and owner-operator (AECOO) industry. Many research efforts were dedicated to the subject and addressed some specific aspects of the issue. At least two decades of research suggest the need for a comprehensive BIM teaching and learning framework to provide decision-makers with practical and neutral guidelines on BIM education at tertiary and other competency levels (Mayo et al. 2020; Boton et al. 2018).

BIM in higher education has come through a long and painstaking process. In the early days, BIM was introduced as technical electives and innovative add-ons to existing university curricula, and there were significant external barriers (e.g. availability of textbooks, industry buy-in, and professional support) and internal barriers (e.g. curriculum redesign, faculty qual-ification, and time commitment) to its integration in higher education (Wu et al. 2017). As BIM was taking off in the industry between 2007 and 2012, the educational community was incentivized to expand its footprint in higher education, with multiple strategies (e.g. vertical and horizontal integration) being adopted to adapt university curricula to prepare students for the rising market demand for BIM talent. BIM education has ever since become ubiq-uitous in 2-year or 4-year architecture, engineering, construction, and facility management programmes, as well as graduate programmes in the United States, Europe, Australasia, and around the world. Whilst there is a consistent dedication to BIM education, higher education is facing some major challenges in meeting the market demand for a BIM-competent work-force from their industry partners. Prior scholarly works reveal that current BIM education tends to focus on specific disciplines for practicality reasons, which is indisputably critical as eventually the job tasks in BIM implementation will be largely performed by specialists in each industry sector (Mayo et al. 2020; Wu et al. 2017). This chapter develops and proposes a comprehensive BIM teaching and learning framework to facilitate educators at tertiary and other competency levels with consideration of former prominent attempts and their short-comings. By integrating the state of the art in such a way, the authors hope this integrated framework will abruptly facilitate BIM education with the desired industry uptake which is a key aspect of the proposed framework.

The use case for education

A lack of a comprehensive BIM teaching and learning framework has influenced edu-cators using different approaches to introduce BIM in construction-related programmes.

From the construction management point of view, an integrated framework should consist of BIM at a fundamental level integrated within various subjects, at an application level where BIM is applied to solve real-world problems, and at an advanced level that focuses on the latest and advanced topics of BIM (Huang 2018). More importantly overall, the strategy of integrating BIM is based on the specific skills that students are expected to acquire by targeting a specific position outcome. A comprehensive framework integrating position outcomes with competencies appropriate for all types of design, construction, and operation (DCO)–related studies is necessary to provide the greater common understanding mandatory in the curriculum development. This environment facilitates the development of streamlined programmes regardless of the educational establishment, university, industry, or country. Employing a widely acceptable integrated framework, decision-makers can define the appropriate teaching approaches. An effective implementation strategy should be gradual in order to progressively raise community awareness through collaboration. Consideration of the local industry needs to be supported through the delivery of the courses, and this should be enhanced by integrating students within the local industry and supporting both parties (Boton et al. 2018).

Limited exposure to empirical BIM knowledge has established a competency gap between industry performance expectations and the actual capacities of vocational and university graduates. The flexibilities associated with digital integrated project delivery and the transformation of roles between construction industry sectors encourage educators to focus on the interdisciplinary perspective of BIM education. The nature of the digital integrated project delivery methodology provides the ability for educators to take a holistic approach in the development of future BIM curricula (Wu et al. 2017).

This integrated framework (other related frameworks: NATSPEC Construction Information Systems 2020; Boton et al. 2018; Huang 2018; Bush & Robinson 2018; Lee et al. 2019; Leite 2016; Macdonald 2012) is especially valuable for education programmes that are dedicated to aligning the student learning outcomes of their BIM curriculum with the career-specific BIM competencies desired by industry partners. Specifically, using a backward design model, educators may utilize these resources to establish and prioritize the end results of the student learning outcomes of their BIM curriculum, to determine metrics for performance assessment, and, lastly, to plan for pedagogy design. The backward BIM curriculum design model encourages outcome-driven, competency-based learning and facilitates a partnership among academia, industry, and subject-matter experts in defining future workforce development strategies with local and regional priorities (Wu et al. 2017).

BIM education within university and vocational education and training

As detailed in previous works (Succar et al. 2012), there are many stakeholders involved in the provision of BIM education, as within academia, universities and TAFE institutions have started delivering a range of BIM course offerings. Whilst many universities are offering BIM integrated curricula worldwide, it is apparent that most universities operate without a real consensus amongst them. BIM has a potentially significant impact on the vocational education and training (VET) sector – the sector responsible for training and retraining the construction industry's tradespeople (carpentry, plumbing, painting, electrical, etc.) and para-professionals (architectural technology, building design, surveying, etc.). VET covers a range of qualifications from Certificate II (AQF 2) up to Advanced Diploma (AQF 6). All of these VET qualifications, the industry roles they represent, and the education they require are significantly affected by BIM technologies and workflows (Succar et al. 2012; Mihindu & Arayici 2008).

First, para-professional qualifications (AQF 4–6) have traditionally generated technician-level graduates within a wide range of discipline areas. With the advent of BIM technologies, many have taken the opportunity to develop into BIM specialists within their respective fields and currently play diverse and increasingly important roles. A great number of modellers, model managers, BIM project coordinators, and BIM managers have their roots in para-professional education. Also, in addition to the knowledge and skills para-professionals need to operate within their chosen specialities, VET trainees must now learn how to use data-rich models and other technologies to collaborate with their peers, tradespeople, and AEC professionals. New BIM-focused courses now need to be developed to ensure para-professionals are "industry-ready" at graduation – courses which necessarily include hands-on, collaborative, and multidisciplinary project work (Succar et al. 2012).

Second, at trade qualification levels (AQF 2–3), the increasing availability of accurate, information-rich models is starting to impact the construction site. While many in the steel detailing and mechanical ducting professions have been using 3D CAD for many years (especially for offsite prefabrication), BIM has highlighted the need for tighter coordination between many trades and specialities. This, in turn, has dramatically increased the need for highly trained technicians with additional experience in model interrogation, clash detection, construction sequencing, and quantity take-off. These skills should now be included in trade certificate courses so trainees benefit from available BIM technologies and can collaborate efficiently with professionals, para-professionals, product suppliers, and others within the construction supply chain (Succar et al. 2012).

BIM education within non-academic bodies

Professional associations can play an important role in promoting BIM education within both academia and industry, where some of them provide course accreditation, certification, and/or continuing professional development (CPD) programmes (e.g. BSI group, ABAB). BIM training within organizations is another important aspect; driven by immediate business benefits, many DCO companies offer their staff the necessary training to generate and share data-rich models with their project partners. The training offered – whether on-the-job or through registered training organizations – is mainly technically focused on cultivating the skills necessary to use BIM's ever-expanding repertoire of tools and workflows (Succar et al. 2012).

Frameworks for BIM education

Over the last two decades, many dozens of frameworks been developed with different viewpoints. Promotion and implementation of them in various academic institutions in multiple countries were facilitated with valuable recommendations. Various authors have analyzed these frameworks at length for their strengths, weaknesses, improvements, and vital results of the curriculum implementation circumstances. Whilst curriculum implementation based on some frameworks appears somewhat straightforward, others certainly require a comprehensive strategy and common understanding between faculties, departments, and the industry. A brief description of two specific frameworks provided shows this diversity in the following sections. Yet again, the important aspect of encapsulation of the full life cycle of the infrastructure DCO has been overlooked.

Framework for construction engineering from Korea

Researchers from Korea developed this framework with the goal of making students industry-ready and employable. This assisted in setting learning goals for the systematic curriculum development process, as such goals provide milestones for the students, as well as invariably represent the knowledge, skills, and behavioural standards that have been set by the curriculum. Figure 20.1 shows this simplified but elaborative BIM-based framework for construction curriculum. The learning topics and the schedule of implementation are detailed in their work (Lee et al. 2019).

BIM knowledge and skills framework from Australia

As a result of the joint partnership of the Australasian Procurement and Construction Council (APCC) and Australian Construction Industry Forum (ACIF), the BIM knowledge and skills framework was published to provide guidance about the required skills and education relevant to BIM for a broad range of industry stakeholders. This wider approach is to enable a consistent upskilling of the construction market sector to improve productivity through collaborative and integrated project delivery. The focus revolved around a three-part model, linking the framework with accredited courseware and a certified qualification procedure. Knowledge areas for the framework (Figure 20.2) reflect the pattern of the Project Managers Body of Knowledge (PMBOK) as given below (ACIF & APCC 2017).

The complete BIM knowledge and skills framework is provided as a spreadsheet for industry use, detailing five levels of proficiencies from fundamental awareness to expert level and

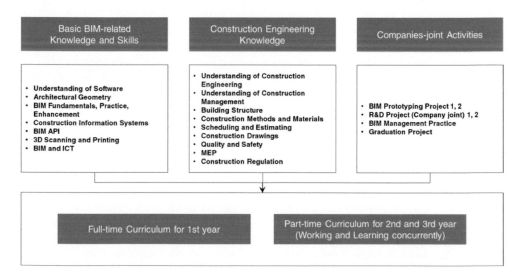

Figure 20.1 Framework of the BIM-based construction engineering education curriculum (Lee et al. 2019).

1.000 INTRODUCTION

2.000 STARTUP

3.000 INITIATION

4.000 PLANNING

5.000 EXECUTION / OPERATION

6.000 MONITORING AND CONTROLLING

7.000 CLOSEOUT / HANDOVER / COMMISSIONING

Figure 20.2 Knowledge areas of the BIM knowledge and skills framework (ACIF & APCC 2017).

applicability for various stakeholders. While this framework is very detailed, the implementation within academic curricula has not been one of its main focuses (ACIF & APCC 2017).

An integrated framework for BIM teaching and learning

As presented at the beginning, there is no comprehensive BIM teaching and learning framework to provide academic decision-makers with practical and neutral guidelines covering the full life cycle of the infrastructure DCO. The framework details here identify the main challenges to address for AECOO departments and academia in general (Figure 20.3). In the future, case studies will be conducted within academia to evaluate for accuracy and to provide further details of implementation support, as well as to illustrate apparent challenges. The strategy of integrating BIM in AECOO should be based on the specific skills the students are expected to acquire. It is then possible to define the appropriate teaching approaches and assessment procedures, which leads to an industrial recognition system (Boton et al. 2018). An effective implementation strategy is a gradual process in order to progressively raise community awareness, learn from mistakes, and identify best practices. Particular emphasis should be placed on the needs of local industry during the curriculum delivery and on integrating students within real-life construction projects where they can apply the acquired knowledge and skills in a collaborative, multi-stakeholder environment. The high-level view of the proposed six-stage Australian Consolidated Framework (ACF) to facilitate BIM teaching and learning depictured here incorporates a clear strategy to encapsulate the full life cycle of DCO. The section development of the integrated framework, discusses the consolidation aspect of ACF development.

Stage 1 and stage 2 are devoted to the recognition of current and future positions within applications of the BIM landscape, and on the analysis of learning, designing, implementing, and improving on the competencies that are necessary to successfully perform on these positions. A learning schedule for attaining these identified competencies is to be delivered progressively, building over a period of study combined with adequate real-life integrated project delivery (IPD) experience. In other words, stages 1 and 2 are more relevant to the academic community, technology developers, and industry partners, attaining the vendor-neutral best possible collaboration amongst them. Stage 3 is devoted to learning, comprehension, and practising all necessary competencies for one or more targeted positions. During stages 4,

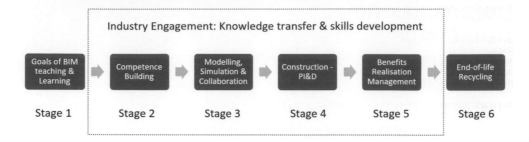

Figure 20.3 ACF: Six stages of BIM teaching and learning, the integrated framework.

5, and 6, students gain real-life experience on BIM in use, while industry-ready graduands acquire further real-life experience with the ambition of attaining their graduate positions. The industry engagement element is the most important of all, covering stages 2 to 5. The experienced graduates collaborate with academia on a particular focus of improving stage 2 and other processors within the competency development. This inevitably assists in the process of gaining industry recognition for curricula and the capability enhancement of all graduates.

BIM teaching and learning, the connotation of ACF

Summing up from all previous chapters, BIM in its most simple terms is the utilization of a database infrastructure to encapsulate built facilities with the specific viewpoints of stakeholders. The BIM process facilitates creating and mapping digital facility life-cycle information. BIM incorporates a methodology based on the notion of collaboration between stakeholders, using ICT to exchange valuable information throughout the life cycle. Utilizing the IPD methodology, it integrates people, systems, and business practices in a process that collaboratively harnesses the talent and insights of all participants to reduce waste and optimize efficiencies through all phases of design, sustainability analysis, fabrication, construction, maintenance, and recycling (Lee et al. 2019).

The application of BIM within the industry has developed several BIM-related positions associated with strategic, operational, or technological aspects within construction development, operation, and maintenance projects. Industry analysis has recognized over 40 BIM-related positions. Amongst the 40 different positions, nine prominent positions were identified and discussed in this chapter. The nine BIM position types are the BIM project manager; director; BIM manager; BIM coordinator; BIM designer; senior architect; BIM mechanical, electrical, and plumbing (MEP) coordinator; BIM technician; and benefits and sustainability manager. Small organizational groups or consortiums can reduce these nine BIM positions to four (see Figure 20.4). Specific positions require specific competencies, and these competencies are grouped into 15 competency sets, which are directly and indirectly associated with the noted nine position types. The integrated framework (i.e. ACF) discussed here encapsulates 15 competency sets advising on curriculum development targeting tertiary and other levels of training programmes (Miyoung et al. 2017). In an ideal situation, when and if a qualification prepares a graduate to be e.g. a BIM designer, during their period of study and practice they must attain all the associated competencies to the required standard targeting this position.

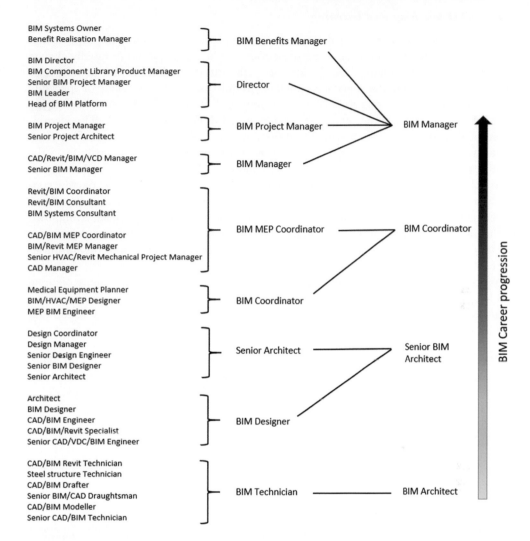

Figure 20.4 Positions within BIM landscape.

BIM competency is divided into several BIM competency sets which are, in turn, subdivided into BIM competency topics. These competency sets, their topics, and their granular subdivisions (competency items) represent all the measurable abilities, outcomes, and activities of individuals who deliver model-based products and services. Importantly, this representation of abilities accurately identifies an individual's competency profile using a broad spectrum of topics. It is driven by the notion that an individual cannot be recognized as either competent or incompetent as a whole but may be an expert in one competency item due to their level of experience and theoretical knowledge, whilst at the same time being a novice in a competency of which they have no experience or background knowledge (Succar et al. 2013).

The next stage details the process of recognizing BIM roles that satisfies a BIM competency hierarchy within the complete BIM landscape.

Goals: BIM teaching and learning

Identifying a set of BIM-specific roles is an important process for recruitment purposes that fulfils current and future position roles; these role definitions are bound to rapidly change to reflect the relentless technological and procedural transformations from which the roles are derived. Identifying the specific competency requirements of a discipline or speciality requires clarity about responsibilities, but this approach does not lend itself to identifying the BIM competencies common across specialities (BIM ThinkSpace 2012). The identification of BIM competencies specific to an organization – the approach taken by specialist consultants – is useful for that organization; however, it does not contribute to the identification of competencies across the wider industry. A more pertinent and persistent approach would be to avoid rigid delimitation of BIM roles within arguably narrow contexts and to focus more intently on identifying industry-wide BIM competencies that shape current roles and affect emergent ones. The significance of this wider approach is amplified by the need to facilitate multidisciplinary collaboration, encourage integrated practices and workflows, and reduce inefficient interdependency between teams and organizations (Succar et al. 2013).

As pointed out in the literature (Succar et al. 2013), the process of industry-wide, as opposed to role-, organization-, or discipline-specific, identification of BIM competencies requires a multi-pronged approach. There are several complementary ways to identify, refine, and validate individual BIM competencies, and a prominent few are listed here:

- Dissecting BIM-specific roles as defined within BIM guides, BIM management plans, and similar documents;
- Adopting and adapting formal skill inventories, competency pools, and accreditation criteria similar to those described by HR-XML-Consortium (https://hropenstandards .org/);
- Harvesting competency requirements from industry associations, organizations, and subject-matter experts through interviews, focus groups, and dedicated surveys;
- Reviewing academic literature and industry publications focused on BIM workflows, deliverables, and their requirements;
- Analyzing "job advertisement descriptions" crafted by recruitment sites.

At least once every two years, researchers are required to use many available resources, established methods, and accessible means of identifying BIM competencies across the DCO industry. Through these multiple sources, newly identified BIM competencies can be collected at an industrial scale, conceptually filtered to isolate those which satisfy the integrated definition, and classified using a specially developed tiered taxonomy. The BIM competency hierarchy is a taxonomy organizing BIM competencies into meaningful, exhaustive, and mutually exclusive clusters. This clustering is goal-driven and aims to simplify a large system by decomposing it into smaller sub-systems (Succar et al. 2013).

From an individual's point of view, one of the major goals of BIM education is to acquire and retain a related position within applications of the BIM process. Through the last three decades of applying BIM in the DCO industry has created various positions that are related to this process, the maturity of the position descriptions has become evident. As education providers, the institutions are responsible for providing a pathway for the individual to achieve this goal. Clear recognition of individual position title has led to the identification of the competencies one must acquire to perform that role successfully. Nevertheless, some competencies can only be gained to the required standard through applying the knowledge and skills ascertained within real-life situations; simulated environments which are very close to real-life situations can provide an initial pathway for this process. If individuals are to acquire these competencies at an appropriate time within the study period, the opportunity for suitable industry engagement must be established. This aspect is clearly identified in ACF.

Competence building

The ACF consists of 105 BIM competency topics, including 15 competency sets depicted under foundation, primary, secondary, execution, and industry engagement themes (see Figures 20.5 and 20.6). Many previous works recognize a few dozen competency topics to satisfy the development of qualifications for preparing industry-ready graduates. However, the consideration of the life-cycle aspect of BIM has not yet been properly integrated into a BIM education framework. A BIM Excellence project has provided a detailed competency inventory for primary and secondary competency sets (8) describing 57 competency topics (BIMe Initiative 2019).

Therefore, moving on, the sustainability competency set is defined as a foundation theme to capture product sustainability over the life cycle: development, maintenance, and recycling. The infrastructure operation set is defined to cover from infrastructure handover to demolition; utilization of BIM during this lengthy period is extremely valuable and inevitable. For example, during safety evacuation scenarios, not only infrastructure managers have access to the infrastructure BIM information but also external emergency services teams. The primary and secondary themes are detailed in the literature well reasonably. Execution theme is included to capture BIM use during construction and industry engagement project theme is included to capture knowledge transfer and skills development through live industry projects covering stages 3, 4, and 5 in ACF. The execution theme has three competency sets and the industry engagement theme has two competency sets. Further, three extra topics are included in each of these stages as core components, ES01 and ES02 – applications of

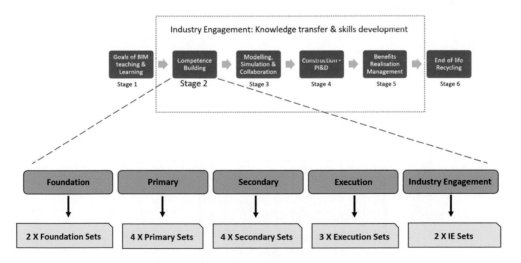

Figure 20.5 Competence building through 15-competency sets.

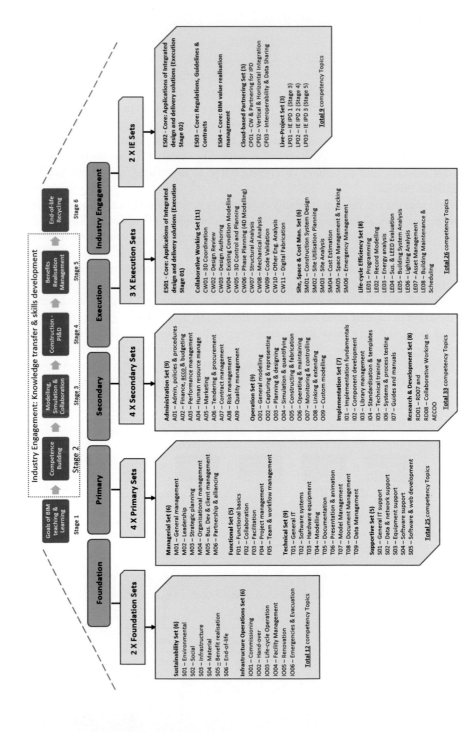

Figure 20.6 BIM teaching and learning: Competency chart.

integrated design and delivery solutions (Execution Stage 01/02) to facilitate applying competencies in each stage, i.e. during execution and knowledge transfer. The second and third extra core topics within the execution stage cover policy issues associated with construction, ES03 – Regulations, Guidelines & Contracts, and, finally, ES04 BIM value engineering/realization management. Further, within industry engagement, three knowledge transfer and integration projects (LP01, LP02 & LP03) are defined to cover stages 3, 4, and 5 in ACF.

To the previous work, primary and secondary competency sets and competency topics, one competency topic has been added to the research and development set. The BIM Excellence competency table describes 57 standard competency topics (BIMe Initiative 2019). At a higher level, they are separated into eight competency sets, comprising four primary competency sets and four secondary competency sets. The BIM Project Execution Planning Guide – Version 2.2, has informed the design of execution competency sets covering the practicality of BIM use in the construction phase (Messner et al. 2019). By referring to these documents (available as online websites), readers can comprehend further details of competency topics.

The industry engagement projects at different levels of competency development are considered as the most valuable consolidated learning and development experience, assessing the participants' comprehension and ability to perform in projects. As needed, the number of live projects can be added to suit the delivery of the training programme covering various stages of the ACF. The overall design of the competency sets and competency topics is given in Figure 20.6.

Modelling, simulation, and collaboration

Modelling, Simulation and Collaboration, stage 3 of this integrated framework (ACF), encapsulates one of the earlier frameworks developed through the CodeBIM project. The CodeBIM project focused on developing a curriculum framework to support the implementation of collaborative design in AEC education utilizing BIM tools. The framework developed is made up of two components: a benchmarking tool and an implementation guide. The four stages of this framework – illustration, manipulation, application, and collaboration (IMAC) – have been mapped onto levels of the taxonomy of learning. These four stages relate to different levels of achievement. The IMAC framework (Figure 20.7) identifies the way in which the students are gaining deeper levels of learning at each stage as they progress through their education. It supports the development of both technical (IT and discipline-specific) and

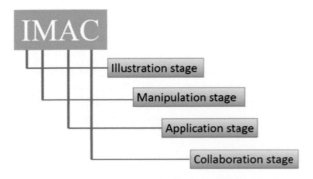

Figure 20.7 IMAC framework (Macdonald 2012).

interpersonal (collaborative and teamwork) skills. The students from the different AEC disciplines study courses of varying lengths, and some skills are introduced earlier in some courses than others (Mills et al. 2013; Macdonald 2012).

The IMAC framework does not dictate in which academic year each stage should be introduced. For example, students of architecture tend to be introduced to modelling tools from first year, whereas students of structural engineering might only be introduced to them in the third year. It may also be possible to progress between stages within one academic year (Macdonald 2012). The details of the IMAC framework are described below.

Illustration (knowledge/comprehension and receiving/responding)

This is an introductory stage. Building information models are used to illustrate key concepts to students, and students will typically be taught in their separate disciplines at this stage. Models will have sufficient detail to allow lecturers/tutors to highlight different components/connections to show how buildings are constructed, insulated, and waterproofed, for example (Macdonald 2012).

Manipulation (comprehension/application and responding/valuing)

At this stage, students start to interact with and manipulate existing models themselves. They will be required to make simple changes to and/or create basic elements within the models in relation to their disciplines. They are also continuing to develop their teamwork and basic IT literacy skills, in addition to developing discipline-specific knowledge (Macdonald 2012).

Application (application/analysis and valuing/organizing)

At this stage, students have acquired basic theoretical knowledge in their disciplines and are starting to apply this knowledge to solve discipline-related problems. For architecture students, they will start to build building information models from scratch and learn how to set the models up for effective interdisciplinary collaboration. Engineers will start to use tools to analyze models using exports from building information models. Construction managers will develop 4D and 5D schedules and plan logistics and materials ordering using models from other disciplines. All disciplines will be taught principles of value engineering and sustainable design and how BIM tools can be used to assist these. They will also be introduced to the roles that the other disciplines play in a construction team, and how models are set up to facilitate information exchange (Macdonald 2012).

Collaboration (synthesis/evaluation and characterizing)

At this stage, the students from the different disciplines come together to work on joint projects. Ideally, this will involve groups containing a student from each AEC discipline. Learning through teaching others can be encouraged by pairing senior engineering and construction students with junior architecture students, for example. Ideally, real-world problems will be given to the students to solve. To ease students into the process, they can be given partly finished models to start with, and will then be asked to make some changes to these models due to "new project information" arising. The students will also learn about the types of contract that facilitate BIM and collaborative working and will continue to learn about group dynamics and improving teamwork (Macdonald 2012).

Construction: Network-based project integration and delivery

BIM is widely accepted as a revolutionary technology and potentially revolutionary socio-technical approach for collaboration in construction projects. BIM adoption and implementations and the associated benefits for construction projects have been widely discussed and researched. Active collaboration between the project participants and smooth data exchange between the tools they utilize is considered a key to successful BIM implementation. The adaptive use of BIM not only requires but also supports multidisciplinarity. BIM technologies, their adoption, and their implementation require collaborative teamwork and processes. Development of new roles and the need for new competencies for disciplines suggest new approaches and requirements for multidisciplinary collaboration amongst the project participants. As an example, the job of BIM coordinator/BIM manager has become a common role only in the past approximately 5–10 years to support coordination and management of multidisciplinary team and activities spread among different disciplines. Therefore, BIM requires multidisciplinary skills and knowledge at individual, team, project, and stakeholder levels (Suwal et al. 2016). The development of some of these competencies must be performed through industry engagement, i.e. through engaging and working with live projects. One of the previous attempts to analyze BIM use during the construction stage and the ranking of achievement of benefits based on the frequency of BIM use is given in Figure 20.8. The benefits realization is an important stage (5) of this framework, and hence, further research in quantifying benefits needs enhancing.

Benefits: Realization management

This stage involves the organization and management of the effort to achieve and sustain envisaged benefits, i.e., DCO benefits realization management (BRM). BRM and the enhancements provided through the application of BIM over the life cycle of the infrastructure are

The rank of BIM Use	Frequency
1. 3D coordination	60%
2. Design reviews	54%
3. Design authoring	42%
4. Construction system design	37%
5. Existing conditions modelling	35%
6. 3D control and planning	34%
7. Programming	31%
8. Phase planning (4D modelling)	30%
9. Record modelling	28%
10. Site utilisation planning	28%
11. Site analysis	28%
12. Structural analysis	27%
13. Energy analysis	25%
14. Cost estimation	25%
15. Sustainability LEED evaluation	23%
16. Building system analysis	22%
17. Space management and tracking	21%
18. Mechanical analysis	21%
19. Code validation	19%
20. Lighting analysis	17%
21. Other engineering analysis	15%
22. Digital fabrication	14%
23. Asset management	10%
24. Building maintenance scheduling	5%
25. Disaster planning	4%

Figure 20.8 BIM use in construction stages (adapted from Kreider et al. 2010).

scrutinized and evaluated. Fundamentally, BRM is a process that is enacted to ensure that the expected benefits of capital investments, such as BIM, are realized. At an early stage of the design (e.g. Sustainability Set, SO5), the development of the BRM plan is to be established, understood by all stakeholders, and acted upon over the life cycle of the project to design life-cycle savvy usable products (Mihindu & Arayici 2008). Over the course, the infrastructure development, commissioning, and operation, the BRM plan informs how and when the benefits of the project are to be delivered or realized.

It includes the metrics to measure the benefits. This plan creates, maximizes, and sustains the project benefits and includes the following elements:

- **Target benefits**: The project benefits expected to be delivered;
- **Strategic alignment**: How well the project benefits align with the organizational strategy;
- **Timeframe for benefits realization**: When the benefits will be realized;
- **Benefit owner**: The person accountable for each project benefit;
- **Metrics**: Measurements for benefits realization;
- **Assumptions**: Assumptions taken;
- **Risks**: Risk assessments for project benefits and the probability of having them;
- **Tracking and reporting**: Processes to record and report the status of benefits.

The above elements assist in the development of a benefits register and benefits dependency map which are inherently facilitate the creation of the benefits realization management plan discussed. The competencies gained during this stage of the framework accomplish more informed knowledge and skills in value-driven assessment and the continual improvement of the implementation of BIM across assets and life-cycle phases. Specific core topic (ES04) is included in the industry engagement; to achieve the best possible outcome of this topic, the notion of value engineering is to be included in any of the topics taught, when and where this is relevant to the competency topic. This assist graduates to recognize the true value of BIM and promotes the value engineering concepts to all stakeholders. BRM in BIM, as a developing research topic, can be a topic in the research and development set on an as-needed basis.

The value of BIM is realized through its benefits for different stakeholders. Benefits arise because BIM, as an advanced information technology collaboration system, enables people to carry out tasks more efficiently and effectively. Stakeholders perform this by allowing and shaping new ways of working through the redesign of intra- and inter-organizational processes or by facilitating new work practices (Mihindu & Arayici 2008). The benefits which have already been identified are to be included in the benefits dictionary with a description of the benefit profile, such that they can be realized, recorded, and assessed for any future issues in realizing them. The benefit profile includes

- a general definition as well as phase-specific descriptions when applicable;
- a list of main beneficiaries;
- a list of tools and processes that enable and maximize the profiled benefit;
- a list of benefits that can flow on from each profiled benefit;
- a list of metrics that can be used to monitor progress towards achieving the benefit;
- examples of projects where they have been achieved.

This dictionary aims to serve as a basis for stakeholders to develop their own value realization management strategy based on those benefits which they identify as being most relevant to their organizational goals. Stakeholders have the added value of the ability to share this knowledge between different infrastructure development projects that utilize BIM (Sanchez et al. 2016; PMI 2019). The key resource in attaining BRM through BIM is not the technology itself, but the processes involved in creating information and knowledge that will be distributed throughout the asset owner's business. As a part of the BIM process, the benefits realization manager must work with the asset owner and BIM manager in developing an asset and facilities management (A&FM) execution plan that specifies their "data needs" for the operation and maintenance of the facility at the outset of the infrastructure development project. The expertise of the BIM manager can provide specific domain knowledge with regard to the execution and integration of the A&FM building information model into existing and future systems for infrastructure operation (Love et al. 2014).

End-of-life: Recycling

The world has changed rapidly towards a sustainable/green world, and the construction industry has been widely known as a major source of pollution, environmental degradation, and natural resource depletion. It can be assumed that the construction industry suffers not only in terms of productivity or economical disaster, but also has to mitigate and maintain the triple bottom line of sustainability. In this regard, the use of BIM in infrastructure development is inevitable, and the momentum of the utilization of efficient tools such as BIM has exponentially increased around the globe. Early collaboration and open information sharing amongst all team members and stakeholders are among the key factors for achieving sustainable infrastructure development in construction projects. The basic focus is on developing aspects such as sustainable sites, environmental pollution, water efficiency and reusability, energy and atmosphere conservation, infrastructure energy modelling, on-site renewable energy and storage, materials and resources efficiency and recyclability, indoor environmental quality, innovation and design quality, amongst many other new themes. BIM plays a vital role in organizing the processes involved within the infrastructure life cycle up to the end-of-life or recycling of the infrastructure through enhanced construction and operational efficiencies and active collaboration between the stakeholders involved through the sharing of intelligent data between various disciplines (Rahman & Suwal 2013).

The tight integration of BIM and design for demolition (DfD) is an important effort in the right direction, and all stakeholders must act upon this theme during planning for effective construction, operation, and end-of-life management of infrastructures. Hence, the associated competencies should be gained during the foundation (foundation sets) where this knowledge and acquired skills can be implemented right from the design stage. Infrastructure design and build is somewhat a short-term aspect compared to its operational stage. Real owners of the infrastructure up to the demolition of it may not be known. However, the processes must be in place for new owners to acquire all the as-built models, renovations carried out and other associated data and information, such that they possess full knowledge and understanding of how to utilize this information in a sustainable manner in the operation or demolition of the infrastructure. BIM will empower DfD tools for improved document management and improved lifecycle management. A deconstruction plan could, therefore, be developed and embedded within a BIM federated model to support end-of-life deconstruction of a building (Akinade et al. 2017).

Industry engagement

In preparation for industry engagement (See Figure 20.6), project-based learning coupled with corporative learning, working with course projects collaboratively in small groups, is to be facilitated. Industry engagement and project-based learning provide students with real-world problems and active learning experiences by encouraging self-directed learning and critical thinking throughout their industry engagement practice. A combination of lectures and live demonstrations with industry practitioners, team-based collaborative learning, and individual learning not only provides students with well-structured knowledge but also enables them to practice working and learning in a collaborative environment supplemented by self-reflection (Leite 2016). During industry engagement, competencies are acquired through nine topics in total: two industry engagement sets and three core topics. Two sets contain topics associated with cloud-based partnering and live projects. As stated previously, this stage allows graduands to get fully immersed in real-industry practices and to maintain close collaboration with project stakeholders. Graduands not only acquire the necessary skills and experience but are also able to pass new knowledge to other project stakeholders as they work in collaboration with them.

The development of the integrated framework, ACF

ACF, the integrated framework, encapsulates six stages of competency development process facilitating the complete life cycle of infrastructure development. Over two decades of various research in competency development associated with the integration of BIM curriculum into the mainstream has somewhat influenced this comprehensive development of the integrated framework. ACF itself consists of other established works by various teams around the globe. Figure 20.9a depicts the association with some of the previous works that influenced this development, and Figure 20.9b shows the resulting framework. Therefore, with the implementation of the BIM teaching and learning, the integrated framework pictured (Figure 20.9b) within curricula will not be an onerous task but will endeavour to facilitate the ongoing effort of curriculum designers and implementers, and it provides an expansive view of the competencies involved, detailing their breakdown considering the infrastructure life cycle to all stakeholders. This has been lacking within previous work.

Framework implementation

The integrated framework (ACF) developed consists of 105 competencies for educating BIM use throughout the life cycle of infrastructure development. As detailed in stage 1, the competencies are associated with specific positions. Most of the current qualifications taught at the universities and VET are closely linked with position outcomes. Adherence to the following 10 points influence greater implementation success in achieving the fundamental focus of BIM teaching and learning through ACF. Further work and case studies are to be published in the future. BIM curriculum designers and consultants, during the process of designing a qualification,

1. Identify all associated position outcomes within the qualification;
2. Identify all competencies that are relevant in performing each position;
3. Identify competency clusters that are appropriate in delivering progressively;
4. Organize competency clusters within semesters or year of delivery;
5. Assure cluster of competencies consist of a balance in theory, practice, and real-use;
6. Focus on conducting a bigger part of the competency assessment during real-use;
7. Plan industry engagement during the second semester of the year or during the final year;
8. Practice goal-driven teamwork, partnering between teams with different responsibilities;
9. Emphasize BIM as a process for achieving the greatest possible sustainability outcomes for all;
10. Link BRM to infrastructure operations during foundation studies and apply throughout IPD.

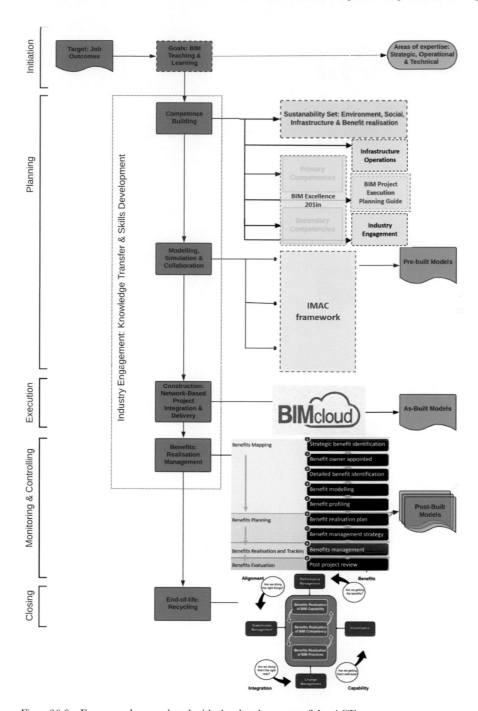

Figure 20.9a Frameworks associated with the development of the ACF.

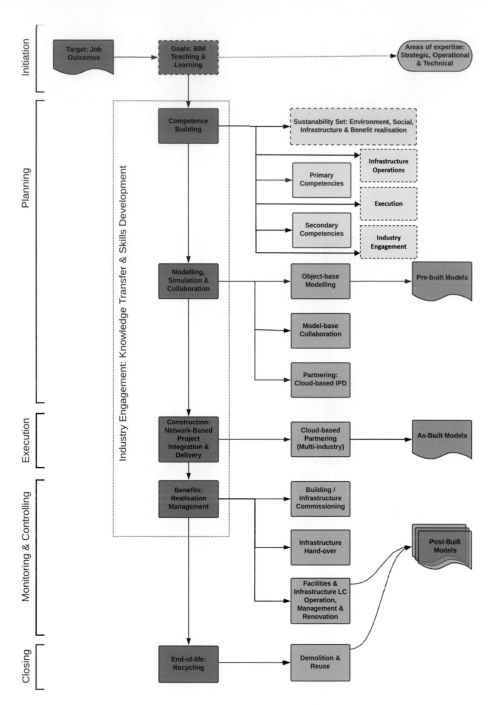

Figure 20.9b BIM teaching and learning: An integrated framework – ACF.

Curriculum enhancement with the ISO 19650 Series

Embedding the use of international standards (e.g. ISO 19650 series) within ACF facilitates easy adoption of this framework universally, enabling better information management and structured construction process. These standards advocate, use of the right amount of information throughout DCO, using the appropriate tools by everyone involved in the project delivery as well as asset management. More importantly, ISO 19650 can apply to any kind of infrastructure and the systems and components within them during the whole life cycle. Inadvertently, any curriculum designed through ACF indeed encompasses a similar scope, i.e. life-cycle focus. The integration of ISO standards in appropriate BIM competency topics and the inclusion of a new topic within a life-cycle efficiency set would be a recommendation.

Conclusion

In summary, although much progress has been made over the past few years, further effort is needed to extend BIM education across professional boundaries and to encourage stakeholders to embrace a more collaborative approach to BIM learning and teaching and integrated project delivery over the infrastructure life cycle. A lack of a consistent BIM learning and teaching framework acceptable to all academics, curriculum developers, and industry counterparts, and which is able to capture the infrastructure life-cycle aspect into BIM teaching and learning, is evident from the literature to date. The described BIM teaching and learning integrated framework, ACF, fulfils the above gap but consolidates major efforts taken towards achieving this goal in recent times.

References

Australian Construction Industry Forum and Australasian Procurement and Construction Council 2017, BIM knowledge and skills framework, Australian Construction Industry Forum (ACIF) and Australasian Procurement and Construction Council (APCC), Canberra, ACT, viewed 15 Apr 2021, < http://buildingsmart.org.au/wp-content/uploads/BIM-Knowledge-and-Skills-Framework-Introduction-Document-MAR2017.pdf>

Akinade, OO, Oyedele, LO, Omoteso, K, Ajayi, SO, Bilal, M, Owolabi, HA, Alaka, HA, Ayrisf, L & Looney, JH 2017, BIM-based deconstruction tool: Towards essential functionalities, *International Journal of Sustainable Built Environment*, vol. 6, no. 1, pp. 260–271, DOI: 10.1016/j.ijsbe. 2017.01.002.

BIM ThinkSpace 2012, *Episode 17: Individual BIM Competency*, BIM ThinkSpace, viewed 10 Jan 2020, <https://www.bimthinkspace.com/2012/08/episode-17-individual-bim-competency.html>.

BIMe Initiative 2019, *201in Competency Table (v2.1)*, BIM Excellence, viewed 22 Jan 2010, <https://bimexcellence.org/resources/200series/201in/>.

Boton, C, Forgues, D & Halin, G 2018, A framework for building information modelling implementation in engineering education, *Canadian Journal of Civil Engineering*, vol. 45, no. 10, pp. 866–877, DOI:10.1139/cjce-2018-0047.

Bush, R & Robinson, M 2018, *Developing a BIM Competency Framework: Research & Key Principles*, Scottish Futures Trust, Edinburgh.

Huang, Y 2018, Developing a three-level framework for building information modeling education in construction management, *Universal Journal of Educational Research*, vol. 6, no. 9, pp. 1991–2000, DOI: 10.13189/ujer.2018.060918.

Kreider, R, Messner, J & Dubler, C 2010, Determining the frequency and impact of applying BIM for different purposes on building projects, In Proceedings of the 6th International Conference on Innovation in Architecture, Engineering and Constructions (AEC), University Park, PA, pp. 1–10.

Lee, S., Lee, J., & Ahn, Y. 2019, Sustainable BIM-based construction engineering education curriculum for practice-oriented training, *MDPI Journal Sustainability, Innovative Practices in Engineering Education: Concept and Implementation*, vol 11, no. 21, 6120, DOI:10.3390/su11216120.

Leite, F 2016, Project-based learning in a building information modeling for construction management course, *Journal of Information Technology in Construction*, Special Issue: 9th AiC BIM Academic Symposium & Job Task Analysis Review Conference, vol. 21, pp. 164–176.

Love, P, Matthews, J, Simpson, I, Hill, A, & Olatunji, O 2014, A benefits realization management building information modeling framework for asset owners, *Automation in Construction*, vol. 37, pp. 1–10, Elsevier, DOI: 10.1016/j.autcon.2013.09.007.

Macdonald, JA 2012, A framework for collaborative BIM education across the AEC disciplines in professional education, In Proceedings of the 37th Annual Conference of the Australasian Universities Building Educators Association (AUBEA), University of New South Wales, Australia, pp. 223–230.

Mayo, G, McCuen, T, Hannon, J & Smith, D 2020, Development of the BIM body of knowledge (BOK) task definitions and KSAs for academic practice, Proceedings of the 56th Annual International Conference, Associated Schools of Construction, *EPiC Series in Built Environment*, EasyChair, vol. 1, pp.124–132, ISSN: 2632-881X.

Messner, J, Anumba, C, Dubler, C, Goodman, S, Kasprzak, C, Kreider, R, Leicht, R, Saluja, C, & Zikic, N 2019, *BIM Project Execution Planning Guide V2.2*, Creative Commons, Penn State: Computer Integrated construction.

Mihindu, S & Arayici, Y 2008, Digital construction through BIM systems will drive the re-engineering of construction business practices, in 2008 International Conference Visualisation, London, pp. 29–34, DOI: 10.1109/VIS.2008.22.

Mills, J, Tran, A, Parks, A & Macdonald, J 2013, *Collaborative building design education using Building Information Modelling (CodeBIM)*, University of South Australia, Adelaide.

Miyoung, U, Lee, G & Jeon, B 2017, An analysis of BIM jobs and competencies based on the use of terms in the industry, *Automation in Construction*, vol. 81, pp. 67–98, DOI: 10.1016 /j.autcon.2017. 06.002.

NATSPEC Construction Information Systems 2020, AS ISO 19650, in *NATSPEC Building Information Modelling Portal*, viewed 30 Mar 2020, <https://bim.natspec.org/documents/iso-19650-documents>.

PMI 2019, *Benefit Realisation Management*, Project Management Institute Inc, USA, ISBN: 978-1-62825-480-8.

Rahman, Md, A & Suwal, S 2013, Diverse approach of BIM in AEC industry: A study on current knowledge and practice, In Proceedings of the 30th CIB W78 International Conference, Beijing, China.

Sanchez, A.X, Hampson, KD & Vaux, S 2016, *Delivering Value with BIM: A Whole-of-Life Approach'*, *1st Edition*, Routledge, Abingdon, Oxon.

Succar, B, Agar, C & Beazley, S 2012, *BIM Education, BIM in Practice*, Australian Institute of Architects and Consult Australia, viewed 10 Jan 2020, < https://wp.architecture.com.au/bim/wp-content/ uploads/sites/40/2014/08/E3-BIM-Learning-Spectrum.pdf>.

Succar, B, Sher, W & Williams, A 2013, An integrated approach to BIM competency assessment, acquisition and application, *Automation in Construction*, vol. 35, pp. 174–189, DOI: 10.1016/ j.autcon.2013.05.016.

Suwal, S, Singh, V & Shaw, C 2016 Towards a framework to understand multidisciplinarity in BIM context: Education to teamwork, In Proceedings of the CIB World Building Congress, 20th CIB World Building Congress, Tampere, ISBN: 978-952-15-3740-0.

Wu, W, Mayo, G, McCuen, R, Issa, RR & Smith, D 2017, The BIM body of knowledge (BOK): A delphi study, *Academic Interoperability Coalition*, viewed 22 Jan 2020, <https:// aicbimed.com/files/2 019-10-02_16_38_12_aic_bim_bok_-_final_07062017.pdf.

Index

Page numbers in **bold** denote tables, those in *italic* denote figures.

Taylor & Francis Group
an **informa** business

Taylor & Francis eBooks

www.taylorfrancis.com

A single destination for eBooks from Taylor & Francis
with increased functionality and an improved user
experience to meet the needs of our customers.

90,000+ eBooks of award-winning academic content in
Humanities, Social Science, Science, Technology, Engineering,
and Medical written by a global network of editors and authors.

TAYLOR & FRANCIS EBOOKS OFFERS:

A streamlined
experience for
our library
customers

A single point
of discovery
for all of our
eBook content

Improved
search and
discovery of
content at both
book and
chapter level

REQUEST A FREE TRIAL
support@taylorfrancis.com

 Routledge
Taylor & Francis Group

 CRC Press
Taylor & Francis Group